Ionic Polymerizations and Related Processes

NATO Science Series

A Series presenting the results of activities sponsored by the NATO Science Committee. The Series is published by IOS Press and Kluwer Academic Publishers, in conjunction with the NATO Scientific Affairs Division.

A. Life Sciences	IOS Press
B. Physics	Kluwer Academic Publishers
C. Mathematical and Physical Sciences	Kluwer Academic Publishers
D. Behavioural and Social Sciences	Kluwer Academic Publishers
E. Applied Sciences	Kluwer Academic Publishers
F. Computer and Systems Sciences	IOS Press
1. Disarmament Technologies	Kluwer Academic Publishers
2. Environmental Security	Kluwer Academic Publishers
3. High Technology	Kluwer Academic Publishers
4. Science and Technology Policy	IOS Press
5. Computer Networking	IOS Press

NATO-PCO-DATA BASE

The NATO Science Series continues the series of books published formerly in the NATO ASI Series. An electronic index to the NATO ASI Series provides full bibliographical references (with keywords and/or abstracts) to more than 50000 contributions from internatonal scientists published in all sections of the NATO ASI Series.
Access to the NATO-PCO-DATA BASE is possible via CD-ROM "NATO-PCO-DATA BASE" with user-friendly retrieval software in English, French and German (WTV GmbH and DATAWARE Technologies Inc. 1989).

The CD-ROM of the NATO ASI Series can be ordered from: PCO, Overijse, Belgium

Series E: Applied Sciences – Vol. 359

Ionic Polymerizations and Related Processes

edited by

Judit E. Puskas

Department of Chemical and Biochemical Engineering,
The University of Western Ontario,
London, Ontario, Canada

assistant editors:

Armin Michel

and

Shahzad Barghi

(assisted by Christophe Paulo)

Department of Chemical and Biochemical Engineering,
The University of Western Ontario,
London, Ontario, Canada

Photographer: Gabor Kaszas

Kluwer Academic Publishers

Dordrecht / Boston / London

Published in cooperation with NATO Scientific Affairs Division

Proceedings of the NATO Advanced Study Institute on
Ionic Polymerizations and Related Processes
London, Ontario, Canada
August 10–20, 1998

A C.I.P. Catalogue record for this book is available from the Library of Congress.

ISBN 0-7923-5811-2 (HB)
ISBN 0-7923-5812-0 (PB)

Published by Kluwer Academic Publishers,
P.O. Box 17, 3300 AA Dordrecht, The Netherlands.

Sold and distributed in North, Central and South America
by Kluwer Academic Publishers,
101 Philip Drive, Norwell, MA 02061, U.S.A.

In all other countries, sold and distributed
by Kluwer Academic Publishers,
P.O. Box 322, 3300 AH Dordrecht, The Netherlands.

Printed on acid-free paper

Dedication

This book is dedicated to two outstanding scientists of our community who recently passed away.

Professor Ferenc Tüdös* was born in 1931. He studied chemistry at the University of Szeged (where Albert Szent-Györgyi isolated vitamin C, for which he received the Nobel Prize). He received his doctorate on polymerization kinetics from Leningrad Institute of Technology with Professor Smirnov.
He was one of the leading researchers in Hungary. He was merely 39 years old when the Hungarian Academy of Sciences (MTA) elected him to be a corresponding member. He was elected to be a regular member in 1982. He was Head of the Department of Chemical Technology, and Head of the Department of Macromolecular Chemistry at the Central Research Institute of MTA. Professor Tüdös' research concentrated on polymerization kinetics; he developed the theory of "hot radicals", giving a unified approach to the kinetics of radical polymerizations, and is the co-inventor of the Kelen-Tüdös method for the calculation of copolymerization reactivity parameters. His legacy is nearly a thousand publications, patents and lectures. He will live on in his work.

Professor Reimund Stadler was born in 1956. He studied chemistry at the University of Freiburg, and did his doctorate on viscoelasticity and stress crystallization in thermoplastic elastomers with Professor W. Gronski.
He was one of the fastest rising star in polymer science in Germany. From 1989 to 1997 he was a Professor of Organic and Macromolecular Chemistry at the University of Mainz. In 1997 he joined the Faculty at the University of Bayreuth as Chair of the Department of Macromolecular Chemistry II. Professor Stadler's research concentrated on fundamental studies of structure-property relationships in thermoreversible and permanent networks. In his recent work with block copolymers he discovered new morphologies. His work is documented in a large number of publications.
His untimely death shocked the scientific community, but he leaves a legacy that secures his place among the best and brightest.

We sorely missed these two friends during the ASI.

*On a personal note: the editor of this book would like to pay a special tribute to Professor Tüdös, who was one of her Ph. D. advisor.

ACKNOWLEDGMENT

This ASI would not have been possible without the dedicated effort of the Organizing Committee:

Professor Judit E. Puskas;
Professor Oskar Nuyken,
Professor Robson F. Storey
Dr. Ron Commander, VP Butyl Technology

Many thanks to the postdoc and student organizers affiliated with the Macromolecular Engineering Research Centre:

Sumitra Angepat, Shahzad Barghi, Bryan Brister, Nathan Del Veccio, Keri Diamond, Brad Dochstader, Armin Michel, Christophe Paulo, Haihong Peng, Eraclis Tzaras

Acknowledgement should also go to Conference Services of the University of Western Ontario (Karen Millard and Melanie Harvey) for the excellen accommodation, food and facilities.

The Advanced Study Institute was generously supported by the Science Committee of the Scientific Affairs Division of NATO and the editor gratefully acknowledges this sponsorship.

Financial contribution from the following companies were greatly appreciated:

Rubber Division, Bayer Inc. Canada
Bayer AG, Germany
Exxon Chemicals Co., USA
Goodyear Tire&Rubber Co., USA
BASF, Germany, USA
WYATT Technologies, USA

Table of Contents

PART II: ANIONIC AND OTHER RELATED PROCESSES

PREFACE

This ASI was the first ever in the field of ionic polymerizations. The ASI provided an excellent opportunity for internationally recognized academic and industrial researchers to discuss and debate the latest developments in ionic polymerizations. The major focus was on cationic polymerizations, but related anionic and controlled radical processes were also discussed. This ASI also provided the opportunity for graduate students and young scientists to meet and have discussions with international experts, whose names they were familiar with from the literature, but whom they had never met before. The general impression shared by the participants was that this ASI served as a excellent forum for new results never presented before, giving rise to a very exciting two weeks. The structure of the ASI, with an entire day devoted to industrial lectures, made it possible to provide an excellent interface between academe and industry. The poster session was very successful, leading to interesting discussions and new ideas. True to the tutorial spirit of the ASI, the lectures included background information on the specific areas discussed; this feature of the ASI was particularly appreciated by the graduate students and young researchers. These young people also thrived on the opportunity to serve as moderators of the sessions, a task usually reserved for senior researchers. In sum, the ASI was a very positive experience for all attendees.

This book contains the main lectures. The papers are a combination of review-tutorial and new results. The editors hope that this book will become a resource material for years to come.

Judit E. Puskas

1999 February

PROGRAM

NATO

Advanced Study Institute (ASI)

Ionic Polymerizations and Related Processes

Hosted by:
The University of Western Ontario
Department of Chemical and Biochemical Engineering
London, Ontario, Canada

August 10 - August 21, 1998

Aug. 10, Monday

9:30 am   Welcome – P.Davenport, President, UWO, A. Margaritis, Chair, Chemical and Biochemical Engineering, UWO, Ron Commander, VP, Butyls R&D.

10:15 am   Judit E. Puskas: Greeting and Introduction

10:30 am   Dr. Heinz Greve, Vice president Technology, Rubber Division, Bayer Inc. Canada: Cationic polymerization: industrial processes and applications.

2:15 pm   J.P. Kennedy: *Plenary Lecture:* Carbocationic polymerizations mediated by weakly coordinating counteranions (WCA).

Aug. 11, Tuesday

9:00 am   H. Mayr: Rate constants and reactivity ratios in carbocationic polymerizations.

10:00 am   H. Schlaad, B. Schmitt, W. Stauf, D. Baskaran, H. Kongismann, J. Feldthusen, B. Ivan, A.H.E. Muller: New initiating systems for the living polymerization of acrylates and methacrylates and their use for the synthesis of well-defined polymer structures.

11:00 am	R.F. Storey: Kinetics and mechanism of living cationic polymerizations of olefins.
2:00 pm	J.E. Puskas: Kinetics of living isobutylene polymerizations.
3:00 pm	Z.G. Meszena: Modelling of ionic polymerization reactions; computational fluid dynamics.
4:00 pm	A.F. Johnson: Modelling and control of polymerization process for tailored synthesis.

Aug. 12, Wednesday

9:00 am	M. Gauthier: Arborescent polymers; Highly branched homo- and copolymers with unusual properties.
10:00 am	M. Sawamoto: Living ionic and living radical polymerizations
11:00 am	J. Mays: Recent developments in anionic synthesis of model graft copolymers.
2:00 pm	K. Matyjaszewski: Comparison of ionic and free radical controlled/"living" polymerizations.
3:00 pm	M. Zsuga: Investigation of star-shaped nanomicelles of block copolymers in solution by dynamic light scattering.
4:00 pm	A. Deffieux: Application of living cationic and anionic polymerizations to the synthesis of comb-like and dendritic polystyrenes.

Aug. 13, Thursday

9:00 am	R. Velichkova: Synthesis and properties of amphiphilic and functional block copolymers.
10:00 am	I. Berlinova: Amphiphilic graft copolymers with poly(oxyethylene) side chains: Synthesis and Properties.
11:00 am	R.N. Young: Synthesis, morphology, and rheology of butadiene/n-butyl methacrylate AB diblock and A_2B_2 star copolymers.

2:00 pm   M. Baird: Polymerization of isobutylene and the copolymerization of isobutylene and isoprene initiated by the metallocene derivative $Cp*TiMe_2(-Me)B(C_6F_5)$.

3:00 pm   K. Weiss: On the route to new initiator systems for carbocationic polymerization of isobutylene.

4:00 pm   M. Bochmann: Cationic zirconium complexes as initiators for carbocationic isobutene homo-and -copolymerization.

Aug 14, Friday

9:00 am   Y. Yagci: Photoinitiated cationic polymerizations.

10:00 am   G. Hizal: Photoinitiated cationic and radical polymerization using charge transfer complexes.

11:00am   J. Crivello: Newer aspects of photoinitated cationic polymerization.

Aug. 16, Sunday

2:00 pm   A. Michel, B. Brister: Novel epoxide initiators for carbocationic polymerization.

3:00 pm   R. Faust: Macromolecular design using non-homopolymerizable monomers.

4:00 pm   O. Nuyken: Amphiphilic graft copolymers of multifunctional polyisobutylene.

Aug. 17, Monday- Industry Day

9:00 am   G. Langstein, Bayer AG: Inimers in carbocationic polymerizations.

10:00 am   A. Halasa, Goodyear Tire & Rubber: Preparation and characterization of solution SIBR via anionic polymerization.

11:00 am   T. Shaffer, Exxon Chemical Co. Isobutylene: cationic to insertion polymerization.

2:00 pm   M. Wulkow, CIT: Simulation of polymerization reactions.

3:00 pm   K.Knoll, BASF: A forgotten class of high T_g thermoplastic
                 materials: anionic copolymers of styrene and 1,1-diphenyl-
                 ethylene.

4:00 pm- 6:00 pm Poster Session:

W. Reyntjens, E. Goethals: Networks based on
poly(octadecyl vinyl ether.)

D. Christova: Polymer networks based on polyoxazoline bis-
macromonomers.

G. Hochwimmer, O. Nuyken, U.S. Schubert: 6,6'-
Bifunctionalized-2,2'-bipyridine metal complexes as
initiators for the polymerization of 2-oxazolines.

J. Ismeier, O. Nuyken: Cationic isomerisation
copolymerization of isobutylene and α-pinene in non-
halogenated solvents.

V. Altstadt, T. Krischnik, H. Ott: PPO/SAN blends with
ABC triblock copolymers: processing, morphology and
mechanical properties.

P. Persigehl, P. Kuhn, O. Nuyken: Amphiphilic polymer
based on poly(oxazoline) block copolymers: end group
functionalization for biophysical and catalytic applications.

H. Schimmel, H. Mayr: Determination of the propagation
rate constants in cationic oligomerizations of styrene.

N. Nugay, T. Nugay, R. Jerome, Ph. Teyssie: Ligated
anionic polymerization of methyl, ethyl, n-butyl and n-nonyl
acrylates.

I. Majoros, T. Marsalko, D. Hull, J.P. Kennedy: Well-
defined PDMS-based star polymerization techniques.

T.L. Maggio, R.F. Storey: Coupling of polyisobutylene via carbocationic polymerization techniques.

C.L. Curry, R.F. Storey: Effect of additives on livingness of controlled carbocationic polymerization of isobutylene.

L.B. Brister, E. Tzaras, J.E. Puskas: Novel star-branched PIB/poly(p-*t*-Bu-
styrene) block copolymers from epoxide initiators.

R. Andrew Kee, M. Gauthier: Anionic Synthesis of Telechelic Poly(n-Butyl Methacrylate)s with Sulfonate End-Groups.

L. Cao, M. Gauthier, D. Nguyen, M. Rafailovich, J. Sokolov: Characterization of Amphiphilic Arborescent Core-Shell Polymers at the Air/Water Interface.

A. Carr: New Initiators for Carbocationic Polymerizations.

I. Berlinova: Multifunctional soluble initiators for the synthesis of amphiphilic star copolymers.

H.A. Nguyen, H. Cheradame, W. Buchmann, B. Desmazieres, J.P. Morizur: Isobutene and related model polymerizations initiated by Isothiocyanate/Lewis acid systems: SFC/GC/MS studies.

M. H.Acar: Synthesis of block copolymer by combination of cationic and initer polymerization system 2.

S. Botzenhardt, M. Krombholz, M. Bauer, M. Thuring, K. Weiss: Isobutylene polymerizations with bifunctional initiators and graft polymerizations.

M. Sangermano: Cationic photopolymerization of vinyl ether systems: influence of the presence of hydrogen donor additives.

K. Weiss, U. Neugebauer, S. Botzenhardt, M. Bauer: Polymerisation of olefins with homogeneous and heterogeneous metallocene catalysts.

Aug. 18, Tuesday

9:00 am   U.S. Schubert: Towards telechelics with specific metal binding sites using the metallosupramolecular initiators for the living polymerization of oxazolines.

10:00 am   M. Moreau, J-P Vairon: Ionic polymerization of aldehydes.

11:00 am   S. Slomkowski: Ionic polymerizations in heterogeneous media: Preparation of heterochain macrospheres.

2:00 pm   S. Penczek: Kinetic features of cyclic ester polymerization.

3:00 pm   F. Du Prez: Amphiphilic polymer networks by cationic polymerization: Design and applications.

4:00 pm   A. Graeser: Cationic host-guest polymerization of vinyl ethers in Yzeolite and in MCM-41 channels.

Aug. 19, Wesnesday

9:00 am   N. Manolova: New fullerene containing polymers prepared by ionic processes.

10:00 am   S. Spange: Principles of cationic polymerization occurring on the surfaces of inorganic materials.

11:00 am   P. Hemery: Ionic polymerization of heterocycles in aqueous emulsion.

2:00 pm   J.P Vairon: Kinetics of living cationic polymerizations followed by the stop flow technique.

3:00 pm   B. Charleux: Uses of diphenylethylene (DPE) in cationic polymerizations.

4:00 pm   J. Kriz: Model studies of growth centers in polymerization using molecular spectroscopy and quantum calculations.

Aug. 20, Thursday

9:00 am   P. Vlcek: Synthetic aspects of ligated anionic polymerization of polar vinyl monomers.

10:00 am   G. Deak: Anionic bulk oligomerization of ethylene and propylene carbonate initiated by bispenol-A/base systems.

11:00 am   T. Nugay: Anionic homo-and block copolymerization of primary acrylates.

2:00 pm   H. Cheradame: Cationic polymerization of dienes.

3:00 pm   I. Majoros: Star polymers: State of the art molecular design.

LIST OF PARTICIPANTS

Filip Du Prez

University of Gent
Polymer Chemistry Division
Krijgslaan 281 S4-bis
B-9000 Gent, BELGIUM

Wouter Reyntjens

University of Gent
Polymer Chemistry Division
Krijgslaan 281 S4-bis
B-9000 Gent, BELGIUM

Irina Berlinova

Bulgarian Academy of Sciences
Acad. G. Bonchev St.
Institute of Polymers
BG113 Sofia, BULGARIA

Darinka Christova

Bulgarian Academy of Sciences
Acad. G. Bonchev St.
Institute of Polymers
BG1113 Sofia, BULGARIA

Nevenka Manolova

Bulgarian Academy of Sciences
Acad. G. Bonchev St.
Institute of Polymers
BG1113 Sofia, BULGARIA

Rumiana Velichkova

Bulgarian Academy of Sciences
Acad. G. Bonchev St.
Institute of Polymers
BG 1113 Sofia, BULGARIA

Michael Baird

Department of Chemistry
Queen's University
Kingston, ON, K7L 3N6, CANADA

Shahzad Barghi

Department of Chem.& Biochem. Eng.
The University of Western Ontario
London, ON, N6A 5B9, CANADA

Florin Barsan

Bayer Inc., Rubber Division
P.O. Box 3001
1265 Vidal Street South
Sarnia, ON, N7T 7M2, CANADA

Lan Cao

Department of Chemistry
University of Waterloo
200 University Avenue West
Waterloo, ON, N2L 3G1, CANADA

Ron Commander

Bayer Inc., Rubber Division
P.O. Box 3001
1265 Vidal Street South
Sarnia, ON, N7T 7M2, CANADA

Mario Gauthier

Department of Chemistry
University of Waterloo
200 University Avenue West
Waterloo, ON, N2L 3G1, CANADA

Heinz Greve

Bayer Inc. Bayer Inc., Rubber Division
P.O. Box 3001
1265 Vidal Street South
Sarnia, ON, N7T 7M2, CANADA

Adam Gronowski

Bayer Inc., Rubber Division
P.O. Box 3001
1265 Vidal Street South
Sarnia, ON, N7T 7M2, CANADA

Gabor Kaszas

Bayer Inc., Rubber Division
P.O. Box 3001
1265 Vidal Street South
Sarnia, ON, N7T 7M2, CANADA

Andrew Kee

Department of Chemistry
University of Waterloo
200 University Avenue West
Waterloo, ON, N2L 3G1, CANADA

Carsten Kreuder

Bayer Inc., Rubber Division
P.O. Box 3001
1265 Vidal Street South
Sarnia, ON, N7T 7M2, CANADA

Jing Li

Department of Chemistry
University of Waterloo
200 University Avenue West
Waterloo, ON, N2L 3G1, CANADA

Argyrios Margaritis

Department of Chem&Biochem. Eng.
The University of Western Ontario
London, ON, N6A 5B9, CANADA

Armin Michel

Department of Chem&Biochem. Eng.
The University of Western Ontario
London, ON, N6A 5B9, CANADA

Christophe Paulo

Department of Chem&Biochem. Eng.
The University of Western Ontario
London, ON, N6A 5B9, CANADA

Haihong Peng

Department of Chem&Biochem. Eng.
The University of Western Ontario
London, ON, N6A 5B9, CANADA

Judit Puskas

Department of Chem&Biochem. Eng.
The University of Western Ontario
London, ON, N6A 5B9, CANADA

Jaroslav Kriz

Institute of Macromolecular Chemistry
Academy of Sciences, Czech Republic
Heyrovsky Sq. 2, 162 06 Prague,
CHECH REPUBLIC

Petr Vlcek

Institute of Macromolecular Chemistry
Academy of Sciences, Czech Republic
Heyrovsky Sq. 2, 162 06 Prague
CHECH REPUBLIC

Bernadette Charleux

Universite Pierre et Marie Curie
Laboratoire de Chimie Macromoleculaire
Tour 44, Couloir 44-54, let Etage
4, Place Jussieu
75252 Paris Cedex 05, FRANCE

Herve Cheradame — Laboratoire de Physico-Chimie des
Biopolymeres
2 rue Dunant, 94320 Thiais, FRANCE

Alain Deffieux — Laboratoire de Chimie des Polymeres
Organiques
CNRS-ENSCPB, Ave. Pey Berland – B.P.
108, 33402 TALENCE Cedex , FRANCE

Patrick Hemery — Universite Pierre et Marie Curie
Laboratoire de Chimie Macromoleculaire
Tour 44, Couloir 44-54 let Etage
4 Place Jussieu
75252 Paris, Cedex 05 FRANCE

Michel Moreau — CNRS, Lab. Chimie Macromoleculaire
Case-185, 4 Place Jussieu
75252 Paris, cedex 05, FRANCE

Hung Ahn Nguyen — Laboratoire de Physico-Chimie des
Biopolymeres
2 rue Dunant, 94320 Thiais, FRANCE

Jean-Pierre Vairon — Universite Pierre et Marie Curie
Laboratoire de Chimic Macromoleculaire
Tour 44, Couloir 44-54, let Etage
4 Place Jussieu
745252 Paris Cedex 05, FRANCE

Monika Bauer — Universitat Bayreuth
Anorganische Chemie 1
D-95440 Bayreuth, GERMANY

Martin Bohnenpoll — Bayer AG,ZF-MFE
Geb. Q18,D-51368
Leverkusen, GERMANY

Sandra Botzenhardt — Universitat Bayreuth
Anorganische Chemie 1
D-95440 Bayreuth, GERMANY

Annett Graser

Strabe der Nationen 82
Technische Universitat Chemnitz-Swickau
Lehrstuhl Polymerchemie
09111 Chemnitz, GERMANY

Georg Hochwimmer

Lehrstuhl fur Makromolekulare Stoffe
Institut fur Technische Chemie TUM
Lictenbergstr. 4, Garching, D-85747
GERMANY

Jurgen Ismeier

Lehrstuhl fur Makromolekulare Stoffe
Institut fur Technische Chemie TUM
Lictenbergstr. 4, Garching, D-85747
GERMANY

Konrad Knoll

BASF AG ZKT/I-B1
67056 Ludwigshafen, GERMANY

Markus Krombholz

Universitat Bayreuth
Anorganische Chemie 1
D-95440 Bayreuth, GERMANY

Gerhard Langstein

Bayer AG, ZF-Q18
D-51368 Leverkusen, GERMANY

Herbert Mayr

Institut fur Organische Chemie
Ludwig Maximilians Universitat Munchen
Karistr. 23.D-80333 Munchen, GERMANY

Oskar Nuyken

Lehrstuhl fur Makromolekulare Stoffe
Institut fur Technische Chemie TUM
Lictenbergstr. 4, Garching, D-85747
GERMANY

Werner Obrecht

Bayer AG, ZF-MFE Q18
D-51368 Leverkusen, GERMANY

Harald Ott

Technische Universitat Hamburg-Harburg
FSP5-09, Denickests A5 2A073
Hamburg, GERMANY

Peter Persiegehl

Lehrstuhl fur Makromolekulare Stoffe
Institut fur Technische Chemie TUM
Lictenbergstr. 4, Garching, D-85747
GERMANY

Holger Schimmel

Institut fur Organische Chemie LMU
Munchen, Karlstr. 23
D-80333 Munchen, GERMANY

Helmut Schlaad

Institute of Physical Chemistry
University of Mainz
Welderweg 15, D-55099 Mainz
GERMANY

Ulrich Schubert

Lehrstuhl fur Makromolekulare Stoffe
Institut fur Technische Chemie TUM
Lictenbergstr. 4, Garching, D-85747
GERMANY

Stefan Spange

Str. der Nationen 82
Technische Universitat Chemnitz -Zwickau
D09107 Chemnitz, GERMANY

MartinThuring

Universitat Bayreuth
Anorganische Chemie 1
D-95440 Bayreuth, GERMANY

Karin Weiss

Universitat Bayreuth
Anorganische Chemie 1
D-95440 Bayreuth, GERMANY

Michael Wulkow

CiT GmbH
Olderburger Str. 200
26180 Rastede, GERMANY

György Deak

Department of Applied Chemistry
Kossuth Lajos University
H-4010 Debrecen, HUNGARY

Sandor Keki

Department of Applied Chemistry
Kossuth Lajos University
H-4010 Debrecen, HUNGARY

Zsolt Meszena

Department of Chem. Information Techn.
Technical University of Budapest
H-1521 Budapest, Muegyetem Rkp. 3
HUNGARY

Miklos Zsuga

Department of Applied Chemistry
Kossuth Lajos University
H-4010 Debrecen, HUNGARY

Marco Sangermano

Dipartimento Scienza dei Materialie
Ingegneria Chimica
C.sc Duca degli Abruzzi 24
10129 Torino, ITALY

Mitsuo Sawamoto

Department of Polymer Chemistry
Graduate School of Engineering
Kyoto University
Kyoto, 606-8501, JAPAN

Stanislaw Penczek

Center of Molecular and Macromol. Studies
Polish Academy of Sciences
90-363 Lodz, Sienkiewicza 112, POLAND

Stanislaw Slomkowski

Center of Molecular and Macromol. Studies
Polish Academy of Sciences
90-363 Lodz, Sienkiewicza 112, POLAND

Metin Acar

Istanbul Technical University
Department of Chemistry
Maslak-Istanbul, TURKEY

Gurkan Hizal

Istanbul Technical University
Department of Chemistry
Maslak-Istanbul, TURKEY

Nihan Nugay

Bogazici University
Department of Chemistry
Bebek, Istanbul, TURKEY

Turgut Nugay

Bogazici University
Department of Chemistry
Bebek, Istanbul, TURKEY

Yusuf Yagci

Istanbul Technical University
Department of Chemistry
Maslak-Istanbul 80626, TURKEY

Manfred Bochmann

University of Leeds
Department of Chemistry
Leeds, LS2 9JT, UK

Andrew Carr

School of Chemistry
University of Leeds
Leeds LS2 9JT, UK

Tony Johnson

IRC in Polymer Science and Technology
School of Chemistry
University of Leeds, Leeds LS2 JT, UK

Ronald Young

Department of Chemistry
University of Sheffield
Sheffield, South Yorkshire S10 2TN, UK

Bryan Brister

University of Southern Mississippi
Department of Polymer Science
Box 10076, Hattiesburg, MS, 39406, USA

David Chung

Exxon Chemical Co.
5200 Bayway Drive
Baytown, TX 77520, USA

James Crivello

Department of Chemistry
Rensselaer Polytechnic Institute
Troy, N.Y. 12180, USA

Cristopher Curry

University of Southern Mississippi
Department of Polymer Science
Box 10076, Hattiesburg, MS 39406, USA

Rudolf Faust

University of Massachusetts at Lowell
1 University Avenue
Lowell, MA 01854, USA

Adel Halasa

The Goodyear Tire&Rubber Co.
142 Goodyear Blvd.
Akron, OH 44305-3375, USA

Diana Hull

The Maurice Morton Institute of Polymer
Science
The University of Akron,
Akron, OH 44325, USA

Joseph Kennedy

The Maurice Morton Institute of Polymer
Science
The University of Akron
Akron, OH 44325, USA

Thomas Maggio

University of Southern Mississippi
Department of Polymer Science
Box 10076, Hattiesburg, MS, 39406, USA

Istvan Majoros

The Maurice Morton Institute of Polymer
Science
The University of Akron
Akron, OH 44325, USA

Krzysztof Matyjaszewski

Department of Chemistry
Carnegie-Mellon University
4400 Fifth Avenue
Pittsburgh, PA, 15213, USA

Jimmy Mays

Department of Chemistry
University of Alabama at Birmingham
Birmingham, Al 35294, USA

Timothy Shaffer

Exxon Chemical Co
Baytown Polymer Center
5200 Bayway Drive
Baytown, TX 77522-5200, USA

Rob Storey

University of Southern Mississippi
Department of Polymer Science
Box 10076, Hattiesburg, MS, 39406, USA

LIST OF MAIN CONTRIBUTORS

F. Du Prez (E.J.Goethals)
University of Gent
Polymer Chemistry Division
Krijgslaan 281 S4-bis
B-9000 Gent, BELGIUM

M. Gauthier
Department of Chemistry
University of Waterloo
200 University Avenue West
Waterloo, ON, N2L 3G1, CANADA

H. H. Greve
Vice President Technology
Bayer Inc., Rubber Division
P.O. Box 3001
1265 Vidal Street South
Sarnia, ON, N7T 7M2, CANADA

J.E. Puskas
Department of Chemical and Biochemical Engineering
The University of Western Ontario
London, ON, N6A 5B9, CANADA

A. Deffieux
Laboratoire de Chimie des Polymeres Organiques
CNRS-ENSCPB
Avenue Pey Berland – B.P. 108
33402 TALENCE Cedex – FRANCE

J-P. Vairon
Universite Pierre et Marie Curie
Laboratoire de Chimic Macromoleculaire
Tour 44, Couloir 44-54, let Etage
4 Place Jussieu
745252 Paris Cedex 05, FRANCE

G. Langstein
ZF-Q18
Bayer Ag
D-51368 Leverkusen, GERMANY

H. Mayr
Institut fur Organische Chemie
Ludwig Maximilians
Universitat Munchen
Karistr. 23.D-80333 Munchen, GERMANY

O. Nuyken
Lehrstuhl fur Makromolekulare Stoffe
Institut fur Technische Chemie TUM
Lictengergstr. 4, Garching, D-85747 GERMANY

Y. Yagci
Istanbul Technical University
Faculty of Science
Department of Chemistry
Maslak, Istanbul 80626, TURKEY

A.F. Johnson
IRC in Polymer Science and Technology
School of Chemistry
University of Leeds
Leeds LS2 JT, West Yorkshire, UK

J. Crivello
Professor of Chemistry
Department of Chemistry
Rensselaer Polytechnic Institute
Troy, New York 12180, USA

R. Faust
University of Massachusetts at Lowell
1 University Avenue
Lowell, MA 01854, USA

J. P. Kennedy
The Maurice Morton Institute of Polymer Science
The University of Akron
Akron, OH 44325, USA

K. Matyjaszewski
Department of Chemistry
Carnegie Mellon University
4400 Fifth Avenue
Pittsburgh, PA 15213, USA

J. Mays
Department of Chemistry
University of Alabama at Birmingham
Birmingham, AL 35294, USA

R.F. Storey
University of Southern Mississippi
Department of Polymer Science
Box 10076
Hattiesburg, Mississippi, 39406, USA

NATO ADVANCED STUDY INSTITUTE
IONIC POLYMERIZATION AND RELATED PROCESSES
AUGUST 10-21, 1998
THE UNIVERSITY OF WESTERN ONTARIO
LONDON, ONTARIO, CANADA
DIRECTOR: JUDIT E. PUSKAS

CATIONIC POLYMERIZATIONS AT ELEVATED TEMPERATURES BY NOVEL INITIATING SYSTEMS HAVING WEAKLY COORDINATING COUNTERANIONS. 1. HIGH MOLECULAR WEIGHT POLYISOBUTYLENES

Z. PI, S. JACOB AND J. P. KENNEDY*
Maurice Morton Institute of Polymer Science
The University of Akron, Akron, OH 44325-3909, USA

Abstract. *In-situ* prepared $(CH_3)_3Si^{\oplus} [B(C_6F_5)_4]^{\ominus}$ efficiently produces extremely high molecular weight polyisobutylene (PIB) in close-to-neat systems (15 % toluene or 5 % methyl chloride to dissolve the $Li[B(C_6F_5)_4]$ coinitiator precursor salt) over the -35 to -8° (reflux) range. In contrast, much lower molecular weight PIBs are obtained when the preprepared initiating system is added to liquid isobutylene (IB). The molecular weights produced by the *in-situ* initiating system are very much higher (e.g., $\overline{M}_n \sim 110,000$ g/mol at - 15°C) than those obtained by common systems, e.g., $AlCl_3$, and match those obtained by the use of γ-rays. The molecular weight distributions ($\overline{M}_w / \overline{M}_n \sim 2$) and the slopes of log M vs. 1/T plots ($\Delta H_{DP}^{\ddagger} = -5.9$ kcal/mol (-24.7 kjoule/mol)) indicate chain-transfer-controlled polymerizations. The *de facto* initiating entity is probably the $H^{\oplus}$ arising from adventitious moisture. Due to the extremely nonnucleophilic nature of the weakly coordinating counteranion $[B(C_6F_5)_4]^{\ominus}$, propagation most likely proceeds by unencumbered carbocations and termination is absent. Living polymerization could not be obtained in the presence of $[B(C_6F_5)_4]^{\ominus}$, which suggests that livingness in carbocationic polymerizations requires counteranion assistance.

1. Introduction

The practice of classical (non-living) cationic polymerizations, specifically the processes leading to isobutylene (IB)-based rubbers, such as polyisobutylene and the various isobutylene-isoprene copolymers (butyl rubbers), are beset, since their introduction some 60 years ago [1], by two major problems: 1. Expensive cryogenic cooling and 2. The use of noxious chlorinated solvents. A case in point is the manufacture of butyl rubber, which in order to attain the desired molecular weights ($\overline{M}_n \sim 100,000$ g/mole, with $\overline{M}_w / \overline{M}_n = 2$-3), is produced at $\sim$ -100°C in methyl chloride by a number of global manufacturers [2]. Similarly, PIB is produced at $\sim$ -102°C by BASF [3]. The effect of temperature on the molecular weights of polyisobutylene-based products has been

1

J.E. Puskas et al. (eds.), Ionic Polymerizations and Related Processes, 1–12.

explored by two generations of researchers and is well documented [4,5]. Numerous efforts have been made over the years to produce high molecular weight PIB-based rubbers without cryogenic cooling; success, however, remains elusive in this domain [4,5].

The other vexing problem in the production of butyl rubbers is the use of noxious chlorinated diluents, specifically methyl chloride. This liquid not only functions as the heat sink of the highly exothermic slurry polymerizations, but also the solvent for the coinitiator of choice, $AlCl_3$, and provides the polar milieu for the ionic process.

The objective of this research was to find new initiating systems which would produce highest molecular weight IB-based polymers and copolymers at temperatures much higher than commonly used under environmentally friendlier, essentially chlorine-free conditions. The clue how to approach these dual objectives came by analyzing and combining two totally disparate polymerization techniques: 1. Cationic olefin polymerizations induced by high-energy (γ-ray) radiation, and 2. Polymerizations induced by metallocenes proceeding in the presence of weakly coordinating counteranions (WCA$^{\ominus}$s).

1.1. CATIONIC OLEFIN POLYMERIZATION BY γ-RAYS

In regard to γ-ray induced cationic olefin polymerizations, it was shown that this technique yields highest molecular weight PIBs, albeit rather slowly [4,5]. The propagating sites in these systems are thought to be highly active "free" carbocations; the ejected electron which provides the pro forma counteranion is thought to be immobilized in a "hole" in the condensed phase [4,5]. γ-ray induced polymerizations proceed in neat monomers (bulk systems), which is both an advantage and disadvantage: The advantages are simplicity (solventless conditions) and the absence of noxious chlorinated solvents; however, bulk polymerizations are difficult to control and are slow because the rate of ionic reactions are greatly reduced in the absence of solvating polar diluents. Methyl chloride or other common chlorinated solvents cannot be used in conjunction with γ-rays because of the large shielding effect by the chlorine atoms. γ-ray induced cationic polymerizations are out of favor at the present mainly because of the stigma associated with this technology and the high cost of the installation.

A few unusual solid-phase initiating systems have also been described (for a survey see refs. [4,5]) which purportedly yield very high molecular weight PIBs albeit at low conversions and rates. In these systems propagation is also thought to proceed by counteranion-unencumbered carbocations with the counteranions sequestered in the solid phase [4,5].

Evidently, unencumbered growing cations produce extremely high molecular weight PIBs because of the absence of the counteranion in the vicinity of the propagating species. Counteranions lurking in the proximity of growing cations are apt to mediate chain transfer and thus reduce molecular weights [5]. The effect of counteranions on molecular weights in cationic polymerizations has been the subject of many studies over the years [2,4,5]. Thus the question arises: Can highest molecular weight PIB based products heretofore obtainable only by γ-rays (or by some ill-defined solids) be produced by the use of chemical initiating systems under conventional conditions. On the basis of recent results presented here, we are inclined to reply in the affirmative.

1.2. METALLOCENE-INDUCED OLEFIN POLYMERIZATION

Turning to metallocene-induced olefin polymerizations (systems which currently are revolutionizing the thermoplastic industry [6]), initiation in these systems involves 14 electron cationic metal centers neutralized by WCA$^\ominus$s, such as $[B(C_6F_5)_4]^\ominus$, in hydrocarbon solvents at or above room temperature. The decisive events (coordination, insertion, propagation) in these polymerizations occur at the cationic metal center, while the stable WCA$^\ominus$s are separated from these events. In this sense γ-rays and metallocenes induced polymerizations are similar.

In line with this chain of thought, several years ago we have started a program to explore WCA$^\ominus$-based initiating systems for carbocationic (*not* coordinated cationic) olefin polymerizations. We have speculated that such systems may provide counteranion-unencumbered "free" carbocations and thus may lead to highest molecular weight products, hopefully in chlorine-free systems under conventional conditions at higher temperatures than was possible earlier. Our earliest efforts by the use of various typical metallocene-based systems were sufficiently encouraging to continue the exploration of the lead. In the course of this work several publications have appeared notably by Baird *et al.* [7,8,9,] and Shaffer *et al.* [10,11,12] and very recently by Bochmann *et al.* [13] demonstrating the polymerization of IB and other olefins and vinyl ethers by a great variety of metallocenes in conjunction with WCA$^\ominus$s. Shaffer et al.'s findings were particularly intriguing since the molecular weights of their products were very high and the mechanism they proposed was very insightful.

This paper concerns the preparation of a novel WCA$^\ominus$-based cationic initiator system, *in situ* $(CH_3)_3Si^\oplus[B(C_6F_5)_4]^\ominus$, and demonstrates its use for the synthesis of highest molecular weight PIB at reasonable rates in close-to-neat systems at unusually high temperatures (up to reflux). Another objective was to explore the possibility of living IB polymerization in the presence of $B(C_6F_5)_4^\ominus$ as the WCA$^\ominus$. IB polymerization experiments carried out by the use of highly reactive unencumbered carbocations under conditions which give living polymerization with BCl$_3$- or TiCl$_4$-based initiating systems, were thought to provide increased insight into the mechanism of living polymerizations.

2. Experimental

2.1. MATERIALS

Chemicals were from Aldrich except as otherwise specified. n-Pentane was freshly distilled from lithium aluminum hydride. Me$_3$SiCl was used without further purification. Isobutylene (Matheson, 99%) and methyl chloride (Matheson) were dried by passing the gases through large columns filled with molecular sieves and BaO and collected in the dry box [14]. B(C$_6$F$_5$)$_3$ (Strem Chemicals) was used without further purification. C$_6$F$_5$Br was distilled before use. Li[B(C$_6$F$_5$)$_4$] was prepared according to reference [15].

2.2. POLYMERIZATION

Polymerizations were carried out in 75 mL culture tubes or 250 mL round bottom flasks equipped with a mechanical stirrer in a stainless steel dry box under nitrogen in a cooling bath at different temperatures as described [14]. Representative polymerizations were as follows:

2.2.1. *IB Polymerization in the Presence of 5% CH₃Cl in the -35 – -8°C Range.*
Into a 250 mL three-neck flask containing IB (142.5 mL, 1.71 mol) and Me$_3$SiCl (0.407 g, 3.75 mmol) charges was added Li[B(C$_6$F$_5$)$_4$] (0.412 g, 0.6 mmol dissolved in 7.5 mL CH$_3$Cl) under stirring. At predetermined time intervals, samples (10 mL) were withdrawn and quenched by pre-chilled methanol. The charges remained colorless throughout the experiments. The polymerizations were quenched before the increases in viscosity would have interfered with agitation (<25%). Above ~ 30% the consistency of the charge becomes like honey and the viscosity of the system increases with increasing conversions. The polymerizations were quenched by methanol and the products were isolated by precipitation in methanol, purified (redissolving in hexane and washing with water and methanol) and dried in vacuum at 40°C. Polymerizations in refluxing charges (-8°C) were carried out in a flask equipped with a cold-finger condenser (filled with a mixture of n-pentane and Dry Ice).

2.2.2. *IB Polymerization in the Presence of 15% Toluene at -15°C.*
Polymerizations were induced by the addition of Li[B(C$_6$F$_5$)$_4$] (0.233g, 0.34 mmol dissolved in 14 mL toluene) to stirred charges of IB (75 mL, 0.86 mol) and Me$_3$SiCl (0.242 g, 2.25 mmol).
Samples (4 mL) were withdrawn at predetermined time intervals; after 90 min. the polymerization was quenched and the products purified as described above.

2.2.3. *Experiments to Test the Possibility for Living Polymerization.*
Two series of AMI (all monomer in) experiments were conducted under the following conditions: Into a series of six culture tubes (75 mL) were placed charges consisting of IB (5 mL), pentane (3 mL), cumyl chloride (CumCl, 1.5 x 10^{-4} mols), dimethylacetamide (DMA, 1.5 x 10^{-4} mols), and DtBP (1.4 x 104 mols) at -50°C.

Li[B(C₆F₅)₄] dissolved in 2 mL CH₃Cl was added to all six tubes at zero time. After 10 min, 5 mL methanol was added to the charge in the first tube to quench the reaction; the same quenching procedure was effected after 20, 30, 40, 50, and 60 min in the balance of the test tubes. The products were worked up as described before.

3. Characterization

Molecular weights were determined by GPC using established procedures [14].

4. Results and Discussion

4.1. THE INITIATING SYSTEM AND THE ROLE OF THE WCA$^{\ominus}$

One of the main objectives of this research was to obtain PIB-based products in the technologically desirable high molecular weight range (>100,000 g/mol) at temperatures substantially higher, and under conditions environmentally more benign than practiced currently. Specifically, we aimed to obtain highest molecular weight PIBs at the highest possible practical temperature, preferably in a refluxing system, in the absence of noxious chlorinated solvents, preferably in neat (bulk) systems. We speculated that these objectives could be attained by the use of novel WCA$^{\ominus}$-containing initiating systems generated *in situ*.

The WCA$^{\ominus}$ was thought to lead to conditions similar to those prevailing in γ-rays induced polymerizations which produce highest molecular weights (see Introduction), while the *in situ* process was expected to produce highly reactive "nascent" cations in sufficient concentrations for efficient initiation. We hypothesized the *in situ* initiating system could be produced from readily available chemicals by:

$$(CH_3)_3SiCl + Li[B(C_6F_5)_4] \longrightarrow (CH_3)_3Si^{\oplus}[B(C_6F_5)_4]^{\ominus} + LiCl \downarrow \qquad (1)$$

The reaction is driven by the high reactivity of (CH₃)₃SiCl toward nucleophilic substitution and by the precipitation of LiCl, and was expected to give very reactive trimethylsilylium cations. The latters would initiate olefin polymerization either directly or by reacting first with adventitious H₂O which would function as the *de facto* cationogen [11],[12],[16],[17],[18] (The brackets indicate intermediates whose existence has not been proven):

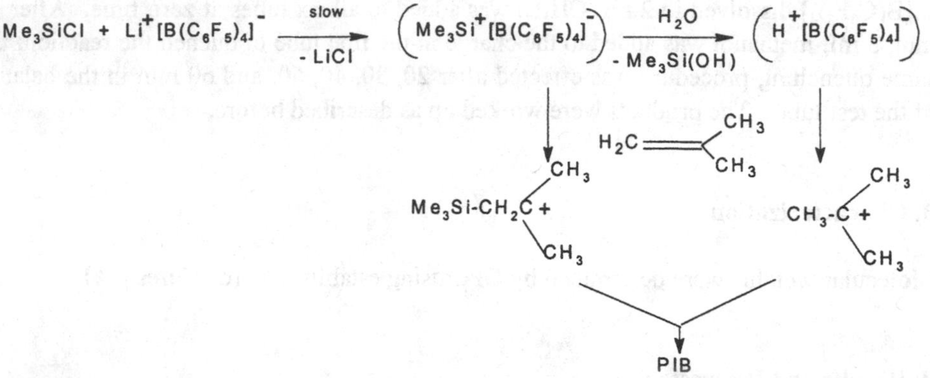

Scheme 1.

Shaffer and Ashbaugh [12] have shown that pre-prepared $(C_2H_5)_3Si^+$ $[B(C_6F_5)_4]^-$ made by Eq 2 [18].

$$(C_2H_5)_3SiH + (C_6H_5)_3CCl + Li[B(C_6F_5)_4] \longrightarrow$$

$$(C_2H_5)_3Si^{\oplus} [B(C_6F_5)_4]^{\ominus} + LiCl + (C_6H_5)_3CH \qquad (2)$$

when added to IB charges produces low molecular weight PIBs. We hypothesized that a similar, even more reactive initiating cation, the $(CH_3)_3Si^{\oplus}$ (the "bulky proton") could be prepared preferably, *in situ* in neat monomer charges in the absence of R_3SiH by Eq 1. The chemistry shown in Eq 2 was deemed unsuitable for this purpose because of the presence of $(C_2H_5)_3SiH$ which we found to be a strong chain transfer agent in IB polymerization in separate experiments (see data in Table 1). Since, in contrast to $(C_2H_5)_3SiH$, $(CH_3)_3SiCl$ was found not to be a chain transfer agent (the Si-Cl bond is stronger than the C-Cl) (see later), we expected the *in situ* $(CH_3)_3Si^{\oplus} [B(C_6F_5)_4]^{\ominus}$, produced by Eq 1, to give highest molecular weight PIBs.

TABLE 1. The effect of Et_3SiH on PIB molecular weight (CH_2Cl_2, -25°C)*

$Li[B(C_6F_5)_4] \times 10^{-3}$ mol/L	$Et_3SiH \times 10^{-3}$ mol/L	$B \times 10^{-3}$ mol/L	Time hr	Conv. %	$\overline{M}_n \times 10^{-3}$ g/mol	$\overline{M}_w / \overline{M}_n$
2.0	0	2.34	0.5	25.1	201	2.1
2.0	2.0	2.34	1	54.3	66.4	1.8
2.0	5.0	2.34	1	55.9	49.2	1.8
2.0	10.0	2.34	1	58.8	36.0	1.7
2.0	50.0	2.34	1	35.6	22.0	1.7
2.0	100.0	2.34	1	33.9	15.3	1.7

*IB = 2 mL; CH_2Cl_2 = 8 mL; total volume = 10 mL

4.2. HOMOPOLYMERIZATIONS IN CLOSE-TO-NEAT MONOMER CHARGES

$Li[B(C_6F_5)_4]$ is insoluble in aliphatic hydrocarbons even at elevated temperatures, however, it is soluble in toluene and methyl chloride. We found that IB polymerizations can be readily initiated by adding toluene or methyl chloride solutions of $Li[B(C_6F_5)_4]$ to quiescent $IB/(CH_3)_3SiCl$ charges. Table 2 shows the results of representative scouting experiments carried out by *in situ* generated $(CH_3)_3Si^{\oplus}[B(C_6F_5)_4]^{\ominus}$ in close-to-neat monomer at -35°C. According to the controls (Expts. 1 and 2) neither $(CH_3)_3SiCl$ nor $Li[B(C_6F_5)_4]$ alone initiate polymerization. In the presence of suitable concentrations of both of these ingredients, however, very high molecular weight PIBs (200,000-210,000 g/mol) can be obtained at reasonable rates (Expts. 3-7 and 8-10). Conversions (rates) could be controlled by adjusting the concentrations of the ingredients (Expts. 3-7) or time (Expts. 8-10) with virtually unchanged molecular weights.

The reasonably low polymerization rates allowed for efficient cooling even in these close-to-neat systems. Expts. 3-7 also indicate that $(CH_3)_3SiCl$ is indeed not a chain transfer agent since the molecular weights did not decrease by increasing the concentration of this reagent (compare these data with those for $(C_2H_5)_3SiH$ shown in Table 1).

TABLE 2. Scouting polymerization experiments at -35°C.*

	$Li[B(C_6F_5)_4] \times 10^{-3}$ mol/L	Me_3SiCl 10^{-2} mol/L	Time min	Conv. %	$\overline{M}_n \times 10^{-3}$ g/mol	$\overline{M}_w/\overline{M}_n$
1	6.0	-	30	0	-	-
2	-	20.0	30	0	-	-
3	3.5	1.0	30	0	-	-
4	3.5	2.0	30	0.4	131	2.2
5	3.5	5.0	30	4.8	201	2.3
6	3.5	10.0	30	13.5	212	2.2
7	3.5	15.0	30	25.0	201	2.2
8	4.0	2.5	30	9.8	210	2.3
9	4.0	2.5	60	21.9	206	2.3
10	4.0	2.5	120	38.0	215	2.3
11‡	4.0	2.5	60	-	-	-
12†	4.0	2.5	60	-	-	-
13(1)°	4.0	2.5	30	14	85	2.1
14(40)°	4.0	2.5	30	13	96	2.2
15(60)°	4.0	2.5	30	13	98	2.2

*$Li[B(C_6F_5)_4]$ was dissolved in MeCl and 0.5 mL of the solution was added to quiescent Me_3SiCl /IB charges; IB = 9.5 mL. ‡$DtBP$ = 4.0 x 10^{-3} M, added with IB. †$DtBP$ = 6.0 x 10^{-3} M, added with IB. °$Li[B(C_6F_5)_4]$ and Me_3SiCl premixed in MeCl before addition. Numbers in parentheses indicate aging time in minutes.

As indicated by Expts. 8, 9 and 10, the rate of formation of the initiating $(CH_3)_3Si^{\oplus}$ ion is slow which is reflected by linearly increasing conversions with time. In contrast, when the components of the initiating system are premixed in MeCl (Expts. 13-15) conversions do not increase as a function of time. The fact that PIB molecular weights are higher by at least a factor of ~2 when using the *in situ* initiating system, is probably due to the low concentration of the active species.

The results of scouting studies were substantiated by larger scale experiments carried out in stirred reactors. Figure 1 shows the results of two series of kinetic experiments carried out with close-to-neat IB systems at -15°C (15% toluene or 5% MeCl were needed to dissolve the $Li[B(C_6F_5)_4]$). After an induction period (30-40 mins most likely due to slow ion generation and/or scavenging of impurities) the polymerizations follow first order kinetics indicating that the number of active species remains constant. The molecular weights show corresponding behavior: After an induction period of rapid growth with increasing conversions, they reach a plateau. This behavior is characteristic of molecular weight control by chain transfer. Termination is most likely absent, however, direct proof for 100% conversions is difficult to obtain in neat monomer because of the increase in viscosities with increasing conversions beyond ~30% (see Experimental).

4.3. THE EFFECT OF TEMPERATURE ON THE MOLECULAR WEIGHTS

To gain insight into the polymerization mechanism, the effect of temperature on molecular weights was investigated by the use of Arrhenius plots (for the background of this method see ref. [4]). Figure 2 shows log $\overline{M}_n$ and $\overline{M}_w$ as a function of 1/T for PIBs prepared by the *in situ* $(CH_3)_3Si^{\oplus}[B(C_6F_5)_4]^{\ominus}$ system together with those obtained earlier by γ-rays [4] and $AlCl_3$ [4] induced polymerizations for comparison. The activation enthalpy difference of molecular weights was found to be -5.9 kcal/mol (-24.7 kjoule/mol), which is within experimental variation of values of the other systems [4]. According to this analysis, IB polymerizations induced by the novel initiating system, just as the earlier systems, are chain transfer dominated [4].

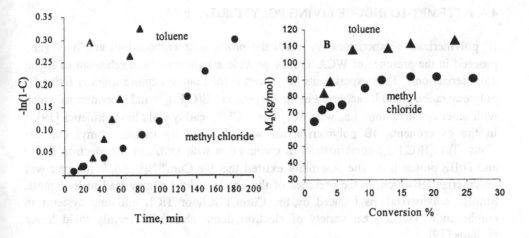

Figure 1. First order rate (A) and M_n vs. conversion (B) plots for close-to-neat IB polymerizations at -15°C (Me_3SiCl = 2.5 x 10^{-2} M; $Li[B(C_6F_5)_4]$ = 4 x 10^{-3} M; 15 % toluene system: IB = 75 mL.; toluene = 14 mL; total volume = 89 mL; 5 % methyl chloride system: IB = 142.5 mL, CH_3Cl = 7.5 mL, total volume = 150mL).

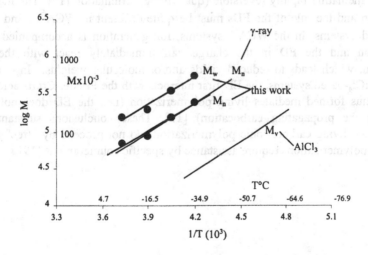

Figure 2. Temperature dependence of PIB molecular weights with different initiating systems in the -8 – - 80°C range.

4.4. ATTEMPT TO INDUCE LIVING POLYMERIZATION

IB polymerization experiments in which the propagating carbocations are 'free" (i.e., proceed in the presence of WCA$^\ominus$s) may provide insight into the mechanism of living polymerizations. Thus experiments have been carried out to explore whether living IB polymerization could be achieved in the presence of $B(C_6F_5)_4^{\ominus}$ under conditions which with other counteranions, i.e., with $BCl_4^{\ominus}$ or $TiCl_5^{\ominus}$, readily yield living kinetics [19].
In the experiments, IB polymerizations were induced by various cumyl chloride (CumCl)/Li[B(C$_6$F$_5$)$_4$] combinations in conjunction with DMA as the electron donor and DtBP proton trap. The possibility existed that the $Cum^{\oplus} B(C_6F_5)_4^{\ominus}$ ion pair will polymerize IB which, in the presence of the above additives, may lead to livingness. Similar polymerizations induced by the CumCl/BCl$_3$ or TiCl$_4$ initiating systems in combination with a great variety of electron donor and DtBP readily yield living systems [19].

Table 3 summarizes the results of two series of experiments. According to the trendless yield (conversion) and molecular weight data, living polymerizations did not occur under the conditions examined.

These results suggest that livingness requires the presence of a suitable counteranion capable of mediating rapidly reversible (quasiliving) termination [19]. The mechanism of initiation and the role of the EDs must be quite different in WCA$^\ominus$ - and BCl$_3$- or TiCl$_4$-based systems: In the WCA$^\ominus$ systems, ion generation is accompanied by LiCl precipitation and the ED in the charge can immediately react with the arising carbocation, which leads to reduced yields and/or molecular weights. In contrast, in BCl$_3$- or TiCl$_4$-based systems, the ED first interacts with the Friedel-Crafts acid and the complex thus formed mediates living polymerization (i.e., the ED does not directly react with the propagating carbocation) [19]. These conclusions substantiate the postulate that living carbocationic polymerizations do not proceed by "free" ions and that living polymerizations require assistance by specific counteranions [19].

TABLE 3. IB polymerizations under living conditions at -50 °C.*

	Li[B(C₆F₅)₄] 10^{-4} mol	Cumyl chloride 10^{-4} mol	Time min	Conv. %	$\overline{M}_n$ 10^{-3} g/mol	$\overline{M}_w/\overline{M}_n$
1	1.5	1.5	20	6.5	40	2.1
2	1.5	1.5	30	18.0	35	2.1
3	1.5	1.5	40	21.1	27	1.9
4	1.5	1.5	50	20.6	29	2.0
5	1.5	1.5	60	16.4	31	2.2
6	0.3	0.3	10	7.0	52	2.1
7	0.3	0.3	25	8.9	64	2.1
8	0.3	0.3	40	4.6	59	2.1
9	0.3	0.3	55	5.3	60	2.0
10	0.3	0.3	70	8.8	54	2.1
11	0.3	0.3	85	9.0	46	2.1

*Li[B(C₆F₅)₄] was dissolved in MeCl and 2 mL of the solution was added to charges. [DMA] same as that of [CumCl]. DtBP = 1.4 x 10^{-4} mols.

5. Conclusions and Comments on the Mechanism

$(CH_3)_3Si^{\oplus}$ $[B(C_6F_5)_4]^{\ominus}$ prepared *in situ* in close-to-neat IB charges is an efficient initiating system that leads to highest molecular weight PIB even under reflux conditions in close-to-neat monomer (see Figure 2).

Pre-prepared $(CH_3)_3Si^{\oplus}$ $[B(C_6F_5)_4]^{\ominus}$, when added to IB under the same conditions, produces lower amounts of much lower molecular weight PIB. According to experiments with 2,6-di-*tert*-butylpyridine, this proton trap inhibits IB polymerization. Thus the initiating system is most likely the hypothetical strong acid $H^{\oplus}$ $[B(C_6F_5)_4]^{\ominus}$ arising from subanalytical amounts of moisture impurities ("H₂O") in the charge. Analysis of our findings together with those generated by others [11],[12] suggest that IB various polymerizations induced by various WCA$^{\ominus}$-based initiating systems (including metallocenes) are predominantly due to $H^{\oplus}$ $[B(C_6F_5)_4]^{\ominus}$. Minor amounts of PIB formed by the use of various metallocenes in the presence of DtBP concentrations about equivalent to that of " H₂O" (~10^{-4}M) may be due to initiation by 14 electron metallocene cations, however, the significance of initiation by the latter species relative to that of "H₂O" is negligible. Hydrocarbons, methyl chloride, and particularly methylene chloride, always contain sufficient "H₂O" for protic initiation which can be removed only by heroic measures (Schlenk techniques are woefully inadequate) [20].

Subanalytical amounts of moisture ("H₂O" <10^{-4}M) are sufficient to initiate efficiently cationic polymerizations since the $H^{\oplus}$ is constantly regenerated by chain transfer and termination is absent due to the WCA$^{\ominus}$. Molecular weights are determined by spontaneous (1st order) $H^{\oplus}$ transfer to monomer, a process less favorable than counteranion-assisted (0 order) transfer who explains the very high molecular weights produced by such systems.

12

In contrast to CumCl/BCl₃ or TiCl₄ combinations which readily lead to living IB polymerization [19], the CumCl/Li[B(C₆F₅)₄] system does not mediate livingness under essentially the same conditions. Evidently, the propagating species in the living systems are strongly affected by the counteranion whereas unencumbered carbocations propagating in the presence (but not under the kinetic influence) of WCA$^\ominus$s lead to chain transfer to monomer which preempts livingness.

Acknowledgement. This material is based on work supported by the NSF under Grant DMR 94-23202. Initial scouting studies with metallocenes by T. Pernecker and Gy. Fenyvesi are gratefully acknowledged.

6. References

1. Purcell, A. T. and Fusco, J. V. (1987) *Butyl The First Fifty Years,* Exxon Chemical Corporation.
2. Vairon, J. P. and Spassky, N. (1996) *Cationic Polymerization Mechanisms, Syntheses, and Applications,* Ed. K. Matyjeszewski, Marcel Dekker, Inc. New York, Chapter 8.
3. Guterbok, H. (1959) *Polyisobutylene,* Springer-Verlag , p. 98.
4. Kennedy, J. P. (1975) *Cationic Polymerization of Olefins: A Critical Inventory,* John Wiley & Sons, New York, Chapter 5.
5. Kennedy, J. P. and Maréchal, E. (1982) *Carbocationic polymerization,* Wiley-Interscience; New York, p. 481.
6. Kaminsky, W. and Arndt, M. (1997) *Adv. Polym. Sci.* **127,** 141-18'2".
7. Wang, Q., Quyoum, R., Gillis, D. J., Tudoret, M. J., Jeremic, D., Hunter, B. K., and Baird, M. C. (1996) *Organometallics* **15,** 693-703.
8. Barsan, F. and Baird, M. C. (1995) *J- Chem. Soc., Chem. Commun. 1065-1066.*
9. Quyoum, R., Q. Wang, Q., Turdet, M.-J., Baird, M. C. and Gillis, D. J. (1994) *J. Amer. Chem. Soc.* **116,** 6435-6436
10. Shaffer, T. D. (1994) U.S Pat. Appl. 08/234,782.
11. Shaffer, T. D. (1997) *American Chemical Society* **665,** Chapter 9, 96-105.
12. T. D.Shaffer and Ashbaugh, J. R. (1997) *J. Polym. Sci., Part A: Polym. Chem.* **35,** 329-344.
13. Carr, A. G., Dawson, D. M., and Bochman M. (1998) *Macromolecules* **31,** 2035-204.
14. Kennedy, J. P. and Smith R. (1980) *J. Polym. Sci., Polym Chem. Ed.* **18,**1523-1537.
15. Massey, A. G. and Park, A. J. (1964) *J. Organomet. Chem.* **2,** 245-250.
16. Lambert, J. B. and Zhao, Y. (1996) *J. Am. Chem. Soc.* **118,** 7867-7868.
17. Lambert, J. B. and Zhao, Y. (1997) *Angew. Chem. Int. Ed. Engl.* **36,**400-40.1.
18. Lambert, J. B. and Zhang, S. (1993) *J. Chem. Soc., Chem. Commun.* 383-384.
19. Kennedy, J.P. and Ivan, B. (1992) *Designed Polymers by Carbocationic Macromolecular Engineering,* Hanser; Munich.
20. Plesh, P. H. (1989) *High Vacuum Techniques for Chemical Syntheses and Measurements,* Cambridge University Press, New York.

Note Added in Proof

A paper coauthored by Baird *et al.* (F. Barsan, A.R. Karman, M.A. Parent, and M. C. Baird, *Macromolecules* **31,** 8439(1998)) has appeared while our manuscript was in press. In this paper the author reassert their earlier contention [7,8,9] that initiation of IB polymerization by the Cp*TiMe₂[μ-Me]B(C₆F₅)₃] system occurs by IB coordination at the Ti cation and that the head group of the polymer is Cp* TiMe₂–CH₂C(CH₃)₂–. The circumstantial arguments for this proposition cannot be accepted unless evidence is presented for the existence of the Ti-CH₂ bond.

CATIONIC POLYMERIZATION: INDUSTRIAL PROCESSES AND APPLICATIONS

R. J. PAZUR, H.-H. GREVE
Bayer Inc. Rubber Division
P.O. Box 3001, 1265 Vidal St. South, Sarnia, ON, N7T 7M2 Canada

Abstract. The wide range of commercial materials that are produced by industrial cationic polymerization processes is presented. Vinyl or ring-opening cationic polymerization methods enable the synthesis of unique polymer structures that are utilized in an array of significant commercial applications. A brief but concise description of the cationic mechanism is necessary in order to enhance current industrial processes as well as to generate novel and useful polymers possessing favorable price/performance characteristics.

1. Introduction

A general overview of the most important commercial polymers produced via cationic polymerization will be undertaken with particular emphasis on the wide scope of applications of these useful materials. The cationic polymerization method provides a unique way to synthesize novel polymer structures possessing important and diverse commercial applications. Cationically produced industrial polymers are obtained by polymerization of either alkene or heterocyclic-based monomers. Time will be spent on reviewing the conventional processes, like butyl rubber, but up-to-date improvements, new products and new trends will also be entertained in this presentation. Given that there exists more than 36 commercial polymers and copolymers, an attempt has been made to only present materials providing important and diverse commercial applications. For the sake of convenience, this treatise will describe each cationic system in terms of the general reaction conditions, the nature and important properties of the products as well as their major end use applications. The main thrust of this summary has been taken from the recent comprehensive synopsis of the field by Vairon and Spassky [1] and the reader is highly recommended to refer to their overview of the field for more detailed accounts of each cationically produced polymer.

The production capacities in kilometric tons for 94-95 are illustrated in Table 1 for all cationically prepared commercially available materials. A good estimate of the market

13

J.E. Puskas et al. (eds.), Ionic Polymerizations and Related Processes, 13–29.

share of cationically produced polymers is difficult to obtain. This is due to small production quantities, to companies withholding production capacity information and to the fact that some cationic polymers can be produced by other means (i.e. polystyrene can be produced cationically, although industrially it is more efficient to produce it radically or anionically). Vinyl polymerizations proceed through carbocation formation of alkene based monomers. As a whole, poly(butene)s, isobutylene based rubbers and hydrocarbon resins are produced in similar quantities, ca. 750 000 metric tons/yr. Poly(butene)s, poly(isobutene)s, butyl elastomers, indene/coumarone and poly(terpene) resins as well as poly(vinyl ether)s will all be looked at in depth. The second series of materials in Table 1 are produced by a cationic ring-opening heterocyclic polymerization. Epoxy and acetal resins both hold a large part of this market. In addition to these two polymers, this paper will look at the manufacture of poly(epichlorhydrin), poly(tetramethylene glycol), poly (alkylene imine) and silicones. The approximate 3.5 million metric tons per year figure, representing about 3% of overall synthetic polymer production, is a good working estimate of the total amount of cationically produced polymers.

TABLE 1. Production capacities of industrial cationic polymers for 1994-95.
(*estimated range)

POLYMER		METRIC TONS (10^3)
Vinyl polymerizations		
Poly(butene)s		650
Poly(isobutene)s: low and high Mw		100
Isobutene based elastomers: butyl, halobutyls		760
Hydrocarbon resins	- C9 + indene/coumarone	350
	- C5 + dicyclopentadiene	274
	- vinyl aromatics	128
	- poly(terpenes)	25-30
Poly vinyl ethers		20-25
Ring-opening polymerizations		
Poly(epichlorohydrin)		10
Epoxy resins		500
Poly(tetramethylene glycol)		150
Poly(ether-b-amide)		6
Acetal resins - copolymers		330
Poly(alkylene imine)s		30-50*
Poly(phosphazene)		25-50
Silicones		60-90
Total		**3 500**

2. Vinyl Polymerizations

2.1. POLY(BUTENES)

The general name poly(butene)s designates a family of copolymers obtained from unsaturated hydrocarbons of petroleum refinery C4 fractions. After butadiene removal, the butane-butene stream consists of isobutene, 1-butene, 2-butenes, isobutane and n-butane as illustrated in Figure 1. The isobutene and butenes are cationically polymerized in their saturated equivalents by way of Lewis acids like $AlCl_3$ or BF_3, which can both initiate the reaction. Coinitiators such as H_2O or HCl for $AlCl_3$ and ethanol for BF_3 must be employed. Industrial processes operate under variable temperature and pressure conditions. The final molecular weight depends on reaction temperature, initiator/coinitiator concentrations and on the composition of the incoming C4 stream.

15 - 45 % isobutene,		
15 - 25 % 1-butene,	-10 to 30 °C (0.4 - 6 bar)	
10 - 20 % Z/E 2-butenes,	⟶	Viscous clear liquid
3 - 40 % isobutane,	$H_2O/HCl/AlCl_3$	($300 < M_n < 2500$)
10 - 15 % n-butane	or ethanol/BF_3	

Figure 1. General reaction scheme for the production of poly(butenes).

The BF_3 based initiator system is favored in the case where a higher amount of terminal unsaturations in the form of reactive vinylidenes is desired. After polymerization, one obtains a viscous clear liquid possessing a number average molecular weight between 300 and 2500. These liquids are non-staining, stable at high shear and demonstrate excellent lubricating ability. Accordingly, they are used as additives in lubricants, fuel, caulks, food wrapping, adhesives, coatings, personal care products, concrete sealers, rubber modifiers and plasticizers.

2.2. POLY(BUTENYLSUCCINIC ANHYDRIDE)

As stated earlier in the polymerization of poly(butene)s, BF_3 based initiations give rise to the formation of terminal vinylidene groups that can be further reacted in order to generate novel materials. In this example, it is possible to end-cap the poly(butene) chain (Figure 2) by reacting the end-chain alkene with a polar group such as maleic anhydride producing poly(butenylsuccinic anhydride) (PIBSA). Good rates and yields are possible thermally when the concentration of terminal vinylidenes is high.

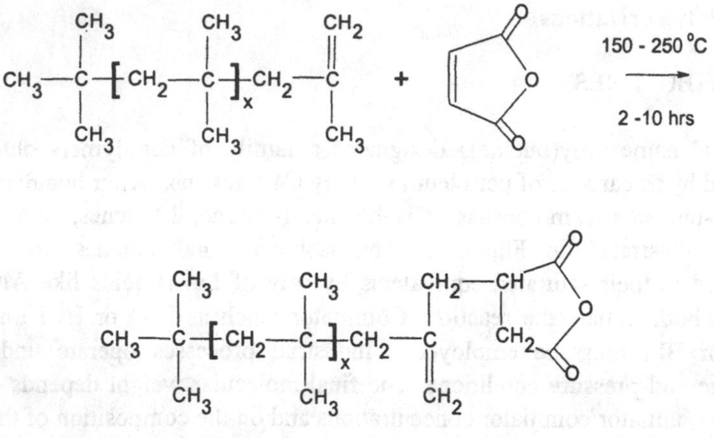

Figure 2. The addition reaction of maleic anhydride produces
poly(butenylsuccinic anhydride).

The addition of a polar end group to the hydrocarbon chain provides unique properties
not observed in poly(butene)s alone. PIBSA is used exclusively as a lubricating oil or
as a fuel active additive with the added advantage that it acts as a corrosion inhibitor. In
addition, PIBSA can be further reacted with oligiopolyalkylenimines to form
poly(butenylsuccinimide)s, which are ash free dispersants used to limit sludge
formation in motor oils.

2.3. POLY(ISOBUTENE)S

In contrast to poly(butene)s, poly(isobutene)s, PIBs, are produced from pure grades of
isobutene monomer. From Table 2, it is observed that the resulting PIB molecular
weight depends directly on the isobutene purity, the solvent/catalyst system as well as
the polymerization temperature. Higher isobutene purites, more polar solvents and
lower polymerization temperatures will systematically increase the molecular weight
thereby producing a solid elastomeric material.

TABLE 2. Summary of poly(isobutene)s commercially available.

ISOBUTENE PURITY (%)	CATALYST/SOLVENT	TEMP (°C)	PRODUCT
≈ 99.5	BF_3/CH_3OH, isobutane	-10	liquid polymers
	$AlCl_3$, hexane	- 10	$M_n < 2300$
> 99.5	BF_3/CH_3OH, isobutane	- 25 to - 40	soft resins
	$AlCl_3$, hexane		$M_n \approx 10^5$
> 99.7	BF_3, ethylene	- 103	HMW elastomer
	$AlCl_3$, MeCl	- 95	$M_n > 10^5$

There are two main industrial processes currently in use with each one depending on the choice of catalyst for the polymerization. BASF uses preferentially the BF_3 catalyst system; a batch polymerization for low to medium molecular weight while a unique belt process is employed for the production of high molecular weight PIB. Exxon uses the $AlCl_3$ catalyst system in a continuous slurry process akin to the manufacture of butyl rubber. Low M_n PIBs are without a doubt the most important product with close to 100 000 metric tons produced annually. Its end use applications are quite similar to those of poly(butene)s. Medium to high M_n PIBs are used as sealants, adhesives, flexibility improvers for waxes/bitumen and as impact additives for thermoplastics.

2.4. ISOBUTENE BASED ELASTOMERS

2.4.1. *Butyl Rubber*
Butyl rubbers are linear, gel-free, random copolymers based on isobutene and relatively small amounts of isoprene [2]. The monomers are introduced in a methyl chloride solution containing dissolved aluminum chloride (see Figure 3). Minute quantities of water or HCl activate the Lewis acid catalyst. The solution polymerization takes place in a continuous reactor at temperatures below – 90 °C. Liquified ethylene is used to remove the heat generated during the vigorous and rapid polymerization reaction. The butyl rubber produced is insoluble in MeCl thereby forming a suspension, which, upon introduction into hot water, causes the organic solvent and unreacted monomers to flash off. At this stage, an antiagglomerant (stearic acid/zinc stearate mixture) is added as well as sodium hydroxide to neutralize the catalyst. The finished and dried product is sold in the form of bales. The individual grades differ with respect to % unsaturations, mooney viscosity and stabilization.

The resulting butyl rubber polymer structure is very much like that of high molecular weight PIB. The polymer chain is flexible leading to rubber-like properties (T_g = -72 °C). The methyl groups along the backbone interfere mechanically with each other reducing the speed with which the molecules respond to deformation. This phenomenon explains its low rebound resilience and excellent damping characteristics.

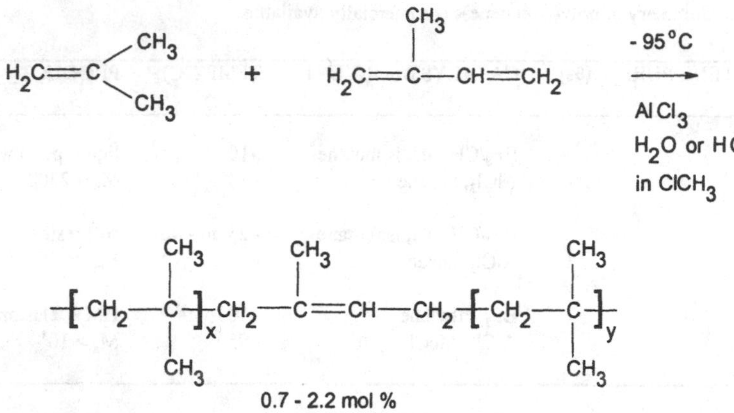

Figure 3. Production of isobutene/isoprene copolymer (butyl rubber).

The low mobility of the chain segments gives butyl its well-known impermeability to small molecules. The low amount of unsaturation in polymer is responsible for its excellent resistance to oxygen/ozone. It can be vulcanized using sulfur based cure systems. The uses take advantage of butyl's unique properties: inner tubes, curing bladders, pharmaceutical stoppers, chewing gum, auto parts, railway pads, wires and cables, tank linings (owing to chemical inertness), hose, belting, packaging film and adhesives.

2.4.2. Halobutyls

Bromobutyl and chlorobutyl are the most important commercial butyl rubber derivatives [2]. In the production of halobutyl rubber illustrated in Figure 4, the halogen is passed through a butyl rubber solution in hexane at temperatures between 40 to 65 °C. Thorough mixing and careful control of halogenation ensures that only the small amount of unsaturation reacts with the halogen by way of a substitution at the allylic position along the polymer backbone. The actual mechanism of substitution is believed to pass by a cationic cyclic intermediate where the halogen attaches itself to the isoprenyl double bond. Thereafter, a series of microstructures, the proportion of which depending on the type of halogen, are produced with the main one being the formation of exomethylene units (70 - 90%) with smaller quantities of endomethylenes (10 - 20%) and minute amounts of the addition product. A stearate based stabilizer and/or epoxidized soybean oil is added after halogenation to mop up any acid (HX) released by the chain.

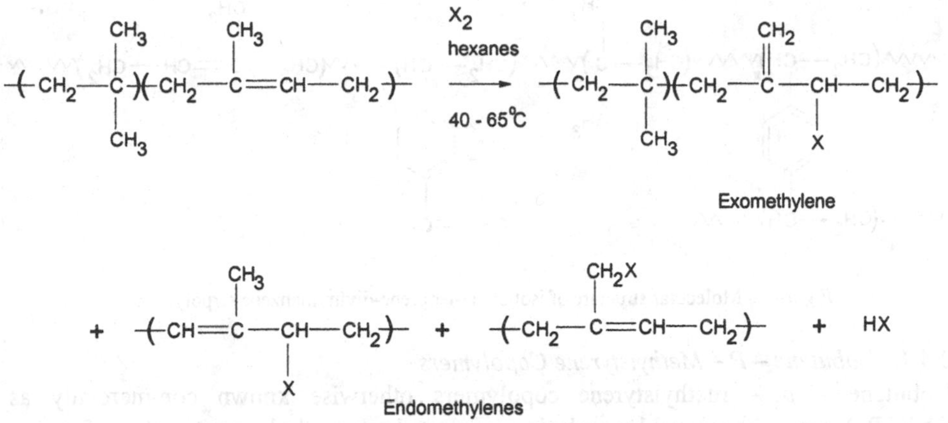

Figure 4. Halogenation reaction of butyl rubber produces halobutyl.

In opposition to butyl rubber curing behaviour, the presence of allylic halogens allows crosslinking by metal oxides such as ZnO thus enhancing the rate of vulcanization over that of regular butyl. Under such vulcanization conditions, only carbon to carbon linkages are formed. Bromobutyl is more reactive than chlorobutyl and can be successfully crosslinked using sulfur alone. A general improvement in compatibility with general purpose rubber is also observed upon halogenating regular butyl rubber. Halobutyls present a wide range of end-use applications: innerliners, innertubes, tire sidewalls, pharmaceutical stoppers, seals and other moulded articles.

2.4.3 *Isobutene-Isoprene-Divinylbenzene Terpolymers*
These terpolymers are prepared like butyl rubber with the major exception that a third monomer, bifunctional divinylbenzene, DVB, is added during the polymerization in the reactor [2]. Its polymer structure is presented in Figure 5. The DVB unit is responsible for precrosslinking during synthesis causing the polymer to become partially insoluble. An increase in green strength, improved dimensional stability, reduced cold flow and enhanced elastic behaviour constitute the main advantages of the terpolymer over butyl rubber. In addition to sulfur/sulfur donor systems, it can be cured by peroxides on the unreacted DVB groups whereas butyl rubber would degrade under the same cure conditions. Isobutene-isoprene-divinylbenzene terpolymers find their major uses in seals and adhesives as well as coatings, cements, tapes for covering pipes and electrical plugs.

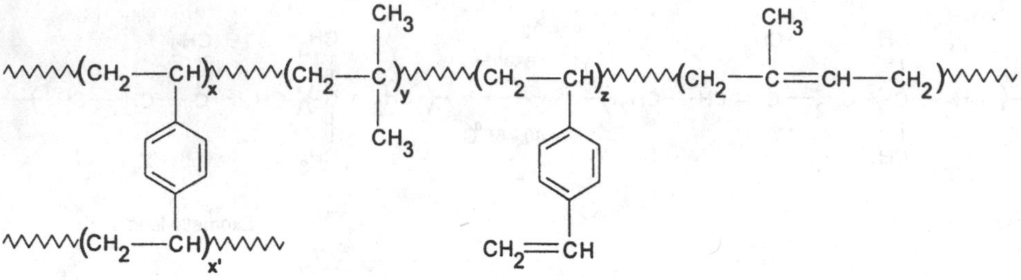

Figure 5. Molecular structure of isobutene-isoprene-divinylbenzene terpolymers.

2.4.4. *Isobutene – P – Methylstyrene Copolymers*

Isobutene – p – methylstyrene copolymers otherwise known commercially as EXXPRO, are polymerized by replacing isoprene by p-methylstyrene and, performing the slurry based $AlCl_3$ initiated polymerization in methyl chloride (see Figure 6). Elastomers are obtained for low p-methylstyrene concentrations (T_g = -45 °C for < 20 wt%) whereas hard glassy theromoplastics (T_g > 100 °C) are achieved at higher p-methylstyrene contents.

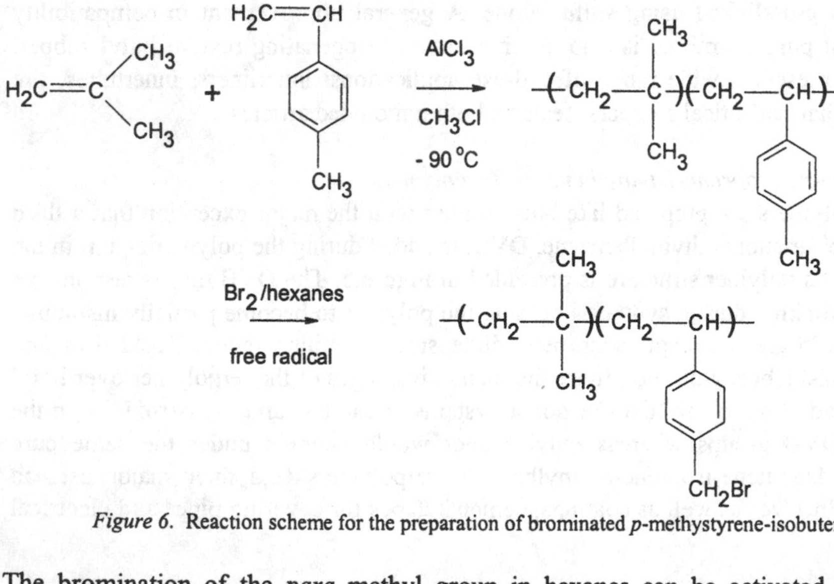

Figure 6. Reaction scheme for the preparation of brominated p-methystyrene-isobutene.

The bromination of the *para* methyl group in hexanes can be activated thermally, photolytically or by use of radical initiator. It is also noted that nucleophilic substitution is possible at the benzylic halide for the addition of other functional groups and the production of new materials. The absence of backbone unsaturation means that the random copolymer possesses an excellent resistance to ozone. As in halobutyls, vulcanization by metal oxides generates thermally stable C-C crosslinks imparting excellent heat aging and fatigue characteristics to the rubber.

Table 3 summarizes 1997 world wide production capacities for butyl rubbers which includes a recent 50 000 ton capacity shutdown at Exxon's Baytown plant. It is noteworthy that Bayer and Exxon corporations dominate in total production capacities as they generate over 90% of the world's butyl and halobutyl rubbers. A proposed synthetic butyl rubber plant in China possessing a production capacity of 30 000 metric tons has not been included in these statistics [3].

TABLE 3. Production Capacities of butyl/halobutyl rubbers in metric tons as of 1997 [4] (JSR = Japan Synthetic Rubber Co.).

PRODUCER	LOCATION	IIR + XIIR
Bayer AG	Sarnia, ON	120 000
	Antwerp, BELG	110 000
Exxon Chem. Co.	Baytown, TX	71 000
	Baton Rouge, LA	125 000
	N.D. de Gravenchon, FR	56 000
	Fawley, UK	75 000
Exxon + JSR	Kawasaki, JP	80 000
	Kashima, JP	25 000
Russian	Nizhnekamsk, RU	50 000
	Togliatti, RU	35 000
Totals		747 000

2.5. HYDROCARBON RESINS

2.5.1 Indene-Coumarone

Hydrocarbon resins are comprised of indene-coumarone and petroleum based resins. They generally have a molecular weight less than 2000 and the resulting resins range from clear viscous liquids to dark brittle solids. Indene-coumarone or coal tar resins are complex copolymers of which indene (50%) is the major component. The initial mixture of monomers is presented in the reaction scheme of Figure 7. The polymerization is performed in aromatic naptha in the presence of a Lewis acid system. The product type (composition, molecular weight, properties) depends directly on feedstock content, the type of initiator and the temperature of the operation. These resins have no commercial applications alone and must be used to modify other materials as processing aids, extenders and plasticizers for resins/rubbers. They are also used in paints, coatings, adhesives and rubber compounding.

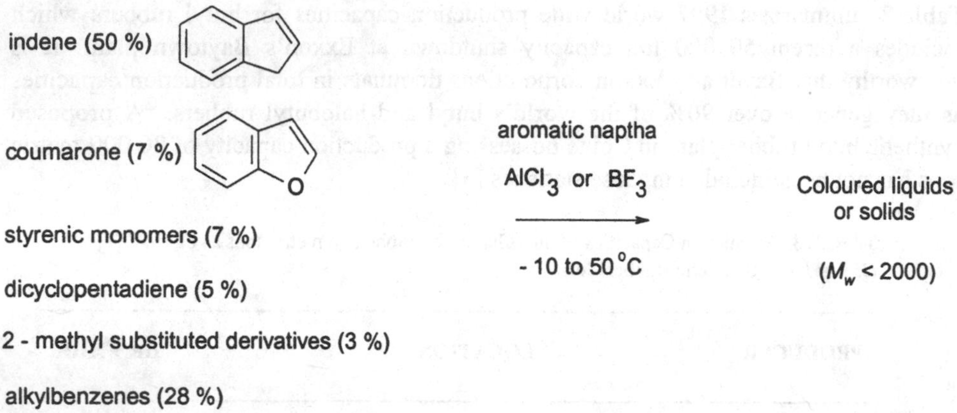

indene (50 %)

coumarone (7 %)

styrenic monomers (7 %)

dicyclopentadiene (5 %)

2 - methyl substituted derivatives (3 %)

alkylbenzenes (28 %)

aromatic naptha

$AlCl_3$ or BF_3
→

- 10 to 50 °C

Coloured liquids
or solids

$(M_w < 2000)$

Figure 7. Preparation of indene-coumarone hydrocarbon resins.

2.5.2. Poly(terpene) Resins

Commercial poly(terpene) resins are synthesized by cationic polymerization of a) Beta and alpha pinenes from turpentine, b) *d, l* - limonene or dipentene from kraft-paper manufacture, or *d* - limonene extracted alone from citrus peels. In general, the mechanisms of polymerization and resulting chain structures are quite complex. The cationic polymerization of beta pinene as seen in Figure 8, will serve as an example of the preparation of these resins.

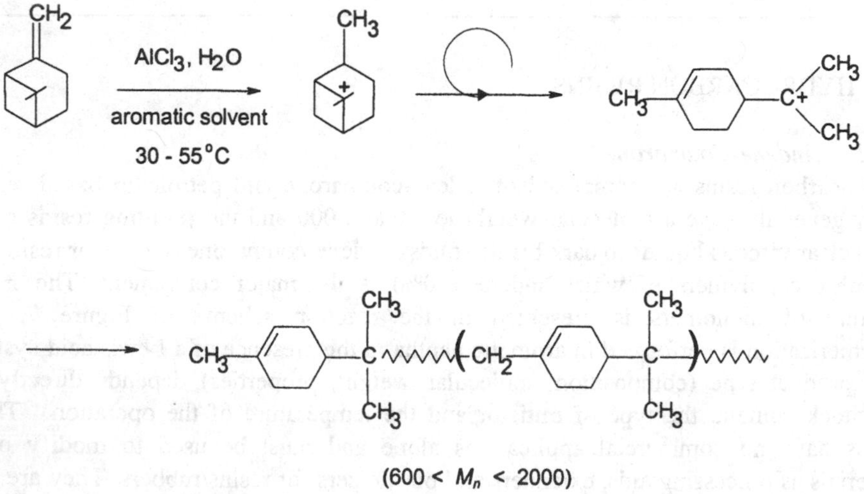

CH_2

$AlCl_3$, H_2O
→
aromatic solvent
30 - 55°C

CH_3

CH_3

$(600 < M_n < 2000)$

Figure 8. Reaction scheme for the preparation of β- pinene.

The polymerization reaction takes place in an aromatic solvent like mixed xylenes from 30 to 55 °C in the presence of an $AlCl_3/H_2O$ initiation system. The purified monomer (72 to 95 %) is mixed in a reactor together with the solvent and catalyst and then stirred between 30 to 60 min. Once polymerization is complete, the catalyst is deactivated by hydrolysis and HCl is neutralized by alkaline washes. Next, the organic solution is dried and solvent removed from the resin by distillation. Chemically, the protonic initiator creates a cyclic tertiary carbenium ion, which rearranges into a more stable species. Successive addition-isomerization leads to a chain structure consisting of isobutyl and cyclohexenyl units. The observed M_n is low due to important chain transfer and termination processes. The final resin is light colored, soluble in most solvents, compatible with polymers and resins and possesses outstanding tackifying characteristics. It is primarily used as a light color tackifier in the production of pressure-sensitive and hot melt adhesives.

2.6. POLY(VINYL ETHER)S

Polymerization of vinyl ether monomers by radical initiation is slow and produces oligomers whereas their cationic polymerization initiated by Lewis acids is much more efficient and preferred for homopolymer production. Vinyl ether monomers are made by the Reppe method, which reacts acetylene in basic conditions with an alcohol (Figure 9).

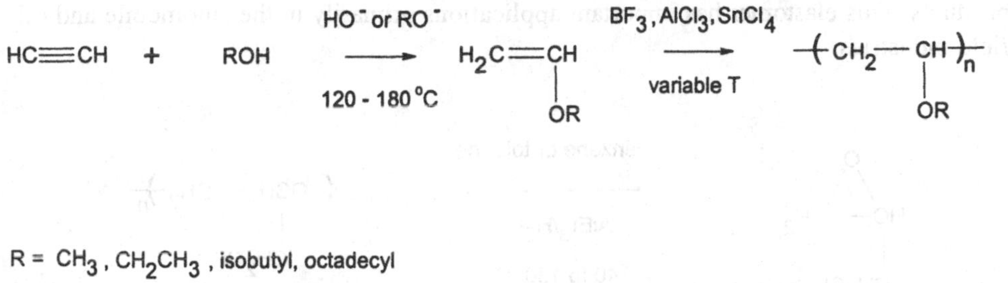

R = CH_3, CH_2CH_3, isobutyl, octadecyl

Figure 9. Preparation of vinyl ether monomers and their polymerization.

Thereafter, a cationic polymerization is performed preferentially in the bulk using Lewis acids at variable temperatures. The molecular weight of the end polymer depends essentially on the purity/dryness of medium, the initiator concentration and the reaction temperature. Higher temperatures generate lower molecular weight materials. The homopolymers are amorphous and atactic with the chain backbone containing a majority of isotactic triads. PVEs exist as viscous oils, soft and tacky resins and elastomeric solids. They are used as non-migrating tackifiers in adhesives, as viscosity index improvers in lubricants and as plasticizers in coatings, films and elastomers.

3. Ring-opening Polymerizations

3.1. POLY(EPICHLOROHYDRIN)

In the case of poly(ethers) such as poly(epichlorohydrin), the presence of oxygen atoms along the chain skeleton contributes greatly to flexibility and thus enhances elastomeric behavior. The cationic polymerization of epichlorohydrin takes places in a benzene or toluene based solution, between 40 to 130 °C in the presence of a so-called Vandenberg catalyst, $AlEt_3/H_2O$ (Figure 10). Stabilizers such as hindered phenols are added prior to polymer isolation by aqueous coagulation. The final homopolymer, consisting of more than 97% head to tail sequences, is completely atactic and amorphous possessing an average molecular weight around 450 000 g/mol. Copolymers with the comonomers being ethylene oxide or allylglycidylether are also synthesized in addition to a terpolymer of the former two together with epichlorohydrin. The approximate 38% chlorine content renders the polymer fuel resistance and promotes flame retardancy. The polar chloromethyl group is the crosslinking site during vulcanization. The polymer also possesses good ozone and heat resistance as well as excellent resistance to vapour permeation. Chemical modification by nucleophilic substitution at the chloromethyl group is possible leading to the production of a wide array of new products. This elastomer has important applications primarily in the automobile and oil field industries.

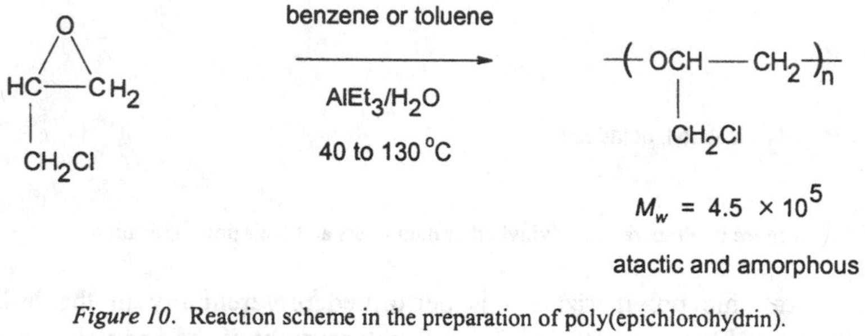

Figure 10. Reaction scheme in the preparation of poly(epichlorohydrin).

3.2. EPOXY RESINS – CURING REACTIONS

Epoxy resins contain epoxy or oxirane groups on an aliphatic, cylcoaliphatic or aromatic backbone. The reaction of these rings by curing agents leads to crosslinked insoluble materials. Crosslinking can be carried out cationically by way of Lewis acid complexes such as boron trifluoride monoethylamine, $BF_3NH_2C_2H_5$ at temperatures from 80 to 100 °C. A more rapidly growing way of crosslinking involves a photoinitiated cationic cure. In this case, one uses aryldiazonium (ArN_2X), aryliodonium, triarylsulfium (Ar_3SX) or ferrocenium salts. In the example of Figure 11, a triarylsulfium salt (not shown) will produce a strong acid, HPF_6, upon UV radiation.

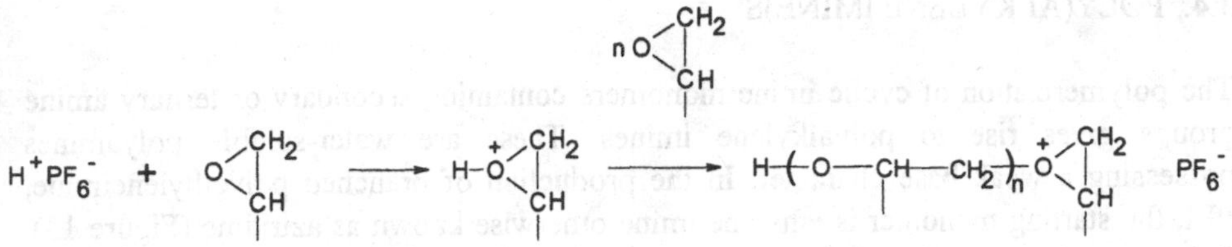

Figure 11. Crosslinking reaction in epoxy resins.

Curing of the epoxy resin takes place through an oxonium intermediate. Photoinitiated cationic curing has the advantage of lower volume shrinkage and oxygen inhibition as compared to radically induced crosslinking. The reaction rate is quite high. The most important area of application for this method is in surface coatings as well as photoresists and printing plates for silk screen printing inks.

3.3. POLY(TETRAMETHYLENE GLYCOL)

In this case, the starting monomer is tetrahydrofuran, a cyclic ether that is produced via the acetylene-formaldehyde route. Fluorosulfuric acid is used to initiate the cationic polymerization shown in Figure 12. The primary polymerization products (not illustrated) are poly(tetramethylene ether) chains possessing sulfate ester groups along the backbone. The sulfate esters are hydrolyzed and the acid is removed by a water extraction, which removes at the same time the shorter glycol chains thereby narrowing the molecular weight distribution of the recovered polymer. Poly(tetramethylene glycol)s, PTMEGs, have excellent elastomeric properties as well as very good fungus and microbial resistance. However, like other monomeric ethers, they are sensitive to oxidation and thermal degradation and must be protected by addition of antioxidants. They are also hydroscopic and must be stored in closed containers under nitrogen atmosphere. PTMEGs are useful as the soft segment in the production of elastomers such as, polyesters and polyamides. Polyurethanes are the most important use of PTMEGs as they find their applications in packaging, tubing, rollers, wheels, automotive parts, cable coating and encapsulating compounds. PTMEG prepared elastomers are beneficial in medical applications on account of their biocompatability and hydrolytic stability: encapsulants for pacemakers, catheter tubing, components for artificial heart devices.

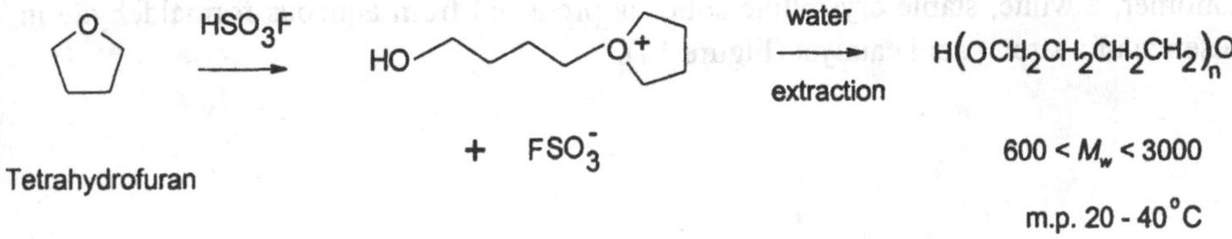

Figure 12. Reaction scheme in the production of poly(tetramethylene glycol)s.

3.4. POLY(ALKYLENE IMINE)S

The polymerization of cyclic imine monomers containing secondary or tertiary amine groups gives rise to polyalkylene imines. These are water-soluble polyamines possessing a weak base character. In the production of branched polyethyleneimine, PEI, the starting monomer is ethylene imine otherwise known as aziridine (Figure 13). Polymerization is initiated by protonic acids, Lewis acids or onium salts. In this particular case, proton transfer from the initial active species together with tertiary amine termination cause the formation of branched structures producing a final polymer structure in which the ratio between primary, secondary and tertiary amine groups is about 1:2:1. PEI has unique colloid chemical properties due to its high affinity for anionic dissolved material. Accordingly, it is used as a flocculent of anionic colloids in the manufacture of paper and board. It enhances dewatering of the pulp and also improves retention of additives and fillers during paper processing. This highly branched macromolecule also has applications in the areas of adhesives and coatings.

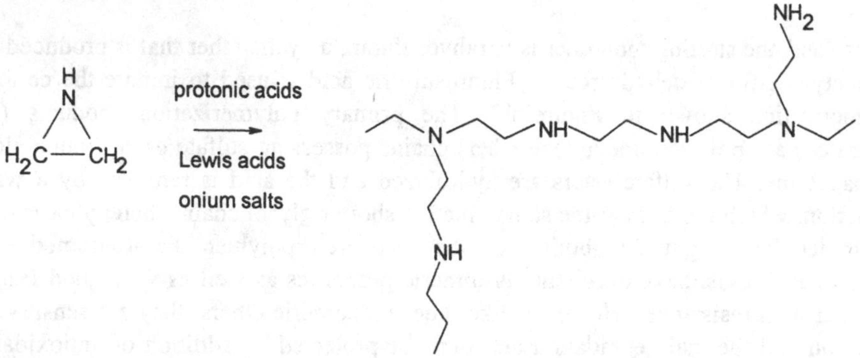

Figure 13. Preparation of branched poly(ethylene imine) from aziridine.

3.5. POLYACETALS– POLY(OXYMETHYLENE ETHYLENE) COPOLYMERS

Acetal resins are high molecular weight polymers and copolymers derived from formaldehyde. Homopolymers are produced by the anionic polymerization of formaldehyde whereas copolymers are obtained preferentially by the cationic polymerization of trioxane with small quantities of cyclic ethers or esters. Trioxane monomer, a white, stable crystalline solid, is produced from aqueous formaldehyde in presence of a strong acid catalyst (Figure 14).

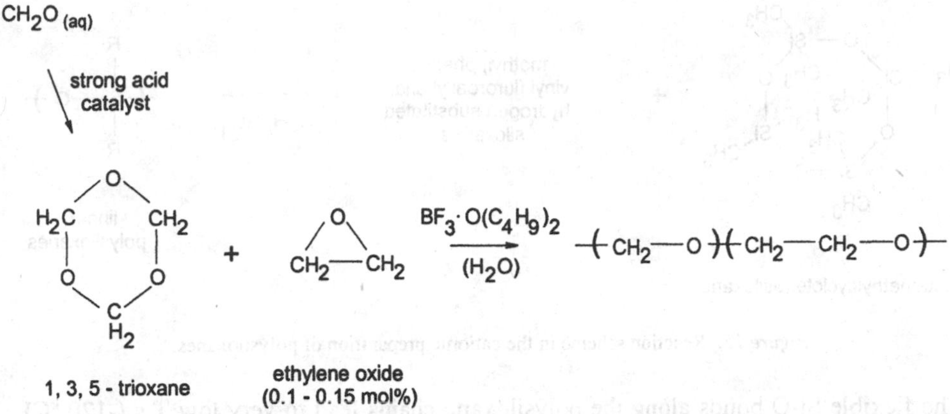

CH$_2$O $_{(aq)}$

strong acid
catalyst

H$_2$C $\diagdown$ O $\diagup$ CH$_2$ +   O / CH$_2$—CH$_2$ $\xrightarrow{\text{BF}_3 \cdot \text{O(C}_4\text{H}_9)_2}{(\text{H}_2\text{O})}$ $+$CH$_2$—O$)($CH$_2$—CH$_2$—O$)+$

1, 3, 5 - trioxane

ethylene oxide
(0.1 - 0.15 mol%)

Figure 14. Reaction scheme for the preparation of poly(oxymethylene ethylene) copolymers.

Copolymerization is conducted thermally (70 °C) in the bulk producing a rapid polymerization reaction. Cationic initiators such as BF$_3$ dibutyl etherate can be used with small amounts of water as a coinitiator. The raw copolymer is heated to remove unstable terminal hemiacetal and formal groups. A random copolymer is produced typically with 95% or more oxymethylene units. Acetal resins are semi crystalline possessing the physical properties that crystallinity confers: high toughness, hardness, rigidity and stiffness. On account of their favourable physical properties, they can be used to replace metals and alloys in many areas; automotive industry (fuel pumps, gauges and caps), domestic and sanitary equipment, mechanical engineering applications (valves and pistons), machinery (conveyors and gears) and the electrical and electronic industry. Blends with polyurethanes generate impact-resistant poly(oxymethylene)s. These versatile copolymers can also be superdrawn in order to form fibres for the textile industry.

3.6. SILICONES

Although the most important industrial processes in silicone technology make use of anionic polymerization, cationic methods have been demonstrated to play a significant role (10 to 15%) in their total production. Under these circumstances, the most usual starting cyclic material is octamethylcyclooctetrasiloxane which is copolymerized with other substituted siloxanes (see Figure 15), in the presence of strong acids or inhomogeneous catalysts like clay activated by sulfuric acid. The actual mechanism of cationic polymerization is quite complex and is not illustrated. Materials ranging from silicone fluids to elastomers are created depending on the R composition and the molecular weight.

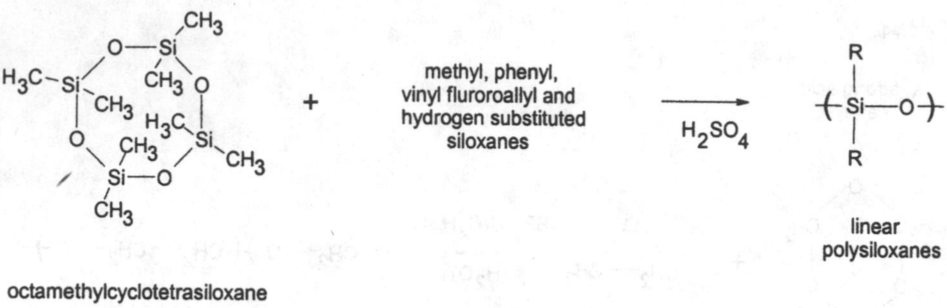

octamethylcyclotetrasiloxane

+ methyl, phenyl, vinyl flurorallyl and hydrogen substituted siloxanes $\xrightarrow{H_2SO_4}$ linear polysiloxanes

Figure 15. Reaction scheme in the cationic preparation of polysiloxanes.

The flexible Si-O bonds along the polysiloxane chains lead to very low T_gs (-120 °C). With respect to physical properties, these materials are stable at high and low temperatures, hydrophobic, act as lubricants, chemically inert, soluble to gases, surface active and are insulators on account of their dielectric properties. Their major uses are found in the automotive and aerospatial industries. More specifically, they are used as refrigerants, heat transfer media, waterproofing agents, protective coatings, antifoaming agents, antiflocculents, lubricants, paint additives, dielectric coolants and as additives in the formulation of elastomers.

4. Conclusions

This survey has shown that there exist a large number of cationically prepared industrial polymers possessing favourable price/performance characteristics as well as important and quite diverse applications. The cationic method is often the only route for the production of these materials, for example, isobutylene can only be polymerized by this route for the production of PIBs and butyl rubber.

It stands to reason that a better understanding of the cationic processes is needed at the fundamental level. Unlike stereoregular or living polymerizations, cationic-based mechanisms are generally less understood and need to be continuously studied at the scientific level with the aid and cooperation of our scientific colleagues in universities and governmental research centers. More cooperation between industry and universities is needed. Only when the detailed mechanism of cationic processes are better comprehended can we hope to control and finally exploit the polymerization reaction in order to create new and desirable materials.

Finally, recent improvements and new processes based on these scientific approaches have been bearing fruit with the advent of new polymer structures like the BF_3-based poly(butene)s with reactive exomethylene chain ends and new benzyl halide based halobutyl rubbers to name a few. The use of new comonomers and catalysts together with innovative living cationic techniques increases the chances for the successful commercial development of more high performance polymers in the near future.

5. References

1. Vairon, J-P. and Spassky, N. (1996) Industrial Cationic Polymerizations: An Overview, in K. Matyjaszewski (ed.), *Cationic Polymerizations: Mechanisms, Synthesis and Applications*, Marcel Dekker, New York, pp. 683-750.
2. Puskas, J.E., Wilson, G., and Duffy, J. (1998) *Synthesis of Butyl Rubber by Cationic Polymerization.* *Ullman's Encyclopedia of Industrial Chemistry,* Bender, H., Berghus, K., Friemann, H., Harmsworth, N., Humme, G., Kempermann, T., Italiaander, E.T., Kuhn-Grund, E., von Langenthal, W., Mezger, M., Pieroth, M., Rohde, E., Steinbach, H.-H., Sumner, A., Thomas, H.D., Warth, M. (1993) Polysar Butyl/Bromobutyl/Chlorobutyl in Th. Kempermann, S. Koch and J. Sumner (eds.), *Bayer Manual for the Rubber Industry*, Bayer AG Rubber Business Group, Leverkusen, pp. 213-235.
3. (1998) *Eur. Rub. J.*, **180**, pp. 10.
4. (1998) *Worldwide Rubber Statistics*, International Institute of Synthetic Rubber Producers, Inc..

NEW BRANCHED POLYISOBUTENES AND BUTYL RUBBERS BY THE INIMER METHOD

G. LANGSTEIN, W. OBRECHT
Bayer AG, Zentrale Forschung, 51368 Leverkusen, Germany
J. E. PUSKAS
The University of Western Ontario, London, ON, N6A 5B9, Canada
O. NUYKEN, M. GRASMÜLLER
Technische Universität München, 85747 Garching, Germany
K. WEISS
Universität Bayreuth, 95440 Bayreuth, Germany

Abstract. This paper reviews the synthesis of branched and hyperbranched polymers from an industrial point of view. The synthesis of branched polyisobutene and butyl rubber using the *inimer* method is presented. Branched copolymers were prepared in solution and bulk, using p-chloromethylstyrene (vinylbenzyl chloride) as *inimer* and MAO and diethylaluminum chloride as coinitiators. The polymers were characterized by SEC-viscometry.

1. Introduction

Butyl rubber, poly(isobutene-co-isoprene) is produced industrially via carbocationic polymerization at 183 K (-90 °C) in methyl chloride [1]. Isoprene is introduced as a comonomer in order to provide curing sites in the resulting polymer. The incorporation of isoprene proceeds in a *trans*-1,4-enchainment, resulting in essentially linear copolymers. Isoprene also acts as a strong chain transfer agent, and copolymerizations yield lower molar mass polymers than homopolymerizations under identical conditions. Very low polymerization temperatures are necessary to produce useful high molar mass materials; low molar mass polmers exhibit undesirably low green strength and low creep resistance. An alternative to such low temperature polymerization would be to produce lower molar mass copolymers at higher temperature, and subsequently connect these chains to produce branched copolymers. However, this is not trivial since the simple addition of a bifunctional monomer, such as divinylbenzene, results in partially crosslinked materials. A novel, alternative way of producing branched butyl rubbers is presented in this paper.

31

J.E. Puskas et al. (eds.), Ionic Polymerizations and Related Processes, 31–44.
© 1999 *Kluwer Academic Publishers. Printed in the Netherlands.*

32

2. Methods for the Synthesis of Branched Polymers

2.1. TRADITIONAL METHODS

Four principal routes are available for preparing branched polymers:

a) Grafting from
b) Grafting onto
c) Polymerization of Macromonomers
d) Linking

These methods are discussed in the literature – some examples are listed in the Reference section [2-24]. The "grafting from" method (a) is based on a polymeric initiator which has to be prepared first (Figure 1).

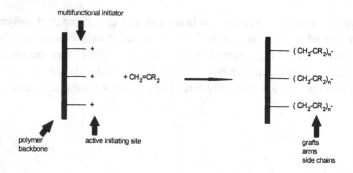

Figure 1. „Grafting from" method for the preparation of branched polymers

The "grafting from" techniques have several disadvantages for the preparation of well-defined polymers, especially in the case of carbocationic isobutene copolymerizations. To be useful as a carbocationic initiator the backbone polymers need to contain easily accessible functionalities which can be activated as a starting point for polymerization. As a consequence, this method is limited to backbones capable of generating tertiary, allylic or benzylic carbocations. If the backbone polymer is considerably shorter than the side chains, true star-branched polymers can be produced. A drawback of the „grafting from" method is the limitation to living polymerization methods; non-living polymerizations lead to a mixture of the desired product plus undesired byproducts. From the industrial point of view, the availability and price of multifunctional initiators also appear to be a limiting factor.

The "grafting onto" method (b) is based on multifunctional termination reagents to built highly branched macromolecules (Figure 2) "Grafting onto" requires that the growing polymer branches are

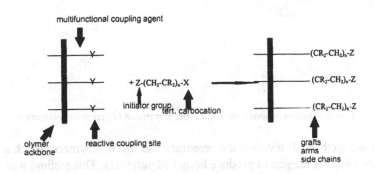

Figure 2. „Grafting onto" method for the preparation of branched polymers

terminated by a multifunctional terminating reagent. Considerable work in this field has been done by EXXON researchers [8]. They have used unsaturated (co)polymers like polybutadiene, SBR rubbers etc. as terminating agents and obtained a mixture of high molar mass star-branched fraction and a low molar mass linear fraction. This mixed product was reported to have improved processability compared to linear butyl rubbers. One problem is the solubility of the polymeric termination reagents which limits the amount of branched fractions in the final product.

The third method for the preparation of branched polymers is the polymerization of macromonomers (c). Macromonomers (macromers, Figure 3) belong to the general class of reactive oligomers [11]. The polymerizable function can be an unsaturation or a heterocycle. Various approaches have been used to prepare macromers for radical, anionic, group transfer as well as for cationic homo- and copolymerizations. Kennedy *et al.* [5] and Nyuken *et al.* [10] have described the synthesis of isobutene macromers.

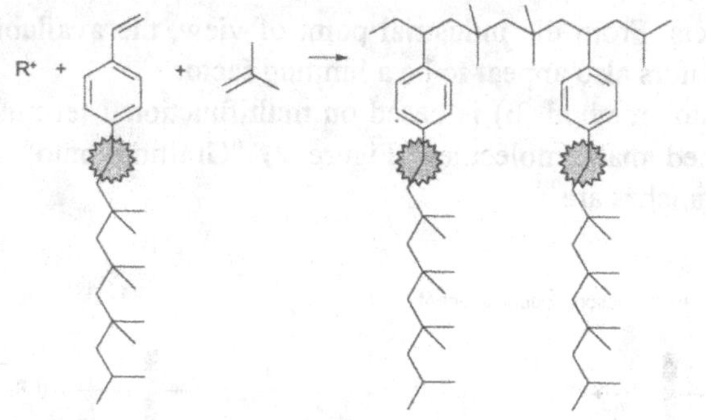

34

Figure 3. Macromonomer method for the preparation of branched polymers

The linking method (d) involves the preparation of linear polymers, which then are reacted with a linking reagent to produce branched polymers. This method was successfully used for the synthesis of star-branched polyisobutenes. In the first step living cationic polymerization of isobutene was used to generate the arms; the addition of the linking monomer (e.g. divinylbenzene) resulted in star-branched polymers [13,15]. However, this method can lead to crosslinking and gelation. In fact, partially crosslinked butyl rubber is produced commercially by the terpolymerization of isobutene, isoprene and divinylbenzene [1,25].

2. 2. INIMER METHOD

A simple method for the one step synthesis of branched polymers is to use an initiator which can also function as a monomer; the *inimer* method (Figure 4):

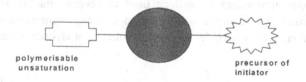

polymerisable
unsaturation

precursor of
initiator

Figure 4. Inimer

This is the extension of the so-called „self-condensing vinyl polymerization" [26] used for the preparation of dendritic structures as shown in Figure 5:

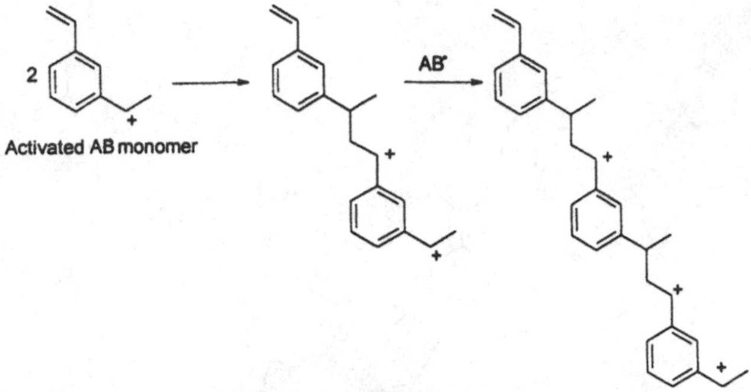

Figure 5. Self-condensing vinyl polymerization

This reaction shows the characteristics of polycondensation reactions. However, when the inimer is copolymerized with another monofunctional vinyl monomer, a chain-growth reaction occurs (Figure 6).

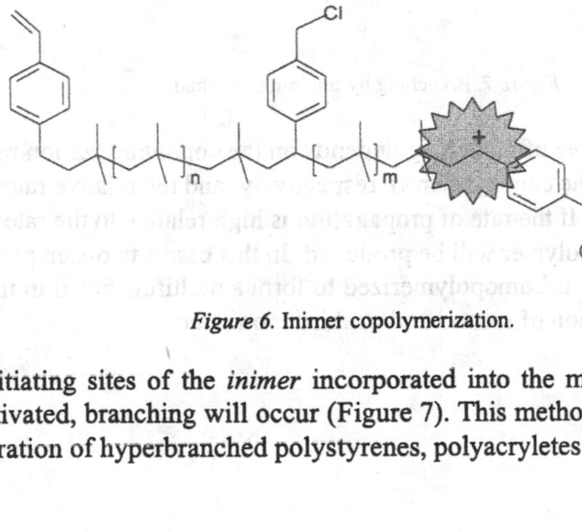

Figure 6. Inimer copolymerization.

When the initiating sites of the *inimer* incorporated into the main chain via the vinyl group are activated, branching will occur (Figure 7). This method was used successfully for the preparation of hyperbranched polystyrenes, polyacryletes and polyisobutylenes

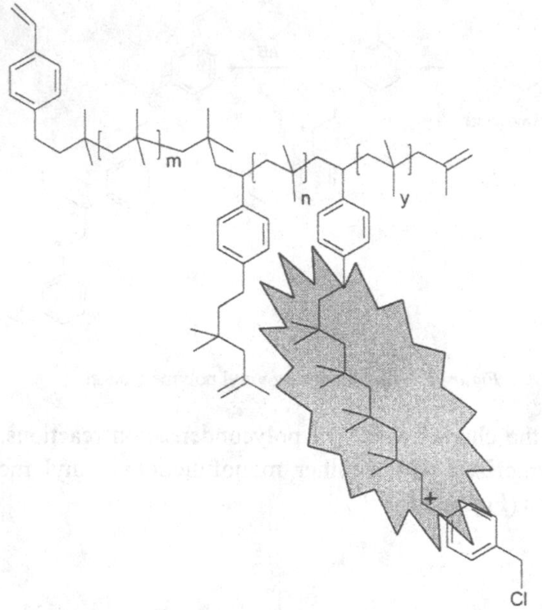

Figure 7. Branching by the inimer method.

[10, 22, 27-30]. The degree of branching depends on the copolymerization reactivity ratios of the inimer and the comonomer(s), respectively, and the relative rates of initiation to propagation. If the rate of propagation is high relative to the rate of initiation, mostly linear polymer will be produced. In this case a two-step process is preferred; first the inimer is homopolymerized to form a multifunctional initiator (Figure 8), followed by the addition of second and/or third monomer.

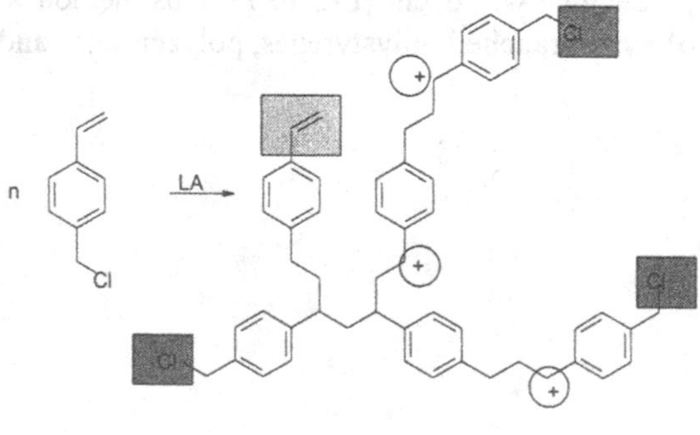

Multifunctional Initiator

Figure 8. Inimer homopolymerization.

If the initiation and propagation rates are comparable, or the initiation is faster than propagation, branching will occur in a one-step process as shown in Figure 7.

3. Synthesis of Highly Branched Polyisobutylene and Butyl Rubber

3.1. INIMER AND COINITIATOR SELECTION

From the industrial point of view, the inimer and coinitiator have to be commercially available, cost-effective compounds. For the synthesis of branched polyisobutylene and butyl rubber by the inimer method, the first functionality of such an inimer has to be active in cationic polymerization, and the second functionality has to be a precursor of an initiator. The incorporation of the inimer in the growing polymer chain depends on reactivity ratios, and initiation and propagation rates. The polymerization can be carried out as one step or two step synthesis. The two step synthesis is based on the homopolymerization of the inimer. The homopolymerization generates a multifunctional initiator which can be used in the second step. The one step synthesis is a simultaneous initiation (co)polymerization reaction of the inimer and the monomers, e.g. isobutene and isoprene.

A readily available potential inimer is p-chloromethyl styrene (vinylbenzyl chloride) Jones et al. [31] described the copolymerization of vinylbenzyl chloride with isobutene in ethyl chloride with boron trifluoride as initiator. They calculated reactivity ratios of 0.7 for vinyl benzyl chloride and 4.5 for isobutene. Powers et al.. [32] described the copolymerization of isobutene and vinyl benzyl chloride in a solution process using a mixture of hexane and methyl chloride. Kennedy et al. [33] described a process for producing isobutene vinyl benzyl chloride copolymers and their use as precursors for graft copolymers. Nuyken et al. [10] investigated the use of vinylbenzyl chloride as an initiator for the polymerization of isobutene. They obtained soluble polymers with broad molar mass distributions and higher molar masses than calculated on the basis of initiator and monomer concentrations. They employed vinyl benzyl chloride as an initiator and trimethyl aluminium/water to yield macromers. Nuyken confirmed the suggestions of Powers and Kennedy that the copolymerization of vinyl benzyl chloride with isobutene leads to complex mixtures of polyisobutylene as well as copolymers of isobutene and vinyl benzyl chloride with highly branched structures.

3.2. POLYMER SYNTHESIS

Table 1. summarizes the experimental conditions used for the synthesis of branched polymers, together with reaction time and conversion data. Polymerizations were carried out under argon in standard glass apparatus using Schlenk techniques. After charging the solvent, monomer(s) and the inimer (except in Exp # 2, when the inimer was also added continuously), the coinitiators (MAO or Et$_2$AlCl) were added continuously by a dropping funnel. The reactions were terminated by methanol, and conversion was determined gravimetrically. SEC/viscometry was performed as reported [21]; absolute molecular mass was measured by laser light scattering (LS).

The first example is a copolymerisation without isoprene; the conversion – time profile is shown in Figure 9. At the beginning there seems to be a accelerating period followed by a slow decrease of polymerization activity.

TABLE 1. Synthesis of branched polymers using vinylbenzyl chloride as inimer

Exp. #	Temp. (K)	Solvent (ml)	Inimer (mmol)	Coinitiator[a] (mmol)	Monomer(s) IB (mol)	IP (mol)	Time (min)	Conv. (%)
1	273	2000	0.1	5[b]	11.4	-	35	56
2	273	250	0.012	0.75[b]	1.34	-	10	20
3	273	250	0.012	0.75[b]	1.34	-	20	59
4	273	2000	0.1[a]	5[b]	11.4	-	40	53
5	213	-	0.1	12[b]	10.7	0.2	40	19
6	213	1000	0.1	12[c]	3.73	0.2	15	44

[a] Continuous addition; [b] MAO; [c] Et$_2$AlCl

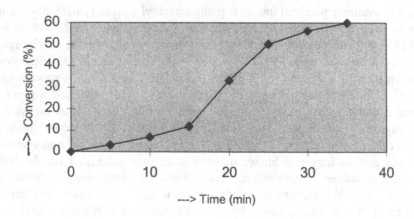

Figure 9. Conversion-time profile of isobutene-vinylbenzyl chloride copolymerization (TABLE 1, Exp. # 1)

The molecular mass increases with conversion in the beginning, with an apparent decrease beyond about 30% conversion, which maybe due to termination and/or transfer reactions (Figure 10).

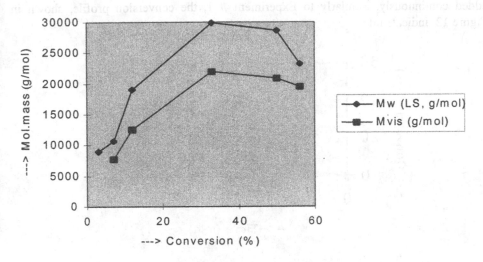

Figure 10. Molecular mass - conversion profile of isobutene-vinylbenzyl chloride
copolymerization (TABLE 1, Exp. # 1)

Experiments # 2 and 3 are similar to # 1, with shorter reaction times. The SEC traces of
the end products demonstrate molecular mass growth with time, and the multimodality
indicates branching.

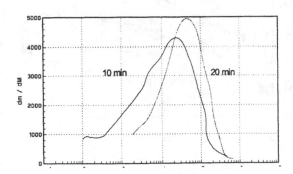

Figure 11. SEC traces of isobutene-vinylbenzyl chloride copolymers (TABLE 1, Exp. # 2 and 3)

40

Experiment # 4 represents a copolymerization where both coinitiator and inimer were added continuously. Similarly to Experiment # 1, the conversion profile, shown in Figure 12, indicate rate

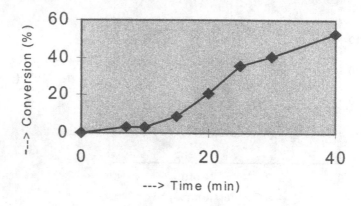

Figure 12. Coversion – time profile of isobutene-vinylbenzyl chloride copolymerization (TABLE 1, Exp. # 4)

acceleration. The molecular mass – time plot (Figure 13) shows a levelling-off at about 40 % conversion.

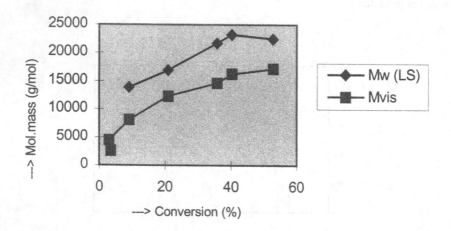

Figure 13. Molecular mass - conversion profile of isobutene-vinylbenzyl chloride copolymerization (TABLE 1, Exp. # 4)

Experiments # 5 and 6 are terpolymerizations (isobutylene, isoprene and vinylbenzyl chloride – see Figure 14) for the preparation of branched butyl rubber. They were conducted at lower temperatures

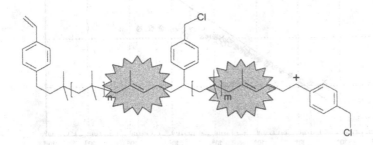

Figure 14. Isobutene-isoprene-vinylbenzyl chloride terpolymerization

to counteract the strong chain transfer effect of isoprene. Figure 15 shows the SEC traces of polymer

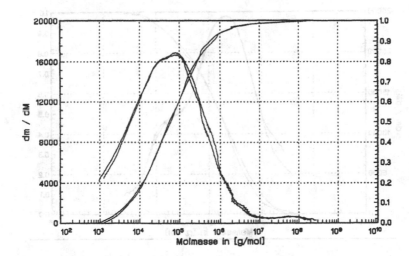

Figure 15. SEC traces of an isobutene-isoprene-vinylbenzyl chloride copolymer (TABLE 1, Exp. # 5)

5. 40% of the polymer has higher than 10^5 molecular mass. Figure 16 shows the Mark-Houwink plot for this polymer. The solid line applies to linear polyisobutylenes [34]. Deviation from the linear demonstrates branching in the high molecular mass fraction (MM > $5x10^5$ g/mol, about 20% of the total polymer). Branching is more pronounced with Et_2AlCl as indicated by the high molecular mass shoulder in the SEC trace (Figure 17). The Mark-Houwink plot for polymer # 6, presented in Figure 18, show a higher amount of branched fraction (about 30%) than for polymer # 5 (Figure 16).

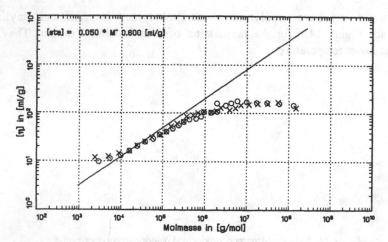

Figure 16. Mark-Houwink plot of an isobutene-isoprene-vinylbenzyl chloride
copolymer (TABLE 1, Exp. # 5).

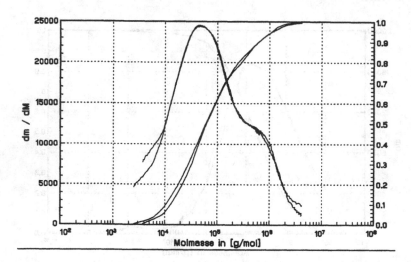

Figure 17. SEC traces of an isobutene-isoprene-vinylbenzyl chloride
copolyme*r (*TABLE 1, Exp. # 6)

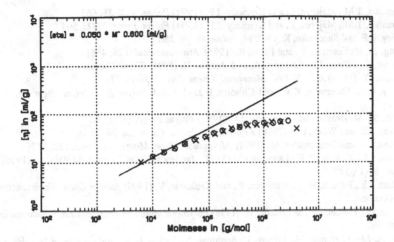

$$[\text{eta}] = 0.050 * M^{0.600} \; [\text{ml/g}]$$

Figure 18. Mark-Houwink plot of an isobutene-isoprene-vinylbenzyl chloride copolymer (TABLE 1, Exp. # 6)

Our results demonstrate that the *inimer* method is suitable for the preparation of high molecular weight branched butyls. Physical and chamical characterization of the new branched butyl polymers is in progress.

Acknowledgement. The authors would like to thank Bayer AG for permission to publish this work.

4. References

1. Puskas, J.E., Wilson, G., and Duffy, J. (1998) *Synthesis of Butyl Rubber by Cationic Polymerization. Ullman's Encyclopedia of Industrial Chemistry*
2. Plesch, H.P. (1958) *Chem. Ind (London),* 954.
3. Kennedy, J.P. (1977) *J. Appl. Polym. Sci.: Appl. Polym. Symp.* **30**, 1-11
4. Kennedy, J.P., Ross, L.R., Lackey, J.E., and Nuyken, O. (1981) *Polym. Bull* **4**, 67
5. Farona, M.F. and Kennedy, J.P. (1984) *Polymer Bulletin* **11**, 359-363.
6. Huang, K.J., Zsuga, M. and Kennedy, J.P. (1988) *Polym.Bull* **19**, 43
7. Kaszas, G., Puskas, J. E., Kennedy, J.P., and Chen, C. C. (1988) *Polym.Bull.* **20**, 413
8. Wang, H-C., Powers, K.W., and Fusco, J. V. (1989) Paper presented at a meeting of the Rubber Division, American Chemical Society, Mexico City, Mexico
9. Kennedy, J.P.and Ivan, B. (1992) *Designed Polymers by Carbocationic Macromolecular Engineering. Theory and Practice.* Hanser Publishers, Munich
10. Nuyken, O., Gruber, F., Pask, S. D., Riederer, A. and Walter, M. (1993) *Makromol. Chem.* **194**, 3415-3432.
11. Mishra, M. K. (Ed.) (1994) *Macromolecular Design: Concept and Practice*, Polymer Frontiers International, New York
12. Storey, R. F., Chisholm, B.J., and Choate, K.R. (1994) *JMS-Pure Appl.Chem.* **A31(8)**, 969

13. Marsalko, T.M., Majoros, I., and Kennedy, J.P. (1993) *Polym. Bull.* **31**, 665
14. Marsalko, T.M., Majoros, I., and Kennedy, J.P. (1994) *Polym. Prepr.* **35(2)**, 504
15. Storey, R.F. and Shoemake, K.A. (1994) *Polym. Prepr.* **35(2)**, 578
16. Wang, L., McKenna, S.T., and Faust, R. (1995) *Macromolecules* **28**, 4681
17. Cloutet, E., Fillaut, J.L., Gnanou, Y., and Astruc, D. (1996) *Chem. Commun.*, 2047
18. Gnanou, Y. (1996) *J. M. S.-Rew. Macromol. Chem. Phys.* **C36(1)**, 77
19. Storey, R.F., Shoemake, K.A., and Chisolm, B.J. (1996) *J. Polym. Sci., Polym Chem.* **34**, 2003
20. Jacob, S., Majoros, I., and Kennedy, J.P. (1997) *Polym. Prepr.* **38(1)**, 208
21. Puskas, J.E. and Wilds, C.J (1998) *J.Polym.Sci. Polym. Chem. Ed.* **36**, 85
22. Puskas, J.E. and Grassmuller, M. (1998) *Makromol. Chem. Macromol. Symp.* **132**, 117
23. Puskas, J.E., Pattern, W.E., Lanzendörfer, M.G., Jamieson, D. and Oraha Al-Rais, L. (1998) *Polym. Prepr.* **39(1)**, 325
24. Puskas, J.E., Pattern, W.E., Wetmore, P., and Krukonis, V. (1999) *Rubber Chem. Techn.*, accepted for publication
25. Kaszas, G., Puskas, J.E., and Baade, W. (1996) *Polymeric Materials Encyclopedia*, J. Salamon Ed., CRC Press, Inc.
26. Frechet, J.M.J., Henmi, M., Gitsov, I., Aoshima, S., Leduc, M.R., and Grubbs, R.B. (1995) *Science* **269**, 1080
27. Langstein, G., Nuyken, O., Obrecht, W., Puskas, J.E., and Weiss, K. (1996) US Patent 5,543,479
28. Gaynor, S. G., Edelman, S., and Matyjaszewski, K. (1996) *Macromolecules* **29**, 1079-1081.
29. Matyjaszewski, K. and Gaynor, S. G. (1997) *Macromolecules* **30**, 7042-7049.
30. Weimer, M.W., Frechet, J.M. J., Gitsov, I. (1998) *J. Polym. Sci: Part A: Polym. Chem.* **36**, 955-970.
31. Jones, D.G., Runyon, J.R. and Ong, J. (1961) *J. Appl. Polym. Sci.* **5**, 452-459.
32. Powers, K.W. and Kuntz, I. (1978) US Pat.ent 4,074,035.
33. Kennedy, J.P. and Frisch, K. C. (1982) US Pat.ent 4,327,201.
34. Mrkvickova, L., Lopour, P., Pokorny, S., and Janca, J. (1980) *Angew. Makromol. Chem.* **90**, 217

NEWER ASPECTS OF PHOTOINITIATED CATIONIC POLYMERIZATION

J. V. CRIVELLO
Center For Polymer Synthesis
Department of Chemistry
Rensselaer Polytechnic Institute
Troy, N.Y. 12180-3590, USA

Abstract. The development of onium salt photoinititors for cationic polymerization has made possible to carry out the rapid, environmentally attractive polymerization of various mono and multifunctional vinyl and heterocyclic monomers. Such monomer-photoinitiator systems can be used in a wide variety of applications including protective and decorative coatings, printing inks and adhesives. Mechanisms for the photolysis of the photoinitiators and their action in the initiation of cationic polymerization are presented.

1. Introduction

In the past few years, there have been many significant advances in the area of the catalysis of cationic polymerization. These developments have substantially changed the way polymer chemists think about this field and have opened the doors to many applications of cationic chemistry which previously were thought to be unattainable. This communication will focus on three specific areas; namely, the development of novel photoinitiators for cationic polymerization which have seen particularly rapid development in the past two decades; the synthesis of specially designed monomers for use in photoinitiated cationic polymerization; and specific applications of photoinitiated cationic polymerization.

2. Historical Background

The cationic polymerization of vinyl and heterocyclic monomers can be initiated by a number of agents including: Lewis and Brønsted acids, carbenium and trialkyloxonium salts and by ionizing radiation [1]. In most cases, polymerization ensues exothermically immediately on mixing the monomer and initiator. Due to the high reactivity of many monomer systems, it is often very difficult to attain good homogeneous conditions before polymerization sets in. For this reason, monitoring of the early stages of the polymerizations is problematical. More significantly, the high reactivity of many cationic polymerizations precludes their use as inks, coating, adhesive, molding,

45

J.E. Puskas et al. (eds.), Ionic Polymerizations and Related Processes, 45–60.
© 1999 *Kluwer Academic Publishers. Printed in the Netherlands.*

encapsulating and reinforced composite applications, which require a reasonable working life and a controlled polymerization rate.

In recent years, several novel photoinitiator systems for cationic polymerization have been developed which exhibit excellent latency under normal conditions of storage but are activated by UV or visible irradiation [2]. The most significant of these are shown in Table 1. These photoinitiators can be grouped into several general classifications. The broadest class of compounds is the onium salt photoinitiators comprising the diazonium, diarylhalonium, triarylsulfonium, dialkylphenacyl-sulfonium, 4-hydroxyphenyldialkylsulfonium, pyrylium, thiopyrylium, and siliconium salts. Ferrocenium salts are a class of organometallic iron-containing photoinitiators. Lastly, α-sulfonyloxyketones and benzyl triarylsilyl ethers are representatives of covalently bonded compounds, which release sulfonic acids and triarylsilanols upon photolysis which can be used initiate certain types of cationic polymerization reactions.

3. Onium Salt Photoinitiators

Onium salts are ionic compounds consisting of an organic cation in which the positive charge resides on a heteroatom. Usually, an inorganic anion accompanies the organic cation. Certain onium salts are photosensitive and the key discovery that these compounds can photoinitiate cationic polymerization has led to their widespread study and use in many applications.

The following general comments with respect to the structure and reactivity in photoinitiated cationic polymerization apply to all onium salt photoinitiators. The first consideration important in the structure of an onium salt photoinitiator is its photosensitivity. Thus, the onium salt structure must possess some light absorbing chromophor, which results in a deep-seated chemical reaction culminating in the cleavage of a primary chemical bond. This photolytic event must also produce either directly or indirectly, a species capable of initiating cationic polymerization. Generally speaking, most onium salt photoinitiators contain either an aromatic or an aryl ketone group, which serves as the primary light absorbing chromophor and which can be further modified by the attachment of various substitutents.

The rate of photolysis of an onium salt and, therefore, the number of initiating species generated per given irradiation time and intensity is related to the structure of the cation, which is the light absorbing species. For this reason, many studies have been conducted on the photoinitiators discussed here in which the absorption characteristics of the photoinitiator cation structures have been systematically varied with the object of increasing their quantum yields or modifying their spectral response. It may be stated, that in general, a bathochromic shift in the absorption bands of an onium salt is observed when electron releasing substituents are introduced into the ortho and para positions of an aromatic rings. Conversely, the absorption bands are shifted to shorter wavelengths when electron withdrawing substituents are placed at these positions.

The mechanism of photolysis of an onium salt also has a profound effect on its efficiency as a photoinitiator. Onium salt photoinitiators may be divided into two basic classes; those which undergo irreversible photolysis and those which undergo reversible photolysis. Typical of those which undergo irreversible photolysis are diarylhalonium

TABLE 1. The Structures of Cationic Photoinitiators

Photoinitiator	Ref.	Photoinitiator	Ref.
$ArN_2^+ X^-$	3, 4	$N^+ - CH_2 \cdot COAr\ X^-$	14
$Ar_2I^+ X^-$	5, 6	$Ar_2 - \overset{O}{\underset{\parallel}{S}}{}^+ - OR\ X^-$	7
$Ar_2I^+O\ X^-$	8	Ar_2S^+ (with Ar groups) X^-	9, 10
$Ar_2Br^+ X$	11	Ar pyrylium O^+ X^-	12, 13
$Ar_2Cl^+ X^-$	14	$[C_5H_5\text{-Fe-Ar}]^+ X^-$	15
$Ar_3S^+ X^-$	16	$Si^+ X^-$ (with 3 groups)	17
$Ar_3Se^+ X^-$	18	$Ar - \overset{O}{\underset{\parallel}{C}} - \underset{R}{\underset{\mid}{CH}} - O\text{-}SO_2\text{-}Ar$	19
$ArS^+Alk_2 X^-$	20, 21	$O_2N\text{-}Ar\text{-}CH_2O\text{-}SiAr_3$	22
$Ar_3P^+\text{-}CH_2COAr\ X^-$	23		

salts and triarylsulfonium salts while examples of those which undergo reversible photolysis are dialkylphenacylsulfonium salts and dialkyl-4-hydroxyphenylsulfonium salts. Diarylhalonium and triarylsulfonium salts are, in general, more efficient photoinitiators of cationic polymerization than dialkylphenacylsulfonium salts and dialkyl-4-hydroxyphenylsulfonium salts and, consequently, have a much broader range of application.

Although the anion of an onium salt plays no role in its photochemistry, it is the dominant factor in the subsequent polymer chemistry since it determines the reactivity of both the initiating and propagating species as well as controlling what kinds of termination processes occur. Among the most useful onium salt photoinitiators are those which bear the very weakly nucleophilic anions such as BF_4^-, PF_6^-, AsF_6^-, SbF_6^- and $(C_6F_5)_4$ B^-. On photolysis, diazonium salts produce the corresponding Lewis acids by loss of a fluoride anion, while diaryliodonium and triarylsulfonium salts generate Brønsted acids. Because in both cases the acids, which are generated are very strong, these photoinitiators are capable of polymerizing almost every known type of cationically polymerizable monomer. This includes not only the polymerization of vinyl compounds such as styrene, α-methylstyrene, acenaphthalene, and vinyl ethers, but also a variety of ring-opening polymerizations among which the polymerization of epoxides, cyclic ethers, lactones, aziridines, oxazolines and many others may be mentioned. Due to the weakly nucleophilic character of these anions, termination is very slow if not absent and in certain cases, such as in the polymerization of tetrahydrofuran, living cationic polymerizations are observed.

In addition to the considerations given to the structural features of the photoinitiator, one must also take into account the reactivity of the monomer. If very highly reactive monomers such as vinyl ethers are to be photopolymerized, correspondingly less reactive onium salts with anions such as trifluromethanesulfonate or perchlorate may be used. In contrast, such initiators will only sluggishly initiate the ring-opening polymerization of epoxide and other heterocyclic monomers. Since the irradiation of onium salt photoinitiators yields acids, it is possible to carry out other types of reactions besides strictly cationic polymerizations. For example, the acid catalyzed condensation polymerizations of phenol-formaldehyde, melamine-formaldehyde and alkoxysilanes can be induced using these same photoinitiators [24].

Among the onium salts, only the diaryliodonium, triarylsulfonium and ferrocenium salts have experienced appreciable commercial development. The high quantum yields and excellent latency in the absence of light have made these the photoinitiators of choice for almost all commercial coating, printing ink and adhesive applications. For this reason, only these three classes of photoinitiators will be discussed here.

3.1. DIARYLIODONIUM SALTS

Diaryliodonium salts with the general structure shown below are a series of stable, colorless, crystalline, ionic salts, which are readily soluble in many organic solvents but nearly insoluble in water. Especially useful is their excellent solubility in a wide variety of polar cationically polymerizable monomers. It was observed that all diarylhalonium salts are photosensitive [25] and can be used to initiate cationic polymerization [5,6] provided that they bear suitably non-nucleophilic anions.

$$Ar-I^+-Ar \quad X^-$$

Within this family of compounds, there are many structural variations possible. For example, salts in which various alternate aryl, heteroaryl and cycloaryl groups are attached to iodine have been prepared. Similarly, X^- can represent virtually any anion.

The chemistry of diaryliodonium salts as it applies to their synthesis and reactions has been reviewed previously [2,26,27,28].

Early studies of the mechanism of the photolysis of diarylhalonium salts pointed to a process involving a photoinduced homolysis of a carbon-halogen bond [2,29,30,31]. In the past few years, there has been a great deal of activity in the field of the study of the mechanism of the photolysis of diaryliodonium salts [32,33]. Recent investigations, have found that the mechanism of the direct photolysis of these compounds is considerably more complex than originally thought. Shown in the following Scheme 1 is the currently accepted mechanism for the photolysis of diaryliodonium salts.

$$Ar_2I^+\ X^- \xrightarrow{\ h\nu\ } [\,Ar_2I^+\ X^-\,]^* \longrightarrow [\,Ar_2I^+\ X^-\,]^1$$

$$[\,Ar_2I^+\ X^-\,]^1 \underset{\;}{\overset{isc}{\rightleftharpoons}} [\,Ar_2I^+\ X^-\,]^3$$

$$\overline{ArI\ \ Ar^+\ X^-}^{\,1} \xrightarrow{\ e^-\text{-transfer}\ } \overline{ArI^{+\cdot}\ Ar\,X^-}^{\,1} \ \rightleftharpoons\ \overline{ArI^{+\cdot}\ Ar\,X^-}^{\,3}$$

		$\downarrow$ RH	$\downarrow$ RH

$$ArI + Ar^+ + X^- \qquad ArArI + HX \qquad ArI + HX + Ar + R\cdot$$

Cage-escape   In-cage recombination   Cage-escape

$$ArR + HX \qquad\qquad\qquad\qquad\qquad ArH + R\text{-}R$$

RH = solvent

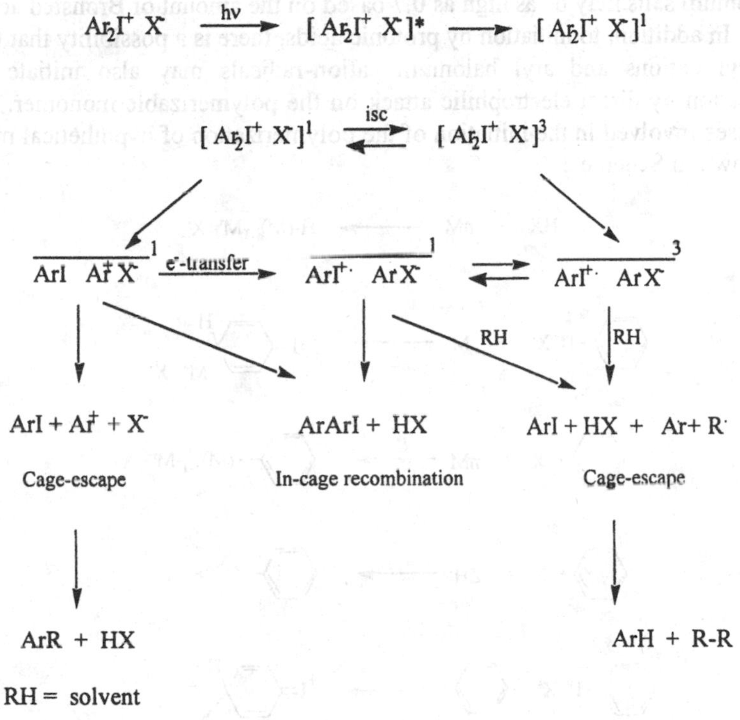

Scheme 1.

All the products of the photolysis can be explained by the multistep mechanism shown in Scheme 1. Excitation by irradiation produces a caged excited singlet diarylhalonium salt. In the scheme, caged species are represented by a bar over the structure. The first process which this species can undergo is a heterolytic cleavage which gives rise to a caged aryl cation-iodoarene pair. Intersystem crossing from the excited singlet leads to the excited triplet which can decay to the ground state or undergo homolysis to a triplet aryl radical-iodoarene geminate pair. The excited triplet may also be produced by the triplet photosensitization. Each of the corresponding caged pairs of reactive intermediates may undergo further reactions depending on whether the reactions occur by an in-cage or cage-escape process. These processes are detailed in the above scheme. It may also be noted that the reactive intermediates can interconvert with one another by equilibration and electron-transfer processes. Thus, the products of the photolysis of diarylhalonium salts arise from radicals, cation-radicals and cations generated respectively, from the homolytic and heterolytic cleavage of a carbon-halogen bond. A detailed product analysis indicates that the heterolytic cleavage reaction is the predominant pathway followed in the irradiation of diaryliodonium salts.

From Scheme 1, it may be concluded that the most ubiquitous initiating species of cationic polymerization are protonic acids. The amount of protonic acid which, is formed from the photolysis of a diaryliodonium salt can be correlated with total amount of iodine containing photoproducts [34]. Based on the work of Pappas and Gatechair [35] and by Timpe and his coworkers [36], the quantum yield of photolysis of diaryliodonium salts may be as high as 0.7 based on the amount of Brønsted acid which is formed. In addition, to initiation by protonic acids, there is a possibility that the cage-escape aryl cations and aryl halonium cation-radicals may also initiate cationic polymerization by direct electrophilic attack on the polymerizable monomer. Some of the processes involved in the initiation of the polymerization of hypothetical monomer, **M**, are shown in Scheme 2.

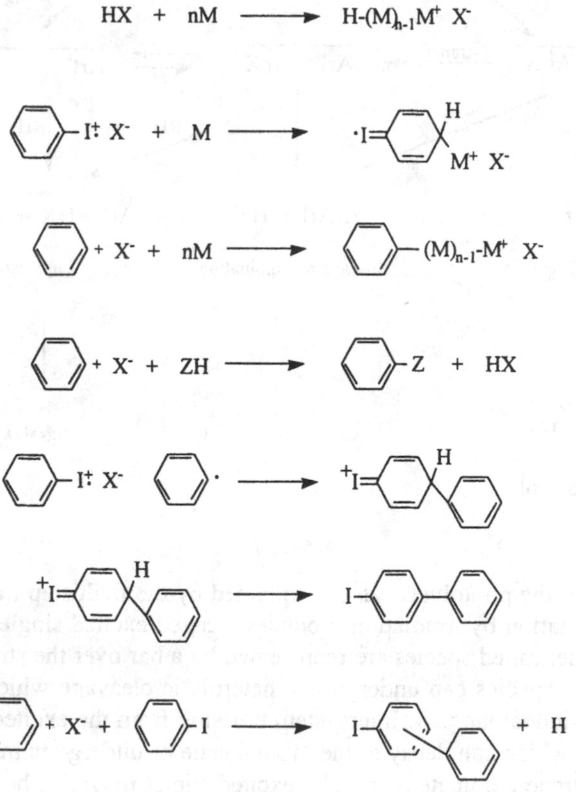

$$HX \ + \ nM \ \longrightarrow \ H\text{-}(M)_{n-1}M^+ \ X^-$$

Scheme 2.

The observation that only a very small portion of the polymer chains which are produced using diaryliodonium salts contain endgroups resulting from initiator residues suggests that processes involving Brønsted acids as the initiating species are dominant. Studies in which the absorption characteristics of diaryliodonium salts have been modified by systematic changes in the cation structures have been reported [37]. In addition, there are further reports of successful attempts to improve the solubility [38,39] and to reduce the toxicity [40,41] of diaryliodonium salt photoinitiators by the attachment of long chain alkyl and alkoxy groups on one or both of the aromatic rings.

3.2. TRIARYLSULFONIUM SALTS

As photoinitiators of cationic polymerization, triarylsulfonium salts are nearly ideal.

$$Ar_3S^+ \ X^-$$

Not only do they have excellent photosensitivity with quantum yields in the range of approximately 0.5-0.7, [2] when irradiated, they also exhibit a high degree of thermal stability. As a result, triarylsulfonium salts are the photoinitiators of choice for many practical applications and are also commercially available from a number of sources. The synthesis of triarylsulfonium salts was reviewed in 1984 as it applies to the preparation of active cationic photoinitiators [2].

Shown in the following scheme of reactions is the currently accepted mechanism for the photolysis of triarylsulfonium salts [42].

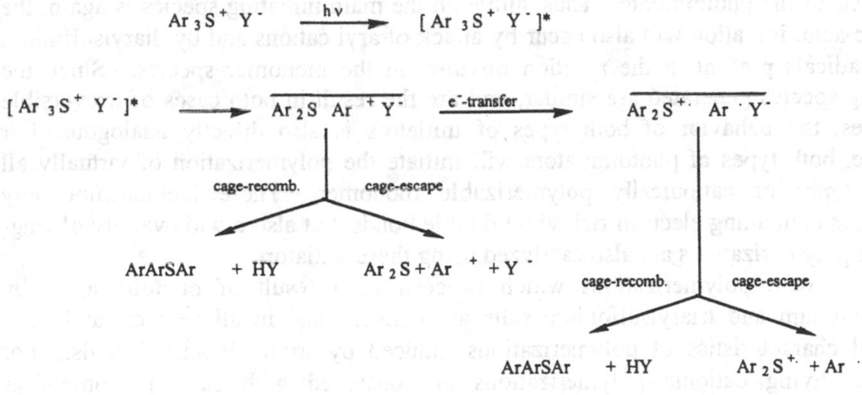

Scheme 3.

Direct irradiation of a triarylsulfonium salt results in the formation of the excited singlet. This species undergoes primarily a heterolytic cleavage of a carbon-sulfur bond to generate an aryl cation and diarylsulfide pair within a solvent cage. This species may undergo decay by either of two modes; either a cage recombination process in which the aryl cation undergoes an electrophilic attack on one of the aromatic rings of diarylsulfide to give the coupled aryl biarylsulfide product or a cage escape in which the highly reactive aryl cation can abstract a proton from or react with other reactants, solvents and monomers to give arylation products. As a result of both of these two processes, protonic acids are generated. The initially generated excited singlet can also undergo interconversion by an electron-transfer process to generate a caged diarylsulfinium cation-radical-aryl radical pair. These fragments can also subsequently undergo caged recombination and cage escape processes as shown in the mechanism. The aryl free radicals and cation-radicals also generate a number of products arising from coupling, and hydrogen abstraction processes.

While the primary absorption of light yields the excited singlet triarylsulfonium salt, there is some evidence that intersystem crossing to the excited triplet state also occurs [43,44]. The triplet state is also accessible by triplet sensitization. The decay products and product distribution from the excited triplet closely resemble those from the excited singlet. The products which have been isolated are in good agreement with the above mechanism. All three biphenyl phenyl sulfides have been isolated from the photolysis of triphenylsulfonium salts and are characteristic of products expected from the reaction of diphenyl sulfide with an aryl cation. In addition to the biphenyl phenyl sulfides, protonic acids are also generated in this reaction. The observed total amount of protonic acid which is produced on photolysis of triphenylsulfonium salts shows a good correlation with the combined amounts of biphenyl phenyl sulfide produced from the heterolytic cleavage together with the diphenyl sulfide derived from the homolytic cleavage [42].

In analogy with what has been described for diaryliodonium salts (Scheme 2), initiation of polymerization takes place not only from the primary products of the photolysis of a triarylsulfonium salt, but also from secondary products of the reaction of those reactive species with solvents, monomers and even other fragments of the photolysis of the photoinitiator. Thus, although the main initiating species is again, the protonic acid, initiation will also occur by attack of aryl cations and by diarylsulfinium cation-radicals present in the reaction mixture on the monomer species. Since the initiating species generated are similar, and are the result in both cases of irreversible processes, the behavior of both types of initiators is also directly analogous. For example, both types of photoinitiators will initiate the polymerization of virtually all known types of cationically polymerizable monomers. These include not only monomers containing electron-rich vinyl double bonds, but also a wide variety of ring-opening polymerizations are also catalyzed using these initiators.

The cationic polymerizations which proceed as a result of photoinitiation by diaryliodonium and triarylsulfonium salts are conventional in all respects and have identical characteristics of polymerizations induced by strong Brønsted acids. For example, living cationic polymerizations are observed with such monomers as tetrahydrofuran [45], while the polymerization of olefins takes place with considerable chain transfer and termination. Figure 1 shows a cartoon depicting some of the more common monomers which can be photopolymerized using these photoinitiators.

It has been shown in the above mechanisms (Schemes 2 and 3) that in addition to protonic acids, aryl cations and cation-radicals, a number of free radical species are also generated from the photolysis of diarylhalonium and triarylsulfonium salts. These free radical species arise from direct fragmentation of the onium salts themselves as well as by reaction of the primary free radicals with solvents and monomers. The free radical species so generated may be used to initiate purely free radical polymerizations [46, 47].

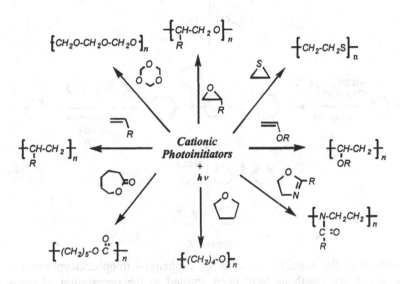

Figure 1. Typical monomers used in photoinitiated cationic polymerizations.

It is also possible to photoinitiate concurrent cationic and free radical polymerizations using these photoinitiators [46]. This, so called, "hybrid" systems lead to simultaneous interpenetrating networks with interesting mechanical and chemical properties. At this time, little has been published concerning these hybrid cationic and free radical polymerizations. Likewise, as has already been mentioned, many acid catalyzed condensation polymerizations can be carried out using these same photoinitiators.

As in the case with diaryliodonium salts, in recent years, much activity has been focused on the development of triarylsulfonium salt photoinitiators having long wavelength absorption bands with high molar extinction coefficients. The object of these efforts is to develop photoinitiators with the highest efficiencies in cationic polymerization possible. The photoinitiators, diphenyl-4-thiophenoxyphenyl sulfonium salts are of particular interest since their UV absorption maxima lies at 300 nm. In contrast, the UV absorption maxima of triarylsulfonium salts possessing unsubstituted phenyl rings or those containing simple substitutents are in the range of 230 to 250 nm. Thus, diphenyl-4-thiophenoxyphenyl sulfonium salts are exceptionally efficient as photoinitiators when used with conventional mercury arc emission sources.

The longer wavelength absorption band in both cases is attributed to the presence of the thiophenoxy group present in both salts, which interacts by resonance through the sulfur atom with the positively charged sulfonium center. Scheme 4 depicts four different methods, which have been employed for the synthesis of these photoinitiators [48,49,50,51].

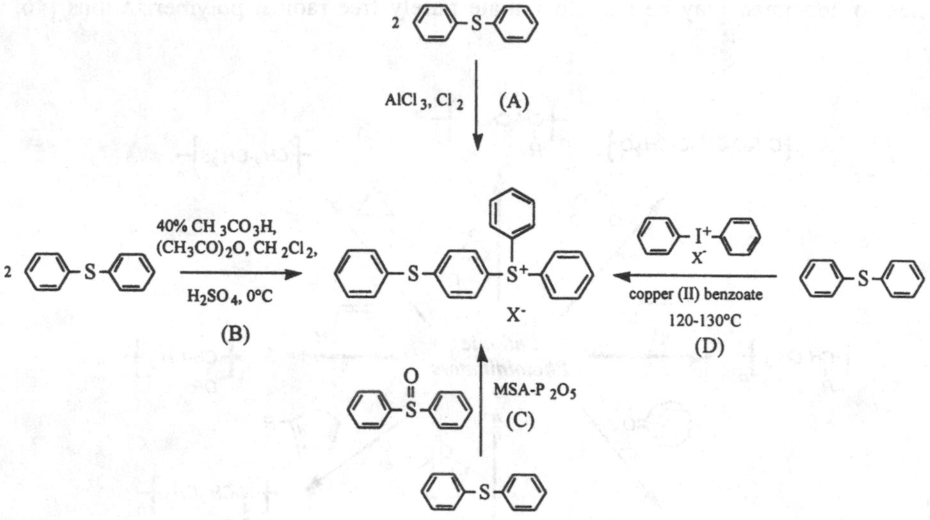

$$2 \; C_6H_5{-}S{-}C_6H_5$$

AlCl$_3$, Cl$_2$ | (A)

40% CH$_3$CO$_3$H,
(CH$_3$CO)$_2$O, CH$_2$Cl$_2$,
$$2 \; C_6H_5{-}S{-}C_6H_5 \xrightarrow{\text{H}_2\text{SO}_4,\ 0°C}$$
(B)

$$\xleftarrow[\text{120-130°C}]{\text{copper (II) benzoate}}$$

(D)

MSA-P$_2$O$_5$

(C)

Scheme 4.

In addition to the specific synthesis of diphenyl-4-thiophenoxyphenylsulfonium salts, the above four methods have been applied to the preparation of many other related triarylsulfonium salt photoinitiators.

3.3. PHOTOSENSITIZATION OF DIARYLIODONIUM AND TRIARYLSULFONIUM SALTS

As an alternative to the synthesis of complex diaryliodonium and triarylsulfonium salts with long wavelength absorbing chromophors, it has been possible to extend the range of spectral sensitivity of those photoinitiators throughout all of the UV and into the visible spectrum. There are three mechanisms by which dye photosensitization occurs in onium salt photolyses [52] they are:

1. Energy transfer (triplet) photosensitization
2. Electron transfer photosensitization
3. Redox photosensitization

Of these three methods, electron transfer photosensitization is the most efficient and useful means by which aryldiazonium, triarylsulfonium and diaryliodonium salts may be photosensitized.
For these reasons, electron transfer photosensitization is commonly employed in such applications as micro- and stereolithography as well as in high speed cationic photopolymerizations used in such applications as coatings. The basic mechanism is depicted in Scheme 5.

In the first step of this mechanism, the photosensitizer is excited by light. The excited photosensitizer then interacts directly with the onium salt to undergo a formal one electron-transfer. This may or may not take place as a result of an intermediate excited state complex. As a result of electron-transfer, two products are produced; the photosensitizer cation-radical and a $Ar_nOn\cdot$ free radical. This latter species undergoes a spontaneous decomposition with the formation of an aryl free radical. The aryl free radical can then couple with the photosensitizer cation-radical to generate an arylated species and a protonic acid. At the same time, the aryl radical can undergo other reactions, such as hydrogen abstraction with the solvent or monomer (Z-H). Polymerization is initiated mainly by the protonic acid, which is generated, although the photosensitizer cation-radical can also participate in the initiation step as an electrophilic species.

$$P \xrightarrow{\text{hv}} P*$$

$$P* + Ar_nOn^+ X^- \xrightarrow{\text{electron-transfer}} P^+ X^- + Ar_nOn\cdot$$

$$Ar_nOn\cdot \longrightarrow Ar_{n-1}On + Ar\cdot$$

$$P^+ X^- + Ar\cdot \longrightarrow Ar\text{-}P^+ X^-$$

$$Ar\text{-}P^+ X^- \longrightarrow Ar\text{-}P + HX$$

$$Ar\cdot + Z\text{-}H \longrightarrow Z\cdot + Ar\text{-}H$$

Scheme 5.

Since the above mechanism involves a redox reaction in which the onium salt is reduced and the photosensitizer is oxidized, electron transfer onium salt photosensitization has been shown to obey the Rhem-Weller relationship [53,54,55].

$$\Delta G = (E^{ox}_{sens} - E^{red}_{onium}) - E*$$

where the (E^{ox}_{sens}) is the energy required to oxidize the photosensitizer, (E^{red}_{onium}) is the reduction potential of the onium salt and $E*$ is the excitation energy of the photosensitizer. Knowing the above values, it is possible to predict with considerable success potential photosensitizers for any given onium salt.

It should be pointed out that technically, that compounds which participate in electron transfer photosensitization are not "true" photosensitizers since they are consumed during this reaction.

It has been found that diaryliodonium salts with lower reduction potentials (-5 k-cal/mol) are easier to electron-transfer photosensitize than triarylsulfonium salts with higher reduction potentials (-28 k-cal/mol). Due to the difference in their reduction potentials, diaryliodonium and triarylsulfonium salts are not usually electron-transfer photosensitized by the same compounds. However, aromatic hydrocarbons such as anthracene, pyrene and perylene with low oxidation potentials will photosensitize both classes of onium salts [54,56].

3.4. FERROCENIUM SALTS

Cyclopentadienyl iron (II) cation salts (ferrocenium salts) having the following general structure have been shown to be very good cationic photoinitiators [57-60].

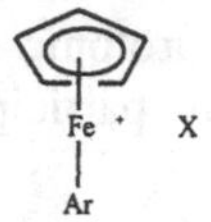

An especially attractive feature of these compounds is the ability to "tune" their absorption characteristics by appropriate choice of the aromatic ligand and the substituents, which it bears. For example, replacing benzene with pyrene as a ligand shifts the major absorption bands from below 300 nm to about 400 nm with the absorption extending into the visible to about 550 nm.

Ferrocenium salts can be prepared as shown below by the reaction of ferrocene in the presence of the desired aromatic ligand with a mixture of aluminum powder, aluminum chloride and titanium tetrachloride [58].

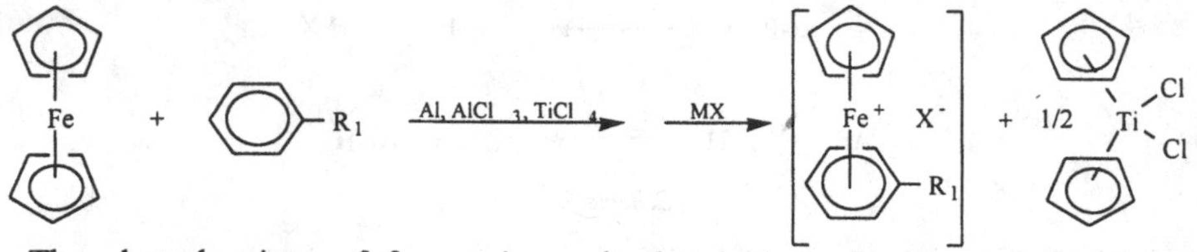

The photochemistry of ferrocenium salt photoinitiators has been the subject of several investigations and the currently accepted mechanism is shown in Scheme 6. On irradiation, a photoinduced ligand transfer reaction takes place with the displacement of the arene ligand. If the photolysis takes place in the presence of an epoxide as a monomer, the positively charged iron atom becomes bonded to three molecules of epoxide. Epoxide polymerization begins in the coordination sphere of the iron atom and then proceeds outward.

Recent experimental evidence using ethylene oxide as a monomer appears to confirm this conclusion [59]. In this experiment, the double crown ether iron complex, iron bis(1,4,7,10-tetraoxydodecane) hexafluorophosphate, was isolated and its formation ascribed to cyclopolymerization by a "template effect" within the coordination sphere of iron.

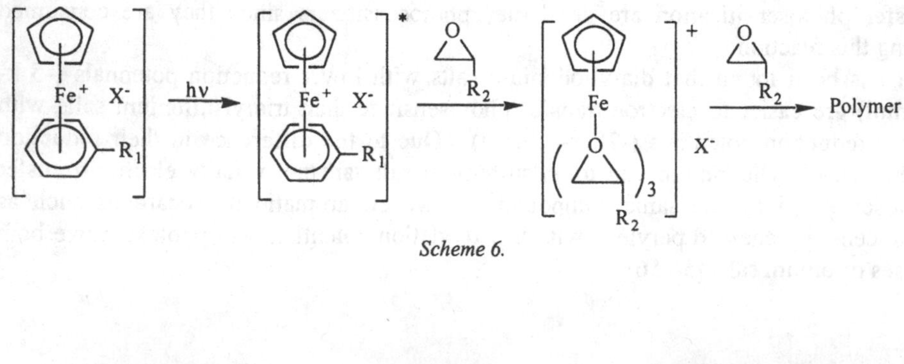

Scheme 6.

Ferrocenium salts have been primarily used for the polymerization of epoxide-containing monomers. Since the photolysis of these compounds produces a relatively weak iron-based Lewis acid, their reactivity in cationic polymerization is considered to be less than the most active triarylsulfonium and diaryliodonium salts. Generally, photolysis is followed by a thermal step required to complete the ring-opening polymerization. More reactive ferrocenium salts systems containing an oxidant such as cumene hydroperoxide or t-butylhydroperoxide have been described [60]. Apparently, a more reactive Lewis acid is produced in these systems by increasing the oxidation state of the iron from +2 to +3. Lastly, it has been reported that the photolysis of ferrocenium salts can be photosensitized [61]. Polynuclear hydrocarbons are active photosensitizers for these compounds and the photosensitization by anthracene was studied.

4. Photoinitiated Cationic Polymerization

It has already been mentioned and shown Figure 1 that the range of monomeric substrates which can be polymerized using cationic photoinitiators is very large and encompasses virtually every known type of cationically polymerizable monomer. This makes photoinitiated cationic polymerizations highly interesting from an academic point of view. However, from a practical standpoint, the range of applicable monomers is limited to a few classes, which either possess outstanding mechanical and chemical properties or have exceedingly high rates of polymerization. Examples of the first class of monomers are epoxides while examples of those comprising the second class are vinyl ether monomers. In both cases, the existing commercial availability of these two types of monomers makes them especially attractive.

For practical reasons, photoinitiated cationic polymerizations are best applied to thin film applications such as coatings, printing inks and adhesives in which the photopolymerization is carried out on a relatively flat, or curved surface easily accessible to irradiation by light. Polymerization is restricted to illuminated areas only and diffusional polymerization into unilluminated regions is slow at best.

Further, photopolymerizations in general are best utilized with multifunctional monomers to produce network or crosslinked polymers. In contrast, linear polymers are more advantageously produced with inexpensive cationic initiators such as Lewis or Brønsted acids. In the production of crosslinked polymers, photoinitiated cationic polymerizations are especially attractive since they provide excellent latency together with high quantum efficiencies.

Epoxide monomers are widely employed for applications such as coatings in which their excellent adhesion, chemical resistance and mechanical toughness are the main attractive advantages. It should be noted, however, that commercially available epoxide monomers are designed and synthesized primarily for use in condensation-type polymerizations in which they are reacted together with stoichiometric amounts of such agents as amines or anhydrides. For this reason, the requirements of purity in the synthesis of these monomers, particularly with respect to the presence of small amounts of basic impurities are not high. Certainly, this can be a problem when photoinitiated cationic polymerizations are carried out in which only catalytic amounts of a photogenerated acid are produced. In many cases, the presence of impurities may be observed by a pronounced induction period or even, in some cases, complete suppression of polymerization. Optimally, monomers should be intentionally synthesized and purified for use in photoinitiated cationic polymerization.

The class of epoxide monomers encompasses several subclasses of monomers with markedly different reactivities in photoinitiated cationic polymerization. Recently, we have completed a study of the comparative reactivity of these monomers [62]. In thin film applications for which photoinitiated polymerizations are most suited, the rate of throughput (i.e. reactivity) of the system is of crucial importance in determining its overall economic viability. Glycidyl ester monomers polymerize most slowly in the presence of diaryliodonium and triarylsulfonium salts. In order of increasing reactivity are: glycidyl ethers, epoxy alkanes, and finally, cycloaliphatic epoxides. The most reactive monomers are those bearing epoxycyclopentyl or epoxycyclohexyl groups. The presence of additional nucleophilic functional groups can also have an influence on reactivity and usually suppresses reaction rates [63]. Acceleration of epoxide photopolymerization rates can be achieved through the use of polyols as chain transfer agents [64].

Photocurable epoxy monomers have found many applications and are currently used in paper release coatings, protective and decorative coatings for metals, pressure sensitive adhesives and in stereolithography. In these applications, the additional factors of low toxicity and reduced shrinkage of the epoxy monomers have been found to be important.

In recent years, multifunctional vinyl ethers have been promoted as photocurable cationic monomers mainly because of their exceedingly high reactivity [65-66]. In this laboratory, work has been proceeding towards the synthesis of a related class of monomers, 1-propenyl ethers, as substrates for photoinitiated cationic polymerization [67,68,69]. These monomers can be prepared by straightforward chemistry as depicted below.

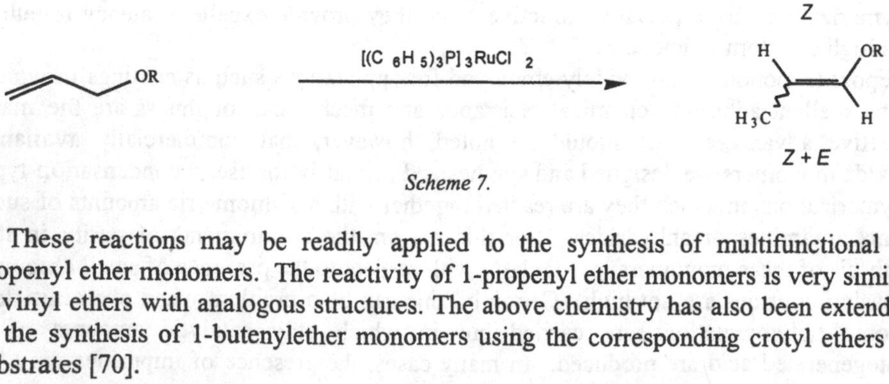

Scheme 7.

These reactions may be readily applied to the synthesis of multifunctional 1-propenyl ether monomers. The reactivity of 1-propenyl ether monomers is very similar to vinyl ethers with analogous structures. The above chemistry has also been extended to the synthesis of 1-butenylether monomers using the corresponding crotyl ethers as substrates [70].

5. Conclusions

Photoinitiation is a highly efficient and practical method of inducing cationic polymerization, which is finding increasingly numerous useful industrial applications. The key to this chemistry has been the development of several classes of onium salt photoinitiators with high quantum yields and equally high polymerization efficiencies. Manipulation of the structure of these onium salt photoinitiators, besides increasing their quantum efficiency has broadened their spectral response as well as increased their compatibility in various monomeric substrates. Employing diaryliodonium and triarylsulfonium salt photoinitiators, a wide variety of cationic vinyl and ring-opening polymerizations can be carried out. Among the most useful polymerizations, which are commonly used are the crosslinking polymerizations of multifunctional epoxides and vinyl ethers. It has been shown that the highest rates of photopolymerizations are achieved when not only the photoinitiator but also the structure of the monomer is optimized. This has presented the opportunity of designing and synthesizing novel monomers expressly for use in photoinitiated cationic polymerization. Multifunctional 1-propenyl ethers are a new class of such monomers, which exhibit extraordinarily high rates of photoinitiated cationic polymerization.

6. References

1. Kennedy, J.P. and Maréchal, E. (1982) *Carbocationic Polymerization*, John Wiley & Sons, New York.
2. Crivello, J.V. (1984) *Adv. in Polym. Sci.* **62**, 1.
3. Licari, J.J., Crepeau, W., and Crepeau, P.C. (1965) U.S. Patent 3,205,257, *Chem. Abstr.* **63**, 17368a.
4. Watt, W.R. (1979) *ACS Symp. Ser. 114*, R.S. Bauer, Ed, American Chemical Society: Washington, D.C., p. 17.
5. Crivello, J.V. and Lam, J.H.W. (1977) *Macromolecules* **10**, 1307.
6. Crivello, J.V. and Lam, J.H.W. (1976) *J. Polym. Sci.*, Symp. **56**, 383.
7. Crivello, J.V. and Lam, J.H.W. (1979) *J. Polym. Sci., Polym. Chem.* **17**, 2877.
8. Irving, E. (1984) U.S. Patent 4,482,679.
9. Crivello, J.V. (1974) U.S. Patent 4,150,988.
10. Ketley, A.D. and Tsao, J.-H. (1979) *J. Rad. Curing* **16**(2), 22.
11. Crivello, J.V. and Lam, J.H.W. (1978) *J. Polym. Sci. Polym. Lett. Ed.* **16**, 563.
12. Green, G.E. and Irving, E. (1980) Eur. Patent 22081; (1981) U.S. Patents 4,299, 938; (1982) U.S. Patents 4,339, 567.
13. Ledwith, A. (1982) *Polym. Prepr.* **23**(1) 323.
14. Crivello, J.V., Lam, J.H.W., Moore, J.E., and Schroeter, S.H. (1978) *J. Rad. Curing* **5**(1), 2.
15. Meier, K. and Zweifel, H. (1985) *Radcure Europe*, Basel, SPE Technical Paper FC 85-417.
16. Crivello, J.V. and Lam, J.H.W. (1979) *J. Polym. Sci. Polym. Chem. Ed.* **17**, 977.
17. Cella, J.A., Schwabacher, A.W., and Schulz, A.R. (1983) *Ind. Eng. Chem. Prod. Res. Dev.* **22**(1), 20.
18. Crivello, J.V. and Lam, J.H.W. (1980) *J. Polym. Sci. Polym. Chem. Ed.* **18**, 2697.
19. Berner, G., Kirchmayr, R., Rist, G., and Rutsch, W. (1986) *J. Rad. Curing* **23**, 10.
20. Crivello, J.V. and Lam, J.H.W. (1980) *J. Polym. Sci., Polym. Chem. Ed.* **18**, 2677.
21. Crivello, J.V. and Lam, J.H.W. (1979) *J. Polym. Sci., Polym. Chem. Ed.* **17**, 1047.
22. Hayase, S., Onishi, Y., Suzuki, S., and Wada, M. (1985) *Macromolecules* **18**, 1799.
23. Crivello, J.V. and Lam, J.H.W. (1980) *J. Polym. Sci., Polym. Chem. Ed.* **18**, 1021.
24. Crivello, J.V. (1983) *Polym. Eng. and Sci.* **33**(17), 953.
25. Khanna, R.K. and Gokel, G.W. (1981) *J. Org. Chem.* **46**, 2153.
26. Banks, D.F. (1966) *Chem. Revs.* **66**(3), 243.
27. Olah, G.A. (1975) *Halonium Ions*, Wiley Interscience, New York.

60

28. Beringer, F.M. and Falk, R.A. (1959) *J. Am. Chem. Soc.* **81**, 2997.
29. Crivello, J.V. (1988) *Makromol. Chem., Macromol. Symp.* **13/14**, 145.
30. Pappas, S.P. (1985) *Prog. Org. Coat.* **13**, 35; (1985) *J. Imaging Technol.* **11**, 146.
31. Pappas, S.P., Pappas, B.C., Gatechair, L.R., and Jilek, J.H. (1984) *Polym. Photochem.* **22**, 69.
32. Devoe, R.J., Sahyun, M.R.V., Serpone, N., and Sharma, D.K. (1987) *Can. J. Chem.* **65**, 2342; (1988) *Can. J. Chem.* **66**, 319.
33. Timpe, H.J. and Schikowsky, V.J. (1989) *J. Prakt. Chem.* **331**, 447.
34. Hacker, N.P. and Dektar, J.L. (1990) *Radiation Curing of Polymeric Materials*; Hoyle, C.E. and Kinstle, J.F., Eds, American Chemical Society: Washington, D.C., p. 82.
35. Pappas, S.P. and Gatechair, L.R. (1982) *Proc. Soc. Photogr. Sci. Eng.* **46**.
36. Baumann, H., Timpe, H.-J., and Böttcher, H. (1983) *Z. Chem.* **23(3)**, 102.
37. Crivello, J.V. and Lee, J.L. (1985) U.S. Patents 4,529,490; (1987) U.S. Patents 4,683,317.
38. Eckberg, R.P. and Riding, K.D. (1989) *Polym. Mat. Sci. and Eng.* **60**, 222.
39. Eckberg, R.P. and LaRochelle, R.W. (1981) U.S. Patents 4,279,717 ; (1983) 4,421,904.
40. Fukuyama, J.M., Lee, J.L., and Crivello, J.V. (1991) U.S. Patent 4,992, 571.
41. Crivello, J.V. and Lee, J.L. (1986) *Polym. Bull.* **16**, 243.
42. Dektar, J.L. and Hacker, N.P. (1987) *J. Chem. Soc., Chem. Commun.*, 1592.
43. Dektar, J.L. and Hacker, N.P. (1990) *J. Am. Chem. Soc.* **112**, 6004.
44. Tilley, M.G. (1988) *Thesis*, North Dakota State University, *Dissertation* Abstr. No. AAC8826980.
45. Crivello, J.V. and Lee, J.L. (1985) *Macromol. Synth.* **9**, 43.
46. Crivello, J.V. and Lam, J.H.W. (1979) *J. Polym. Sci. Polym. Lett. Ed.* **17**, 759.
47. Crivello J.V., Lee, J.L., and Conlon, D.A. (1983) *J. Radiat. Curing* **10**(1), 6.
48. Crivello, J.V. and Lam, J.H.W. (1980) *J. Polym. Sci. Polym. Chem. Ed.* **18**, 2677.
49. Crivello, J.V. (1991) U.S. Patent 5,012,001.
50. Akhtar, S.R., Crivello, J.V., Lee, J.L., and Schmitt, M.L. (1990) *Chem. Mat.* **2**, 732.
51. Crivello, J.V. and Lam, J.H.W. (1978) *J. Org. Chem.* **43**, 3055; (1980) *J. Polym. Sci. Polym. Chem. Ed.* **18**, 2697.
52. Küstermann, E., Timpe, H.-J., Gabert, K., and Schübert, H. (1987) *Wissensch. Zeitschr. TH Leuna Merseburg* **29**(3), 287.
53. Dektar, J.L. and Hacker, N.P. (1988) *J. Org. Chem.* **53**, 1833.
54. Crivello, J.V. and Lam, J.H.W. (1979) *J. Polym. Sci. Polym. Chem. Ed.* **17**, 1059.
55. Gatechair, L.R. and Pappas, S.P. (1982) *Proc. Org. Ctg. and App. Polym. Sci. Division* **46**, 707; Pappas, S.P., Gatechair, L.R., and Jilek, J.H. (1984) *J. Polym. Sci., Polym. Chem. Ed.* **22**, 77.
56. DeVoe, R.J., Sahyun, M.R.V., and Schmidt, E. (1989) *J. Imag. Sci.* **33**, 39.
57. Meier, K. and Zweifel, H. (1986) *J. Imaging Sci.* **30**, 174.
58. Doggweiler, H.O. and Desobry, V. (1986) Eur. Patent Appl., 270490.
59. Meier, K and Rihs, G. (1985) *Angew. Chem.* **97**, 879.
60. Meier, K., Eugster, G., Schwarzenbach, F., and Zweifel, H. (1983) Eur. Patent Appl., 126712, to Ciba-Geigy.
61. Gaube, G.G. (1986) *Proc. Radcure '86 Conf.Proceedings*, p. 15.
62. Crivello, J.V. and Linzer,V. (1995) *Polimery*, manuscript in press.
63. Crivello, J.V. and Varlemann, U. (1995) *J. Polym. Sci., Polym. Chem. Ed.* **33**(14), 2463.
64. Crivello, J.V., Conlon, D.A., Olson, D.R., and Webb, K.K. (1986) *J. Radiat. Curing* **13**(4), 3.
65. Crivello, J.V., Lee, J.L., and Conlon, D.A. (1983) *J. Radiat. Curing* **10**(1), 6.
66. Lapin, S. and House, D.W. (1988) *U.S. Patent 4,751,273.*
67. Crivello, J.V. and Conlon, D.A. (1984) *J. Polym. Sci., Polym. Chem. Ed.* **22**, 2105.
68. Crivello, J.V. and Jo, K.D. (1993) *J. Polym. Sci. Polym. Chem. Ed.* **31**(6); (1993) *J. Polym. Sci. Polym. Chem. Ed.* **31**(6), 1483; (1993) *J. Polym. Sci. Polym. Chem. Ed.* **31**(8), 2143.
69. Crivello, J.V. and Bratslavsky, S.A. (1994) *J. Polym. Sci. Polym. Chem. Ed.* **32**. 2755.
70. Crivello, J.V., Bratslavsky, S.A., and Yang, B. (1995) *Polymer Mat. Sci. and Eng. Preprints* **72**, 473.

CATIONIC MACROMOLECULAR DESIGN USING NON(HOMO)POLYMERIZABLE MONOMERS

Y. C. BAE, S. HADJIKYRIACOU, H. SCHLAAD, AND R. FAUST

Polymer Science Program, Department of Chemistry,
University of Massachusetts Lowell,
One University Avenue, Lowell, MA 01854, USA.

Abstract. The last half decade's development in the cationic macromolecular using non(homo)polymerizable monomers is reviewed. This process involves the intermediate capping reaction of living polyisobutylene with non(homo)polymerizable monomers, such as 1,1-diphenylethylene derivatives or 2-alkylfurans. Kinetic and mechanistic aspects of this capping reaction is discussed, and an overview of capping and coupling reactions is presented as advanced tools for the cationic design of macromolecular structures and properties. Examples are given for the synthesis of functional polymers, block copolymers, star-block copolymers, and macromonomers.

1. Introduction

Over the last half decade, we have made a concerted effort to examine the scope and limitations of non(homo)polymerizable monomers, such as 1,1-diphenylethylene (DPE) or 1,1-ditolylethylene (DTE), in cationic macromolecular engineering [1-3]. The importance of these compounds, from synthetic standpoint, derives from their application in the capping reaction of living cationic polymers. A quantitative addition of DPE or DTE to a living chain end, i.e., a capping reaction, results in a stable and completely ionized cationic living chain end, which was shown to be well suited for quantitative end-functionalization with a variety of nucleophiles [4,5]. The resulting diphenylcarbenium ion was also found to be an efficient initiating species for the polymerization of reactive monomers [6-13]. Therefore this intermediate capping reaction can provide synthetic polymer chemists with advanced tools for the control of macromolecular structures and properties. Our recent progress in the cationic macromolecular engineering is based on this strategy, and block copolymers and chain-end functionalized polymers have been prepared with unprecedented chemical as well as structural integrity. In this paper, on the basis of the last half decade's developments, an overview of capping and coupling reactions is presented as advanced tools for the cationic design of macromolecular structures and properties.

61

J.E. Puskas et al. (eds.), Ionic Polymerizations and Related Processes, 61–73.
© 1999 *Kluwer Academic Publishers. Printed in the Netherlands.*

2. Mechanistic and Kinetic Aspects of the Capping Reaction

The capping reaction of living polyisobutylene (PIB) with DPE (or its derivatives) comprises of two consecutive reactions as shown in Scheme 1; (i) the ionization of dormant PIB (PIBCl) by $TiCl_4$ and (ii) the addition reaction of DPE to ionized PIB (PIB^+). This capping reaction yields a stable and fully ionized diarylcarbenium ion (PIB-DPE^+) but is quantitative only under selected reaction conditions [14,15]. It has been found that the capping reaction of living PIB with DPE is an equilibrium reaction which shifts toward the right-hand side with decreasing temperature, or with increasing Lewis acidity, solvent polarity, electron-donating ability of p-substituents, or concentration of reactants [14].

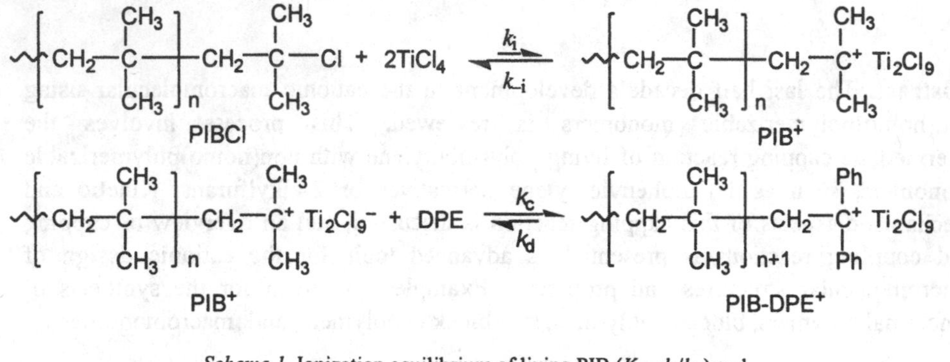

Scheme 1. Ionization equilibrium of living PIB ($K_i = k_i/k_{-i}$) and capping/decapping equilibrium ($K_{cd} = k_c/k_d$).

Thermodynamic ($K_e = K_i K_{cd}$) and kinetic constants ($k_c K_i$ and k_d) of the capping/decapping equilibrium reaction have been determined using spectroscopic methods (NMR and UV/vis) in the temperature range from –80 °C to –40 °C [14,16]. The value of the apparent equilibrium constant, K_e, was found to decrease approximately four orders of magnitude and that of the apparent rate constant, $k_c K_i$, two orders of magnitude upon increasing the temperature from –80 °C to –40 °C as summarized in Table 1. According to the latter finding, the value of the activation energy of the capping reaction is apparently negative which might be attributed to a negative enthalpy of the ionization of PIBCl, already invoked for the polymerization of isobutylene (IB) [17,18]. Interestingly, the kinetics of the capping and the decapping reaction depends substantially on the chain length of living PIB, i.e., the degree of polymerization, indicating that the chain length affects the reactivity/stability of the cationic species. The capping reaction is also applicable to living polystyrene [19]. This reaction, however, is irreversible, i.e., decapping of the chain ends does not occur.

Although it has been known that 2-alkylfurans are highly reactive toward electrophilic attack at C–5 position, no attempt has been made on the synthetic utilization of 2-alkylfurans in the living cationic polymerization of vinyl monomers. Our mechanistic studies revealed that 2-alkylfurans, such as 2-methylfuran and 2-*tert*-

TABLE 1. Apparent equilibrium constants K_e and rate constants $k_c K_i$ and k_d for the capping/decapping reaction of 2-chloro-2,4,4-trimethylpentane (TMPCl) and PIBCl with DPE[a]

T, °C	TMPCl (DP$_n$ = 2)			PIBCl (DP$_n$ = 36)		
	K_e, M^{-3}	$k_c K_i$, M^{-3}s^{-1}	k_d, s^{-1}	K_e, M^{-3}	$k_c K_i$, M^{-3}s^{-1}	k_d, s^{-1}
−80	$4.9 \cdot 10^6$	24.0	$4.9 \cdot 10^{-6}$	$5.2 \cdot 10^6$	170	$3.3 \cdot 10^{-5}$
−70	$7.4 \cdot 10^5$	13.9	$1.9 \cdot 10^{-5}$	$1.2 \cdot 10^5$	35.8	$3.0 \cdot 10^{-4}$
−60	$9.2 \cdot 10^4$	7.0	$8.0 \cdot 10^{-5}$	$2.4 \cdot 10^4$	18.4	$7.7 \cdot 10^{-4}$
−50	$7.0 \cdot 10^2$	5.0	$7.7 \cdot 10^{-3}$	$2.0 \cdot 10^3$	3.0	$1.5 \cdot 10^{-3}$
−40	$4.9 \cdot 10^2$	2.1	$1.5 \cdot 10^{-2}$	$2.2 \cdot 10^2$	1.1	$5.0 \cdot 10^{-3}$

[a][TMPCl]=[PIBCl]=2.0×10^{-3} M, [TiCl$_4$]=$(1.6-80) \times 10^{-2}$ M, and [DTBP]=3.0×10^{-3} M in hexane/methyl chloride (60/40, v/v) solvent mixture.

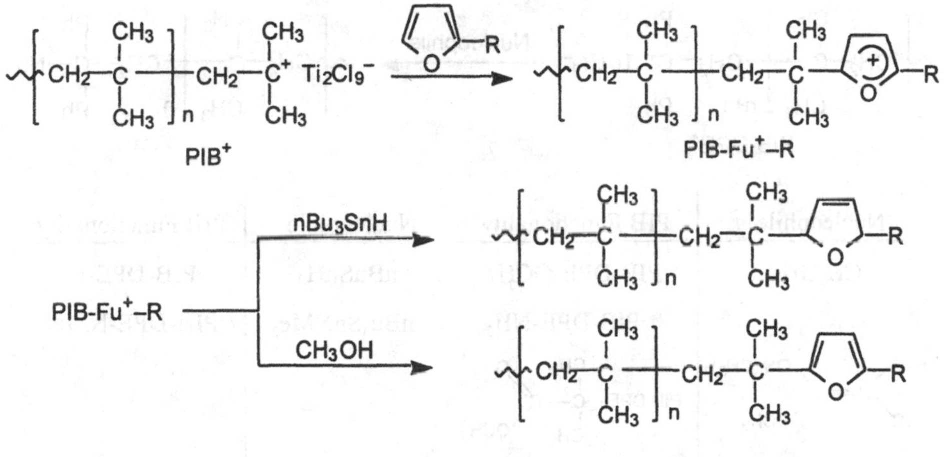

Scheme 2. Capping reaction of living PIB with 2-alkylfurans

butylfuran, quantitatively add to living PIB as shown in Scheme 2 [20]. Addition occurs exclusively at C–5 position and a stable tertiary allylic cation is generated at C–2 position which is further stabilized by the neighboring O-atom. The formation of the stable allylic cation was further confirmed from the trapping experiment of the resulting cation with tributyltin hydride. Quenching the resulting cation with tributyltin hydride yielded PIB with dihydrofuran functionality. Interestingly, quenching with methanol yielded PIB with furan functionality, most probably due to the intermediate formation of an acetal group which eliminates methanol.

One of the promising features in the capping reaction of living PIB with 2-alkylfurans is that the resulting tertiary allylic cation is stable up to −40 °C without

64

decapping [21]. Therefore, a quantitative capping reaction of living PIB could be achieved with 2-*tert*-butylfuran using the BCl$_3$/CH$_3$Cl/–40 °C system. Since the capping reaction of living PIB with DPE or DTE was found to be slow and incomplete under these conditions, it appears that 2-alkylfurans are more suitable capping agents when the subsequent functionalization or block copolymerization should be carried out at elevated temperature.

3. Synthetic Applications of the Capping Reaction

3.1. SYNTHESIS OF CHAIN-END FUNCTIONALIZED POLYMERS

The stable and fully ionized diarylcarbenium ion, obtained in the capping of PIBCl with DPE, is readily amenable for chain-end functionalization by quenching with appropriate nucleophiles as shown in Scheme 3 [1,4,5]. Using this strategy, a variety of chain-end functional PIBs, including methoxy, amine, carbonyl, and ester end-groups have been prepared. While PIBs with terminal-allyl functionality can be

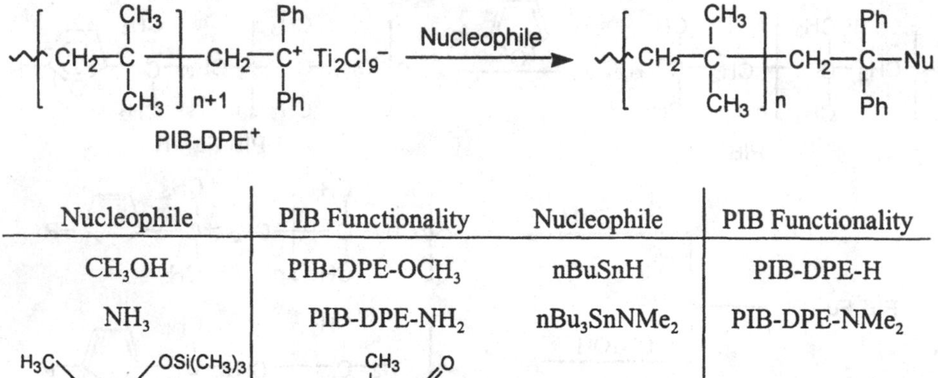

Nucleophile	PIB Functionality	Nucleophile	PIB Functionality
CH$_3$OH	PIB-DPE-OCH$_3$	nBuSnH	PIB-DPE-H
NH$_3$	PIB-DPE-NH$_2$	nBu$_3$SnNMe$_2$	PIB-DPE-NMe$_2$

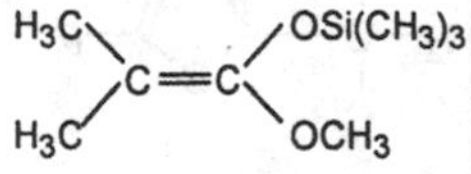

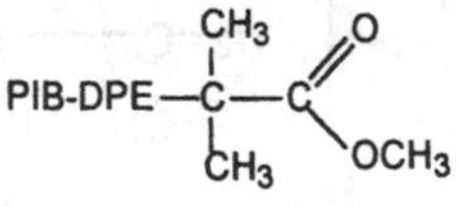

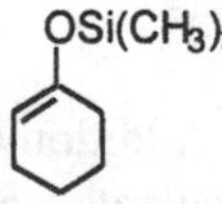

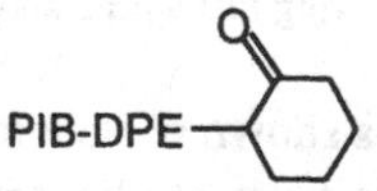

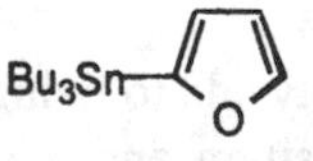

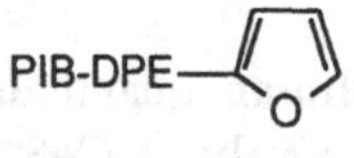

Scheme 3. Synthesis of chain-end functionalized PIBs

prepared by direct quenching of living PIB with allyltrimethylsilane [22], it was found that the capping reaction is necessary for the quantitative functionalization of living PIB with other organosilicon compounds containing Si–O bonds [4]. It is also notable that, when living PIB is capped with DPE, organotin compounds can also be used to introduce new functionalities such as –H, –N(CH$_3$)$_2$, and furan [5].

Although α,ω-telechelic PIBs are readily available when a difunctional initiator is employed in the IB polymerization, the synthesis of α,ω-asymmetrically functionalized PIBs has long been a challenge. By the rational combination of haloboration-initiation and capping techniques, a series of α,ω-asymmetrically functionalized PIBs have been prepared [23-25]. Polymers, prepared by haloboration-initiation, invariably carry an alkylboron head group [26-29] which can easily be converted into a primary hydroxyl group [30] or a secondary amine group [24,25]. It was found that living PIB, prepared by haloboration-initiation, can be capped with DPE, and the same functionalization strategy shown in Scheme 3 is applicable for the chain-end functionalization [23]. Silyl ketene acetals, such as 1-methoxy-1-trimethylsiloxy-2-methylpropene, 1-methoxy-1-trimethylsiloxypropene, and 1-methoxy-1-trimethylsiloxyethene, were employed to incorporate methoxycarbonyl groups as ω-functionality. After oxidation of the alkylboron head group, α-hydroxyl-ω-methoxycarbonyl asymmetric telechelic PIBs were prepared, and further hydrolysis of the sterically less hindered ω-ester group yielded α-hydroxyl-ω-carboxyl asymmetric telechelic PIBs. It should be noted that this procedure can be a general methodology for the synthesis of α,ω-asymmetric telechelic PIBs with a hydroxyl or an amine group as α-functionality and any functional groups shown in Scheme 2 as ω-functionality.

3.2. SYNTHESIS OF BLOCK COPOLYMERS

The synthesis of well-defined block copolymers by sequential monomer addition requires that the rate of a crossover reaction to a second monomer (R_{cr}) be faster than or at least equal to that of homopolymerization of a second monomer (R_p). Therefore, living polymerization is a necessary but not sufficient condition for the block copolymerization by sequential monomer addition. It has been our continuous theme of research to control the relative ratio of R_{cr}/R_p in the block copolymerization of IB with more reactive monomers, such as p-methylstyrene (pMeSt) [6,7], α-methylstyrene (αMeSt) [8,9], isobutyl vinyl ether (IBVE) [10] or methyl vinyl ether (MeVE) [11]. As shown in Scheme 4, a general scheme was developed to increase the relative ratio of R_{cr}/R_p, especially when the second monomer is more reactive than the first one. This process involves the capping reaction of living PIB with DPE or DTE, followed by the Lewis acidity tuning to the reactivity of the second monomer. The purpose of the Lewis acidity tuning is to generate stronger nucleophilic counterions which ensure a high R_{cr}/R_p ratio, as well as the living polymerization of a second monomer. This has been carried out using three different methods; (i) by the addition of titanium(IV) alkoxide [Ti(OR)$_4$], (ii) by the substitution of a strong Lewis acid with a weaker one, or (iii) by the addition of nBu$_4$NCl.

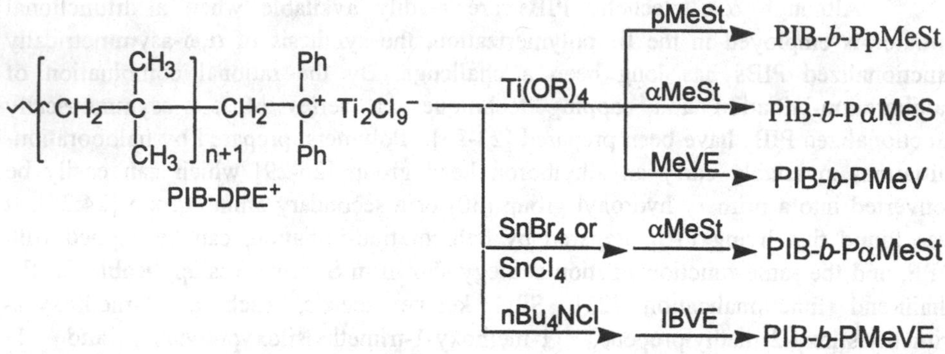

Scheme 4. Synthesis of block copolymers by sequential monomer addition

The first method has been successfully employed in the block copolymerization of IB with αMeSt, pMeSt, or MeVE [6-9,11]. This concept of the Lewis acidity tuning by mixing TiCl$_4$ with Ti(OR)$_4$ was also adapted by Sawamoto et al. in the living cationic polymerization of IBVE and β-pinene [31,32]. The substitution of TiCl$_4$ with a weaker Lewis acid (SnBr$_4$ or SnCl$_4$) was also proved to be an efficient strategy in the synthesis of poly(IB-*b*-αMeSt) diblock and poly(αMeSt-*b*-IB-*b*-αMeSt) triblock copolymers [8,9]. When SnCl$_4$ was employed, it was necessary to keep [SnCl$_4$] equal to or below 0.5[chain end] to increase the relative ratio of R_{cr}/R_p. Mechanistic studies indicated that, when [SnCl$_4$] ~ 0.5[chain end], a double charged counterion, SnCl$_6^{2-}$, is involved during the crossover reaction and converted to a single charged counterion, SnCl$_5^-$, during the polymerization of αMeSt [12]. The block copolymerization of IB with IBVE was also achieved by the Lewis acidity tuning using nBu$_4$NCl [10]. The addition of nBu$_4$NCl reduces the concentration of free and uncomplexed TiCl$_4$ ([TiCl$_4$]$_{free}$), and mechanistic studies indicated that, when [TiCl$_4$]$_{free}$ < [chain end], the dimeric counterion, Ti$_2$Cl$_9^-$, is converted to a more nucleophilic monomeric TiCl$_5^-$ counterion.

In addition to the block copolymerization by sequential monomer addition, the capping reaction of living PIB with DPE also provides a facile route for the preparation of block copolymers by the coupling reaction with living anionic polymers. This approach was employed in the synthesis of poly(IB-*b*-methyl methacrylate) in which the original poly(methyl methacrylate) was prepared by group transfer polymerization [33]. The coupling reaction of two living homopolymers with antagonistic functions can be an expedient alternative to the cumbersome site-transformation technique.

Block copolymerization of IB with MeVE was also carried out using 2-methylfuran or 2-*tert*-butylfuran as a capping agent [20]. While the crossover efficiency of ~66% was obtained using 2-*tert*-butylfuran as a capping agent, the crossover efficiency was slightly higher (~75%) when 2-methylfuran was employed as a capping agent under similar conditions. Structural analysis of the products indicated that the initiation of the MeVE polymerization occurs at C-4 position. This can be accounted for not only by the steric hindrance at C-2 position but by the formation of a more reactive (i.e., less stable) cation at C–4 position. In order to increase the relative ratio of R_{cr}/R_p, it appears that the optimum condition for the fine tuning of the Lewis acidity should be found.

4. Coupling Reaction of Living Cationic Polymers

The synthetic application of non(homo)polymerizable monomers was further extended to bis-DPE compounds, such as 2,2-bis[4-(1-phenylethenyl)phenyl]propane (BDPEP) and 2,2-bis[4-(1-tolylethenyl)phenyl]propane (BDTEP), consummating the living coupling reaction of living PIB [34,35]. It was demonstrated that living PIB

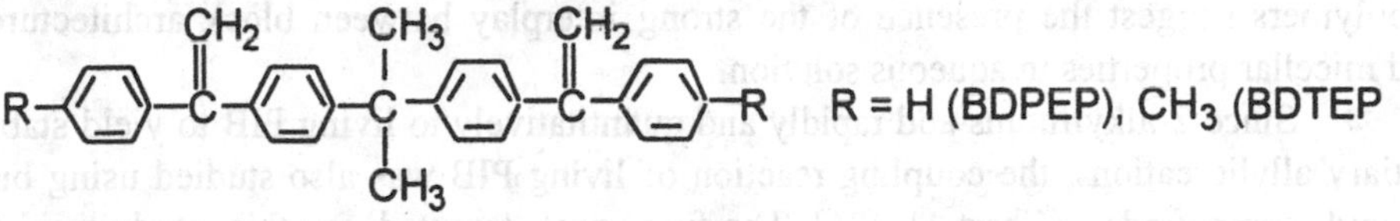

R = H (BDPEP), CH₃ (BDTEP

reacts quantitatively with BDPEP or BDTEP to yield stoichiometric amounts of bis(diarylalkylcarbenium) ions, as confirmed by the quantitative formation of diarylmethoxy functionalities at the junction of the coupled PIB. Since the resulting diaryl cations have been successfully employed for the controlled initiation of styrenic monomers and vinyl ethers, the concept of a living coupling reaction was proposed as a general route for the synthesis of A_2B_2 star-block copolymers. As a proof of the concept, an amphiphilic A_2B_2 star-block copolymer (A = PIB and B = PMeVE) has been prepared by the living coupling reaction of living PIB followed by the chain-ramification polymerization of MeVE at the junction of the living coupled PIB as shown in Scheme 5.

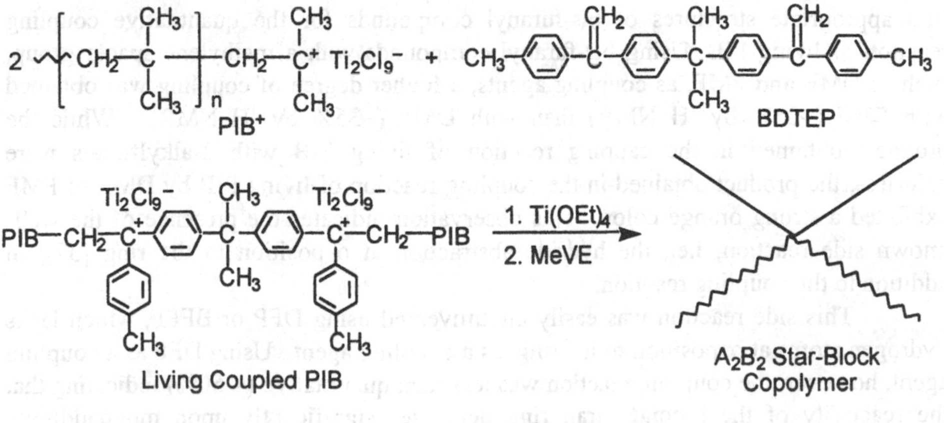

Scheme 5. Synthesis of A_2B_2 star-block copolymers via the living coupling reaction

Architecture/property studies were also carried out by comparison of micellar properties of star-block copolymer and two linear diblock analogues. A_2B_2 star-block copolymers with 80 wt% of PMeVE composition $((IB_{45})_2$-b-$(MeVE_{170})_2)$ exhibited a critical micelle concentration (CMC) of 4.25×10^{-4} M in water, which is an order of magnitude higher than CMCs obtained with linear-diblock copolymers with same total M_n and composition $(IB_{90}$-b-$MeVE_{340})$ or with same segmental lengths $(IB_{45}$-b-$MeVE_{170})$. This suggests that block copolymers with star architectures exhibit less tendency to association than their corresponding linear diblock copolymers. Interestingly, $(IB_{45})_2$-b-$(MeVE_{170})_2$ exhibited an average particle sizes comparable to that of IB_{90}-b-$MeVE_{340}$ in aqueous solution above the CMCs. While no evidence was observed for the architectural effect on the thermal properties such as the glass transition, preliminary results in the solution properties of star-block or linear block copolymers suggest the presence of the strong interplay between block architectures and micellar properties in aqueous solution.

Since 2-alkylfurans add rapidly and quantitatively to living PIB to yield stable tertiary allylic cations, the coupling reaction of living PIB was also studied using bis-furanyl compounds (Chart 1) [36]. The first goal targeted in this study was to

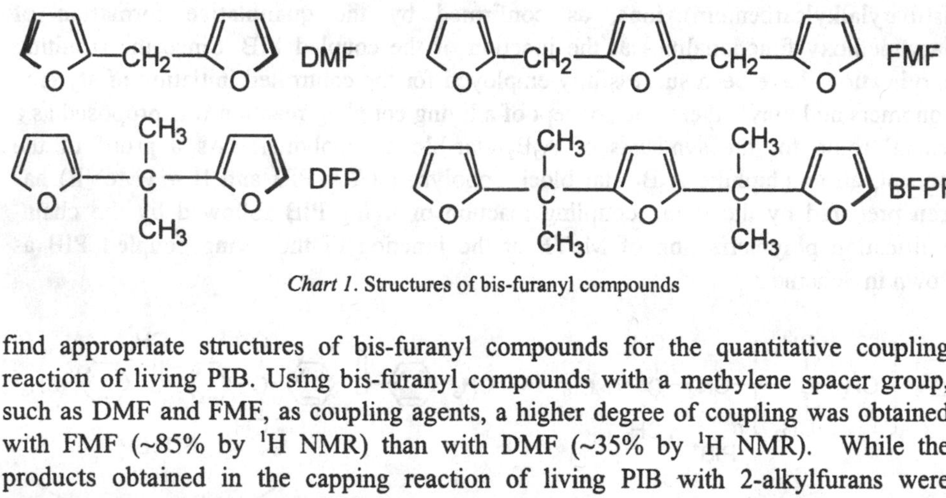

Chart 1. Structures of bis-furanyl compounds

find appropriate structures of bis-furanyl compounds for the quantitative coupling reaction of living PIB. Using bis-furanyl compounds with a methylene spacer group, such as DMF and FMF, as coupling agents, a higher degree of coupling was obtained with FMF (~85% by ^{1}H NMR) than with DMF (~35% by ^{1}H NMR). While the products obtained in the capping reaction of living PIB with 2-alkylfurans were colorless, the product obtained in the coupling reaction of living PIB by DMF or FMF exhibited a strong orange color. This observation indicates the presence of the well-known side reaction, i.e., the hydride abstraction at α-position to the ring [37], in addition to the coupling reaction.

This side reaction was easily circumvented using DFP or BFPF, which lacks hydrogen atoms at α-position to the ring, as a coupling agent. Using DFP as a coupling agent, however, the coupling reaction was less than quantitative (< 50%) indicating that the reactivity of the second furan ring decreases significantly upon monoaddition. When BFPF was used as a coupling agent, the coupling was complete within 30 min in hexane/CH$_3$Cl (60/40 or 40/60, v/v) solvent mixtures on the basis of spectroscopic as well as chromatographic analyses. For example, the ^{1}H NMR spectrum of the final product exhibited the absence of resonance signals for aromatic group as well as for PIB with a terminal-chloro group, indicating the absence of both monoadduct and unreacted PIB. Furthermore, the final product exhibited doubled M_n as well as narrowed M_w/M_n, confirming the quantitative coupling reaction of living PIB by BFPF.

Unique applications of the coupling reaction by BFPF can be found in the BCl$_3$/CH$_3$Cl/–40 °C system in which the coupling reaction of living PIB is not convenient by bis-DPE compounds. For instance, in-situ coupling of living PIB, prepared by haloboration-initiation using the BCl$_3$/CH$_3$Cl/–40 °C system, by BFPF yielded α,ω-telechelic PIB with alkylboron functionality. After oxidation, this telechelic PIB was converted to α,ω-hydroxyl PIB. The synthesis of α,ω-telechelic PIBs with a vinyl functionality was also achieved by the coupling reaction of living PIB, prepared using 3,3,5-trimethyl-5-chloro-1-hexene as an initiator in the presence of TiCl$_4$. Potentials of BFPF as a living coupling agent are under current investigation along with block copolymerization using 2-alkylfurans as capping agents.

5. Synthesis of Precursor Macromonomers for Controlled Architectures

Macromonomers with a terminal non(homo)polymerizable vinylidene group, such as DPE, have unique and appealing applications in the preparation of block copolymers with controlled architectures. They have been used as precursor polymers for a variety of block copolymers with nonlinear architectures such as ABC-type star-block [38-42] or comb-type graft copolymers [41]. Since 2-alkylfurans can be used as cationic capping agents [20], it is readily conceivable that furan-functionalized polymers can also serve as precursor polymers for the synthesis of nonlinear block copolymers.

5.1 SYNTHESIS OF ω-DPE-FUNCTIONALIZED PIB

During the study of the living coupling reaction using bis-DPE compounds, it was found that, for the quantitative coupling reaction of living PIB, two DPE moieties of a coupling agent should be separated by a spacer group which provides independent reactivity of two double bonds towards living PIB. When 1,3-bis(1-phenylethenyl)

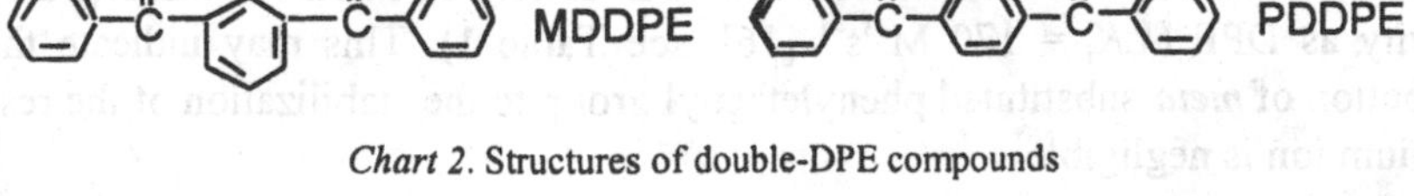

Chart 2. Structures of double-DPE compounds

benzene (MDDPE) (Chart 2) was employed as a potential coupling agent, the coupling reaction of living PIB was not observed indicating that the second double bond is far less reactive than the first one [35]. This was attributed to the formation of the electron deficient α-substituent to the second double bond due to the delocalization of positive charge over the *meta*-substituted aromatic ring. Taking advantage of this monoaddition reaction, the synthesis of PIB macromonomers with terminal-DPE functionality has been achieved using double-DPE compounds such as MDDPE and 1,4-bis(1-phenylethenyl)benzene (PDDPE) as shown in Scheme 6 [43].

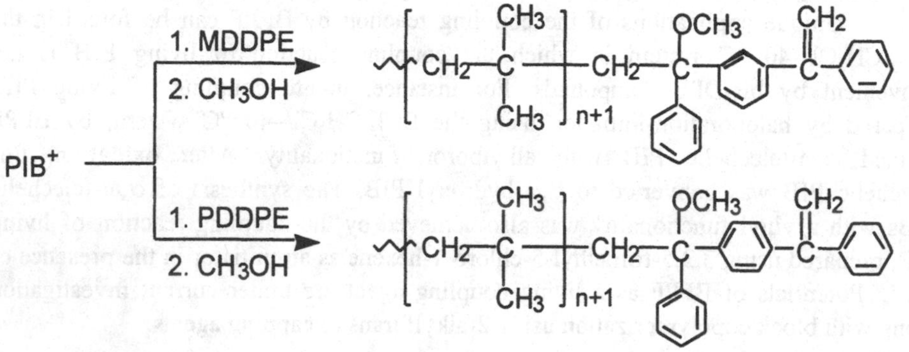

Scheme 6. Synthesis of PIB macromonomers with DPE functionality

Using 2 equiv. of MDDPE, a maximum conversion of ~90% was obtained with negligible but detectable amounts of the coupled product (~3%). However, when a large excess of MDDPE (4 equiv.) was employed, ~100% DPE functionality could be achieved without the formation of the coupled product. Since electron deficient α,α-disubstituents are formed in the second double bond upon monoaddition of living PIB to PDDPE, it was readily conceivable that PDDPE might be a better candidate for the synthesis of PIB-DPE macromonomer without using a large excess. As expected, the addition reaction of living PIB to PDDPE was much faster than to MDDPE, and ~100% functionalization was achieved in 75 min using only 2 equiv of PDDPE. Moreover, no evidence was observed for the formation of the coupled product from the GPC trace of the final product. With 4 equiv of PDDPE, the half-life of the addition reaction was calculated to be 2 min and ~100% functionalization was achieved within 20 min.

The kinetics of the addition reaction of living PIB to double-DPE compounds was further studied using a fiber-optic visible spectrometer. Using a similar kinetic equation which was developed for the capping reaction of living PIB with DPE, the apparent rate constant ($k_c K_i$) in the addition reaction of living PIB to double-DPE compounds was determined to be 370 and 940 $M^{-3}s^{-1}$ with MDDPE and PDDPE, respectively. Although direct comparison is not possible due to the slight difference in solvent polarity, it is also notable that the DPE moiety in MDDPE exhibits a similar reactivity as DPE ($k_c K_i = 170$ $M^{-3}s^{-1}$ [16], see Table 1). This may indicate that the contribution of *meta*-substituted phenylethenyl group to the stabilization of the resulting carbenium ion is negligible.

5.2. SYNTHESIS OF ω-FURAN-FUNCTIONALIZED PIB

As shown in Scheme 2, ω-furan functionalized PIB was prepared by the capping reaction of living PIB followed by functionalization using 2-tributylstannylfuran. While elaborating the synthesis, it was further found that ω-furan functionalized PIB can be obtained by direct functionalization of living PIB using 2-tributylstannylfuran without DPE capping. The resulting ω-furan functionalized PIB can be used as a precursor polymer for the synthesis of AB diblock copolymers or ABC and *AA'B* type star-block copolymers where *A* and *A'* represents PIB segments with different molecular weights and *B* and C represent chemically different block segments. For instance, the strategy for the synthesis of *AA'B* type star-block copolymers, shown in Scheme 7, involves the coupling reaction of PIB with ω-furan functionality (*A*) with living PIB with a different molecular weight (*A'*), followed by the chain-ramification polymerization of a second monomer to yield a chemically different block segment (*B*). The addition reaction of living PIB to PIB with ω-furan functionality was found to be quantitative, and research is under progress for the chain-ramification polymerization of a second monomer to construct *AA'B* type star-block copolymers.

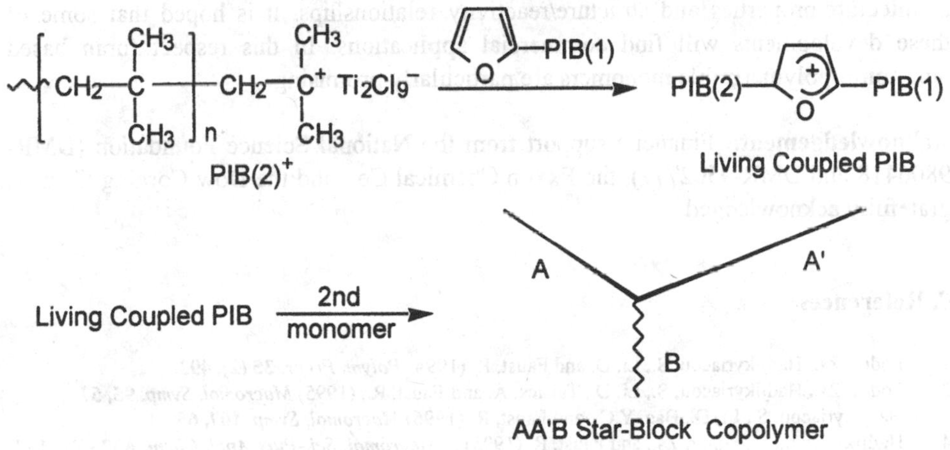

Scheme 7. Synthesis of *AA'B* type star-block copolymers

6. Future Perspectives

It has been show in this review that the methodology of using non(homo)-polymerizable monomers is a powerful new tool in cationic macromolecular design and synthesis. Although extensively demonstrated only for living PIB, this technique should be generally applicable to all living cationic polymers. Thus functional polymers and block copolymers based on styrene and other styrenic monomers (pMeSt, αMeSt, indene etc.) should follow PIB based materials. An especially promising goal easily attainable using this methodology is the design and construction of nonlinear

72

block architectures, such as graft, comb, ring, star, or H-shaped polymers, that self-assemble into ordered nano-structures.

Applications of non(homo)polymerizable monomers in kinetic and mechanistic studies are also promising. A significant and unexpected finding mentioned earlier, is the chain length dependence of the rate constants $k_c K_i$ and k_d and the respective activation parameters for capping PIBCl with DPE in the presence of $TiCl_4$. These findings can be attributed only in part to the effect of the chain length on the position of the ionization equilibrium ($\rightarrow K_i$), and it appears that the chain length might also affect the reactivity/stability of the cationic end group. This hypothesis shall be followed by e.g. ab initio calculations on the transition state. When $k_{-i} \ll k_c[DPE]$, ionization becomes the rate-determining step, and then monitoring the capping reaction by UV/vis spectroscopy should provide the value of the rate constant of ionization, k_i. In order to reach $k_{-i} \ll k_c[DPE]$, the reactivity and the concentration of the capping agent must be sufficiently high. Therefore, kinetic studies of the capping reaction with DTE, or furan derivatives at high concentrations are currently under investigation. These studies shall also be extended to living styrene and αMeSt chain ends that should provide relative electrophilicities of polymer cations.

While the above research will advance our general understanding of architecture/properties and structure/reactivity relationships, it is hoped that some of these developments will find commercial applications. In this respect furan based non(homo)polymerizable monomers are particularly promising.

Acknowledgement. Financial support from the National Science Foundation (DMR-9806418 and DMR-9502777), the Exxon Chemical Co., and the Dow Corning Corp. is gratefully acknowledged

7. References

1.   Fodor, Zs., Hadjikyriacou, S., Li, D. and Faust, R. (1994) *Polym. Prepr.* **35** (2), 492.
2.   Fodor, Zs., Hadjikyriacou, S., Li, D., Takacs, A. and Faust, R., (1995) *Macromol. Symp.* **95**, 57.
3.   Hadjikyriacou, S., Li, D., Bae, Y.C., and Faust, R. (1996) *Macromol. Symp.* **107**, 65.
4.   Hadjikyriacou, S., Fodor, Zs., and Faust, R. (1995) *J. Macromol. Sci.-Pure Appl. Chem.* **A32** (6), 1137.
5.   Hadjikyriacou, S. and Faust, R. (1997) *Polym. Mater. Sci. Eng.* **76**, 300.
6.   Fodor, Zs. and Faust, R. (1994) *J. Macromol. Sci.-Pure Appl. Chem.* **A31** (12), 1983.
7.   Fodor, Zs. and Faust, R. (1995) *J. Macromol. Sci.-Pure Appl. Chem.* **A32** (3), 575.
8.   Li, D. and Faust, R. (1995) *Macromolecules* **28**, 1383.
9.   Li, D. and Faust, R. (1995) *Macromolecules* **28**, 4893.
10. Hadjikyriacou, S. and Faust, R. (1995) *Macromolecules* **28**, 7893.
11. Hadjikyriacou, S. and Faust, R. (1996) *Macromolecules* **29**, 5261.
12. Li, D., Hadjikyriacou, S., and Faust, R. (1996) *Macromolecules* **29**, 6061.
13. Mayr, H., Roth, M., and Faust, R. (1996) *Macromolecules* **29**, 19.
14. Bae, Y.C., Fodor, Zs., and Faust, R. (1997) *ACS Symp. Ser.* **665**, 168.
15. Charleux, B., Moreau, M., Vairon, J.-P., Hadjikyriacou, S., and Faust, R. (1998) *Macromol. Symp.* **132**, 25.
16. Schlaad, H., Erentova, K., Faust, R., Charleux, B., Moreau, M., Vairon, J.-P., and Mayr, H. (1998) *Macromolecules* **31**, 8058.
17. Sigwalt, P. (1998) *Macromol. Symp.* **132**, 127.

18. Fodor, Zs., Bae, Y.C., and Faust, R. (1998) *Macromolecules* **31**, 4439.
19. Canale, P.L. and Faust, R. (1998) *Polym. Prepr.* **39** (2), 400; submitted to *Macromolecules*.
20. Hadjikyriacou, S. and Faust, R. (1998) *Polym. Prepr.* **39** (2), 398.
21. Hadjikyriacou, S. and Faust, R. (1998) *US Patent Application* # 967,002.
22. Ivan, B. and Kennedy, J. P. (1990) *J. Polym. Sci.: Part A: Polym. Chem.* **28**, 89.
23. Koroskenyi, B. and Faust, R. (1998) *ACS Symp. Ser.* **704**, 135.
24. Koroskenyi, B. and Faust, R. (1998) *Polym. Prepr.* **39**(2), 492.
25. Koroskenyi, B. and Faust, R. (1999) *J. Macromol. Sci.*, in press.
26. Balogh, L., Faust, R., and Wang, L. (1994) *Macromolecules* **27**, 3453.
27. Balogh, L., Fodor, Zs., Kelen, T. and Faust, R. (1994) *Macromolecules* **27**, 4648.
28. Wang, L., McKenna, S.T., and Faust, R. (1995) *Macromolecules* **28**, 4681.
29. Koroskenyi, B., Wang, L., and Faust, R. (1997) *Macromolecules* **30**, 7667.
30. Wang, L., Svirkin, J., and Faust, R. (1995) *Polym. Mater. Sci. Eng.* **72**, 173.
31. Kamigaito, M., Sawamoto, M., and Higashimura, T. (1995) *Macromolecules* **28**, 5671.
32. Lu, J., Kamigaito, M. Sawamoto, M., Higashimura, T., and Deng, Y.-S. (1995) *Macromolecules* **30**, 22.
33. Takacs, A. and Faust, R. (1995) *Macromolecules* **28**, 7266.
34. Bae, Y.C., Fodor, Zs., and Faust, R. (1997) *Macromolecules* **30**, 198.
35. Bae, Y.C. and Faust, R. (1998) *Macromolecules*, **31**, 2480.
36. Hadjikyriacou, S. and Faust, R. (1998) *US Patent Application* # 965,443.
37. Gandini, A. (1992) In *Comprehensive Polymer Science, Suppl. 1*, S. L. Aggarwal and S. L. Russo Eds., Pergamon, Oxford, UK, p527.
38. Fujimoto, T., Zhang, H., Kazama, T., Isono, Y., Hasegawa, H., and Hashimoto, T. (1992) *Polymer* **33**, 2208.
39. Quirk, R.P. and Kim, Y.J. (1996) *Polym. Prepr.* **37** (2), 643.
40. Quirk, R.P. and Yoo, T. (1993) *Polym. Bull.* **31**, 29.
41. Quirk, R.P., Hong, D., Kim, Y.J., and Yoo, T. (1996) *Polym. Prepr.* **37** (2), 402.
42. Huckstadt, H., Abetz, V., and Stadler, R. (1996) *Macromol. Rapid Commun.* **17**, 599.
43. Bae, Y.C. and Faust, R. (1998) *Macromolecules*, in press.

SEGMENTED POLYMER NETWORKS BY CATIONIC POLYMERIZATION: DESIGN AND APPLICATIONS

F.E. DU PREZ AND E.J. GOETHALS
University of Gent, Department of Organic Chemistry, Polymer Chemistry Division, Krijgslaan 281 S4, 9000 Gent, Belgium

Abstract. Well-defined acrylate terminated bis-macromonomers of poly(vinylether)s, poly(2-alkyl-2-oxazoline)s and poly(1,3-dioxolane) have been synthesized by making use of macromolecular design concepts of cationic polymerization. The acrylate end groups of the bis-macromonomers have been radically copolymerized with vinyl monomers such as methyl methacrylate, butyl acrylate and hydroxyethyl acrylate for the synthesis of hydrophobic, amphiphilic or hydrophilic segmented copolymer networks. The nature and the ratio of the two components and the molecular weight of the bis-macromonomers determine the morphology, hydrophilic/hydrophobic balance and swelling properties of the network structures. These parameters will be described in terms of the use of these materials for a variety of purposes, such as for pervaporation, shape memory properties or damping applications.

1. Introduction

The properties of polymer materials can be improved and adjusted by physical or chemical combination of two or more structurally different polymers. In the case of intrinsically incompatible polymers, the properties are determined to a great extent by the way of blending [1,2]. It is now well established that the introduction of chemical cross-links between the polymers allows the control of the phase morphology of the end products. One way to obtain a thermoset combination of polymers is by grafting of polymer A to polymer B at both chain ends, which results in the formation of AB-cross-linked polymers (ABCPs), further referred to as segmented networks. The physicochemical properties of these networks are expected to be the result of a combination of the properties of segmented structures such as block or graft copolymers on one hand and of polymer networks on the other hand. For example, in block copolymer systems, the length of the blocks determines the size of the domains. Correspondingly, the cross-link level (length of blocks between the cross-links) will play a major role in determining the domain size of the segmented networks.

In this paper, we will focus on the synthesis and properties of segmented networks, mainly amphiphilic ones. In general, amphiphilic polymers are defined as

75

J.E. Puskas et al. (eds.), Ionic Polymerizations and Related Processes, 75–98.

macromolecular substances consisting of covalently bonded blocks or segments of opposite philicity. In the case of amphiphilic networks, the random assemblage of hydrophilic and hydrophobic chains results in the swelling of these materials in both water and hydrocarbons. The amphiphilic nature of copolymer structures has given rise to their unique properties in the solid state, as well as in selective solvents [3].

For many applications, a precise control of molecular weight, molecular weight distribution and hydrophilic/hydrophobic balance of the copolymer structure is required. Although several methods have been described to prepare segmented networks, only few of them result in the formation of macromolecular network structures with control of the structural parameters such as molecular weight between the cross-links and number of branches at the junction points [4]. Preferred building blocks for the synthesis of well-defined segmented networks include telechelics and bis-macromonomers as they afford the possibility to separate the polymerization process from network formation. These functionalized prepolymers are respectively defined as linear macromolecules bearing reactive functional groups and polymerizable groups at both chain ends. Tezuka *et al.*, for example, proposed a system in which the ion-coupling of telechelics provides a novel cross-linking reaction process to synthesize network copolymers with controlled structures [5]. Another approach, given by Kennedy *et al.*, consisted of the radical copolymerization of hydrophobic bis-macromonomers of polyisobutylene with hydrophilic monomers, such as N,N'-dimethylacrylamide or 2-(dimethylamino)-ethyl methacrylate [6,7]. The term "chameleon" networks was introduced to describe the ability of the networks to change their conformations depending on the medium they are exposed to [8]. It was demonstrated that such conformational adaptability is important for delayed drug release applications.

In our research group, the first stage in the synthesis of the segmented networks is directed towards the preparation of well-defined bis-macromonomers by cationic polymerization procedures. The thus obtained prepolymers are combined with vinyl polymers by free radical copolymerization of their end groups with a variety of vinyl monomers such as methyl methacrylate, hydroxyethyl acrylate or styrene. The bis-macromonomers used for this purpose were mostly derived from hydrophilic polymers such as poly(2-methyl-2-oxazoline), poly(1,3-dioxolane) or poly(N-tert-butylaziridine) [9,10]. The synthesis and characterization of some of the corresponding networks will be reviewed. Recently, bis-macromonomers of the hydrophobic poly(vinyl ether)s (PVE), poly(isobutyl vinyl ether) and poly(octadecyl vinyl ether), have been synthesized. The first results on the properties of the hydrophobic PVE-containing segmented networks will also be presented. The dependence of the morphology and swelling properties of the networks on the nature and ratio of both segments as well as on the molecular weight of the bis-macromonomer will be described. The use of these materials will be discussed for a variety of purposes, such as for pervaporation, shape memory properties or sound and vibration damping.

2. Synthesis of Bis-macromonomers by Cationic Polymerization Procedures

As outlined in the introduction part, bis-macromonomers of different polymers will be used as starting materials for the synthesis of the block copolymer networks. In the following sections, we report on the synthesis of these bis-macromonomers. A variety of cationic polymerization procedures, based on well-known macromolecular design concepts, have been used. Vinyl polymerization has been applied for the synthesis of bis-macromonomers of poly(vinyl ether)s while ring opening polymerization has been applied for the synthesis of bis-macromonomers of polyDXL and poly(alkyl oxazoline)s. Both polymerization processes proceed on electron-deficient active species, being attacked by nucleophiles, respectively double bonds or hetero-atoms. On the condition that suitable initiating systems are used, these polymerizations result in well defined polymers with predetermined molecular weights, low polydispersity and controlled functionality. It is beyond the scope of this publication to describe the mechanistic and experimental details and the roles of the different constituents in the polymerizations. We refer the reader to several reviews dealing with the chemistry of cationic vinyl and ring-opening polymerizations and with the synthesis of telechelic polymers and macromonomers [11,12].

2.1. SYNTHESIS OF BIS-MACROMONOMERS OF POLY(ISOBUTYL VINYL ETHER) AND POLY(OCTADECYL VINYL ETHER)

Since the discovery of the first controlled/living cationic polymerization of vinyl ethers with the HI/I_2 initiating system [13], many vinyl ether monomers have been polymerized using a large number of initiating systems [11,14]. In the last decade, many investigations have been devoted to the optimization of the initiating systems, leading to the design of numerous polymer architectures.

In this paper, the living character of the polymerization is used to synthesize end group-functionalized hydrophobic poly(isobutyl vinyl ether) (polyIBVE) and poly(octadecyl vinyl ether) (polyODVE). PolyIBVE is an amorphous polymer with a T_g of –35 °C, while polyODVE is a highly hydrophobic semi-crystalline polymer with a melting point around 50 °C.

Both vinyl ether polymerizations were initiated with the initiating system acetal/trimethylsilyl iodide (TMSI)/activator [15,16,17]. The reaction of an acetal with TMSI leads to an α-iodo ether derivative, which leads, in combination with the activator ZnI_2, to a controlled synthesis of poly(vinyl ether)s (PVE). For the preparation of the bis-macromonomers, a bifunctional chain precursor was formed in a first step by reaction of a bis-acetal compound, 1,1,3,3-tetramethoxypropane (TMP) (or malonaldehyde bis(dimethylacetal)) with TMSI at –40 °C in toluene (Scheme 1):

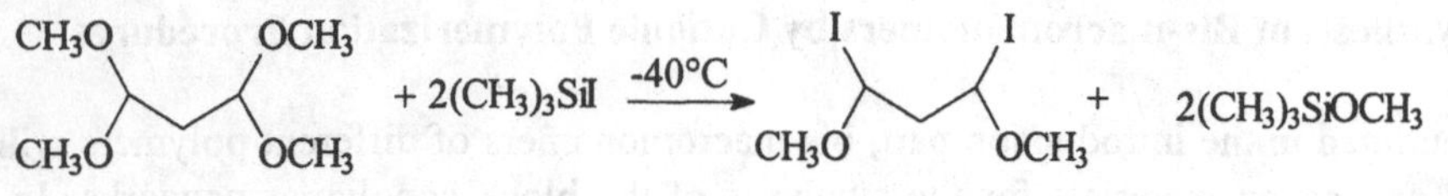

Scheme 1.

Bifunctional initiation of IBVE or ODVE was then performed respectively at –40 °C and 0 °C, in toluene, by adding successively the monomer and ZnI_2 as activator. The polymerization of ODVE is carried out at 0 °C because of solubility problems of the monomer at lower temperatures. Keeping the activator/initiator molar ratios lower than 0.1 could minimize side reactions, such as the transfer of a proton in the β-position of the carbocation to monomer. Finally, the bifunctionally living vinyl ether polymers were end-capped with a functionalized nucleophile, such as hydroxyethyl methacrylate (HEMA) or hydroxyethyl acrylate (HEA) (Scheme 2). It was shown before that, notwithstanding transfer reactions at monomer to initiator concentration ratios ([M]/[In]) higher than 20 (Mn > 6000), it is possible to prepare end group functionalized polyODVE by end-capping with a functionalized alcohol [18].

As shown in Table 1, the molecular weights obtained by end group analyses of the polymers by ^1H NMR spectroscopy are close to those obtained by GPC analyses, indicating that the TMSI/TMP initiation system can be used for the synthesis of bis-macromonomers of polyODVE and polyIBVE with controlled molecular weight and narrow polydispersities.

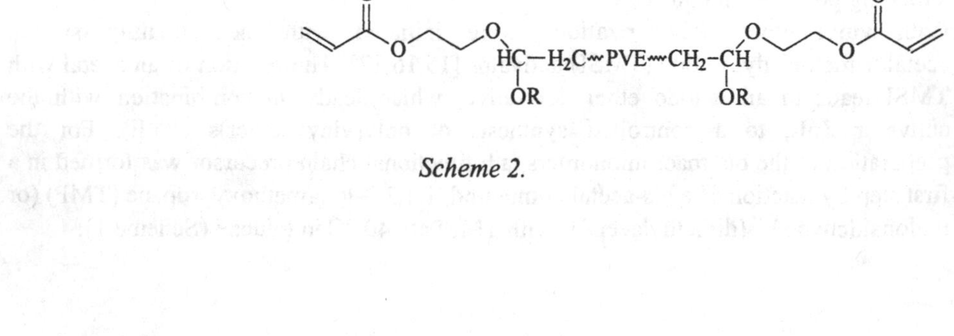

Scheme 2.

TABLE 1. Polymerization of IBVE and ODVE with the TMSI/TMP/ZnI$_2$ initiating system in toluene [a].

Monomer	$[M]_0/[I]_0$	M_n (theo)[b]	M_n (^{1}H-NMR)[c]	M_n (GPC)[d]	M_w/M_n
IBVE	29.6	3300	2800	3300	1.2
IBVE	59.4	6300	6200	6800	1.2
IBVE	80.1	8300	7900	7700	1.2
ODVE	11.2	3600	4800	4500	1.1
ODVE	22.4	6800	5400	5900	1.1

[a] IBVE : in toluene at -40°C; [ZnI$_2$]/[bis-α-iodoether] = 1 : 20; terminated with HEA
   ODVE : in toluene at 0°C; [ZnI$_2$]/[bis-α-iodoether] = 1 : 100; terminated with HEMA
[b] M_n (theo) = $[M]_0/[I]_0$ * M (monomer) + M (initiator) + 2.M (end group)
[c] M_n (^{1}H-NMR) = calculated from ratio of integration of HEMA or HEA end group to polymer side chain
[d] M_n (GPC) = measured by GPC calibrated with polystyrene standards

2.2. SYNTHESIS OF BIS-MACROMONOMERS OF POLY(1,3-DIOXOLANE)

The cationic ring-opening polymerization of the cyclic acetal, 1,3-dioxolane (DXL), is characterized by a reversible termination reaction. Due to the continuous termination followed by re-initiation ("scrambling"), the molecular weight distribution reaches a theoretical value of two. Intramolecular reactions, i.e. reaction of oxygen atoms in the polymer chain on the active center, also lead to the formation of a fraction of cyclic oligomers, the concentration of which is given by the Jacobson - Stockmayer theory.

However, the molecular weight can be controlled by the ratio of monomer to initiator concentration because the concentration of active species remains constant. One approach that has been developed to introduce functional end groups is based on the so-called "end blocker" method. This method is in fact based on an intermolecular reaction that is similar to the "transfer to polymer" but taking place with an added transfer agent, the end blocker. This end blocker consists of a low molecular weight acetal that contains two functional groups.

This kind of transfer reaction has formerly been applied for the synthesis of functionalized polysiloxanes and recently, for the synthesis of functionalized polyacetals [10,19]. PolyDXL α,ω-bisacrylate was prepared by polymerization of DXL in the presence of methylenebis(oxyethyl acrylate). This end blocker was obtained by the reaction of HEA with paraformaldehyde under acid catalysis.

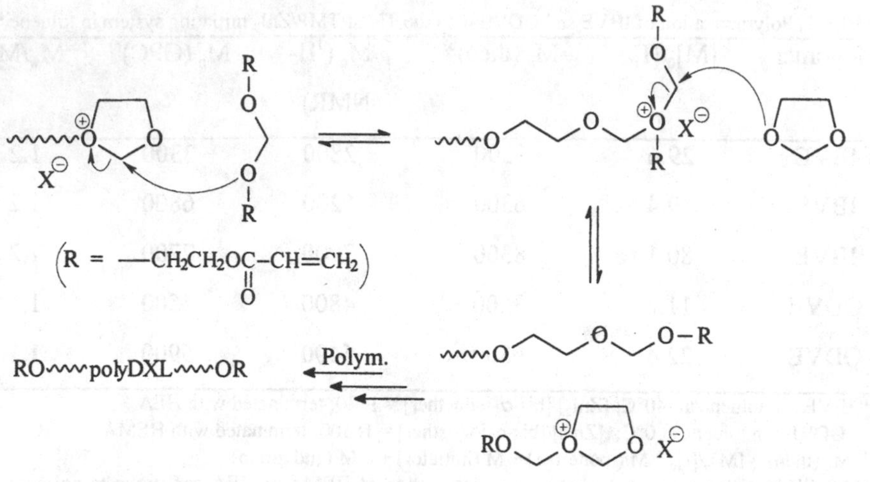

$$\left(R = -CH_2CH_2OC-CH=CH_2\right)$$
$$\qquad\qquad\underset{O}{\|}$$

Scheme 3.

The average degree of polymerization DP_n of the linear fraction of the polymer is given by equation 1:

$$\overline{DP}_n = \frac{[M]_o - [M]_e - [M]_c}{[I]_o + [T]_o - [T]_e} \qquad (1)$$

where $[M]_o$ and $[M]_e$ are the initial and equilibrium monomer concentrations and where $[M]_c$ corresponds to the concentration of cyclic oligomers. $[I]_o$ is the initiator concentration, $[T]_o$ and $[T]_e$ are the initial and equilibrium concentration of the transfer reagent respectively.

2.3. SYNTHESIS OF BIS-MACROMONOMERS OF POLY(2-ALKYL-2-OXAZOLINE)S

A well-known example of polymerizations with a highly living character is that of the 1,3-oxazolines. This type of polymerization differs from the previous one by the fact that chain transfer to polymer is of no importance. During the propagation step, the iminoether function in the monomer isomerizes to an amide function, whose nucleophilicity is much lower than that of the former. The driving force for the polymerization of these five membered cyclic imino ethers, which have little ring strain, is provided by this isomerization to repeating units containing the more stable carbonyl double bond of an alkyl aminoacyl group.

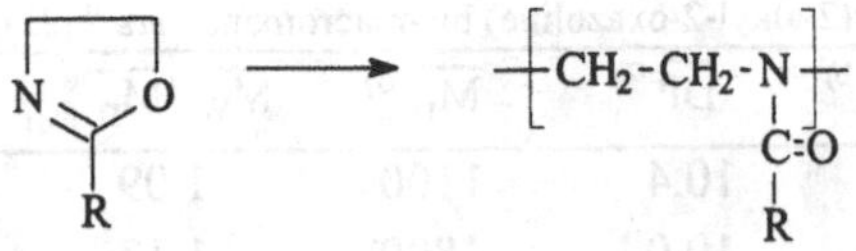

Scheme 4.

Such polymerization allows the optimal control of the polymer structure, including the incorporation of functional end groups by end-capping [20,21]. For the synthesis of the poly(2-oxazoline) bis-macromonomers, 1,4-dibromo-2-butene (DBB) was chosen as bifunctional initiator and acrylic acid (AA) as end-capping agent for the living oxazolinium species (in the presence of a proton scavenger) [22,23].

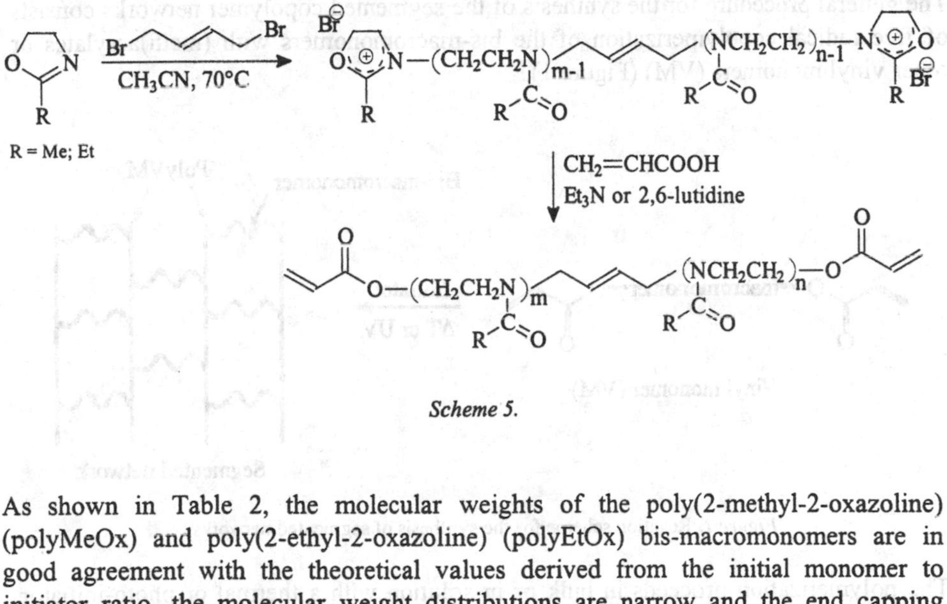

Scheme 5.

As shown in Table 2, the molecular weights of the poly(2-methyl-2-oxazoline) (polyMeOx) and poly(2-ethyl-2-oxazoline) (polyEtOx) bis-macromonomers are in good agreement with the theoretical values derived from the initial monomer to initiator ratio, the molecular weight distributions are narrow and the end capping reaction proceeded quantitatively (functionality of two).

TABLE 2. Synthesis of poly(2-alkyl-2-oxazoline) bis-macromonomers [a] [22].

R	$[ROx]_0/[DBB]_0$ [b]	$\overline{DP}$ [c]	$\overline{M_n}$ [d]	$\overline{M_w}/\overline{M_n}$ [e]	Yield in %	F [f]
Me	9.9	10.4	1100	1.09	97	1.92
Me	19.3	19.0	1800	1.13	99	2.00
Me	52.4	53.7	4800	1.18	96	2.07
Et	19.1	22.8	2500	1.14	98	1.97

[a] Polymerization conditions: solvent CH_3CN, 70°C, 6-8 h; termination: 20 h, 60°C.
[b] Initial monomer to initiator ratio.
[c] Average degree of polymerization, calculated from ^1H NMR spectra.
[d] Number-average molecular weight, obtained by ^1H NMR.
[e] Molecular weight distribution, measured by GPC.
[f] End group functionality, calculated from ^1H NMR spectra.

3. Synthesis and Properties of Segmented Networks

The general procedure for the synthesis of the segmented copolymer networks consists of free radical copolymerization of the bis-macromonomers with (meth)acrylates or other vinyl monomers (VM) (Figure 1).

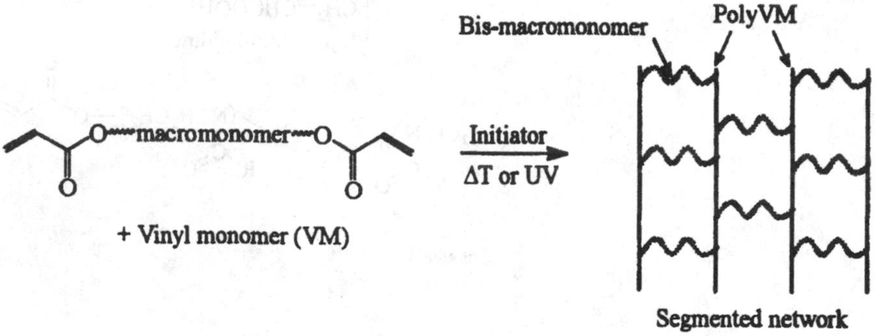

Figure 1. Reaction scheme for the synthesis of segmented networks.

The polymerization proceeds in bulk or in solution with a thermal or photo-initiator, changing the reaction time from respectively several hours to a few minutes [10,24].
For the composition of the segmented networks, the samples are identified by a nomenclature indicating the molecular weight of the bis-macromonomer and the fraction of both components. In the network *net*-poly(MeOx$_{2000}$(30)-*co*-MMA(70)), for example, polyMeOx bis-macromonomer (M_n = 2000) has been copolymerized with MMA in a ratio of 30/70 (w/w).

3.1. POLYIBVE- AND POLYODVE-CONTAINING SEGMENTED NETWORKS

3.1.1. *Swelling Behavior*

Methyl methacrylate (MMA), butyl acrylate (BA) and styrene (St) have been selected as co-monomer for the construction of the PVE-containing segmented networks. All networks are hydrophobic as can be concluded from their swelling behavior in a polar solvents such as toluene or cyclohexane (Table 3). The swelling in water is negligible for all the samples.

TABLE 3. Soluble fraction [a] and equilibrium degree of swelling (Q) [b] of *net*-poly(IBVE-*co*-St) and *net*-poly(ODVE-*co*-BA) in various solvents at room temperature.

Sample	Q (CH_2Cl_2)	Q (Cyclohexane)	Q (Toluene)	Soluble fraction (wt%)
net-poly(IBVE$_{5200}$(20)-*co*-St(80))			830	1
net-poly(IBVE$_{5200}$(40)-*co*-St(60))			450	1
net-poly(IBVE$_{5200}$(60)-*co*-St(40))			360	5
net-poly(IBVE$_{5200}$(80)-*co*-St(20))			290	7
net-poly(IBVE$_{3500}$(20)-*co*-St(80))			310	1
net-poly(IBVE$_{3500}$(40)-*co*-St(60))			190	1
net-poly(IBVE$_{3500}$(60)-*co*-St(40))			120	2
net-poly(ODVE$_{7000}$(20)-*co*-BA(80))	670	380		
net-poly(ODVE$_{7000}$(30)-*co*-BA(70))	560	450		
net-poly(ODVE$_{7000}$(40)-*co*-BA(60))	500			

[a] The soluble fraction was calculated from $100.(W_0 - W_e)/W_0$, where W_0 and W_e, respectively, denote the initial weight and the weight after extraction and drying.

[b] The degree of swelling (Q) was calculated from $100.\rho.(W_e - W_0)/W_0$ where W_0 and W_e, respectively, denote the weight of the dry and swollen network, and ρ is the density of the solvent.

It can also be observed that the degree of swelling (Q) decreases with increasing cross-link density, i.e. increasing fraction and/or decreasing molecular weight of the bis-macromonomer.

The low values for the soluble fractions indicate that the cross-linking reaction between the vinyl monomers and the bis-macromonomers proceeds with high yields. The soluble fraction increases for higher fractions of the bis-macromonomer. This is explained by taking into account the higher viscosity from the beginning of the polymerization, the lower concentration of the polymerizable end groups, the higher segment density around the propagating radicals, etc. [25].

3.1.2. *Phase Morphology*

The dependence of the phase morphology of the polyIBVE-containing segmented networks on the nature and the fraction of the vinyl polymer is illustrated by dynamic mechanical analysis (DMA) (Figures 2 and 3). *Net*-poly(IBVE-*co*-MMA) segmented networks are materials with a phase separated morphology, evidenced by the occurrence of two transitions (maximum of tanδ) at temperatures close to the T_g's of the homopolymers (resp. -20°C and 105°C) (Figure 2).

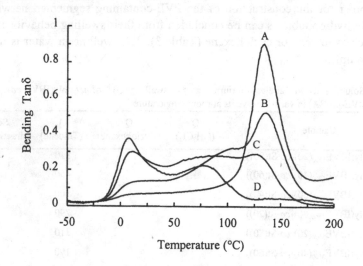

Figure 2. Tanδ versus temperature of *net*-poly(IBVE$_{5200}$-*co*-MMA) with the ratio (w/w) of polyIBVE/polyMMA equal to 20/80 (A), 40/60 (B), 60/40 (C) and 80/20 (D) (Frequency: 1 Hz, heating rate: 2°C/min).

At intermediate compositions of the components (curve B and C), a third transition can be observed between the two other transitions, corresponding to the T_g of domains in which the two components are mixed intimately. It can be supposed that this complex phase morphology results from the fact that both the chemical reactions and the phase separation proceed under nonequilibrium conditions. After a high degree of chemical conversion and cross-linking is reached, microphase separation is impeded and a nonequilibrium structure, which is characterized by incomplete phase separation, is obtained. A similar concept was used to describe the differing degrees of component segregation in interpenetrating polymer networks [26].

The *net*-poly(IBVE-*co*-St) networks, on the contrary, are transparent materials that show only one tanδ-transition at a temperature between the T_g's of the homopolymers (Figure 3). The presence of this single transition indicates a high degree of compatibility of the components. The position of this transition depends on the ratio of the components and is always at a higher temperature than the temperature calculated by the equation of Fox (Eq. 2) for miscible polymer systems:

$$\frac{1}{Tg} = \frac{w_1}{Tg_1} + \frac{w_2}{Tg_2} \quad (w_1 + w_2 = 1) \tag{2}$$

(w_1 (w_2) and T_{g1} (T_{g2}) are the weight fractions and glass transition temperatures of both components).

The latter phenomenon is explained by the presence of cross-links between the components, which reduces the overall segmental mobility of the system compared to the system with the linear homologues.

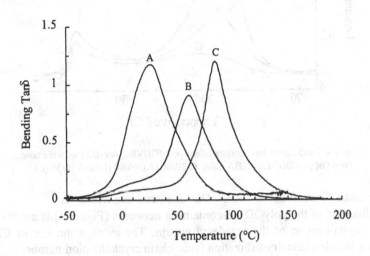

Figure 3. Tanδ versus temperature of *net*-poly(IBVE$_{5200}$-*co*-St) with the ratio (w/w) of polyIBVE/polySt equal to 80/20 (A), 60/40 (B), 40/60 (C) and 80/20 (D) (Frequency: 1 Hz, heating rate: 2°C/min).

In the DMA-curves of the *net*-poly(ODVE$_{7000}$-*co*-BA) segmented networks, a transition around 60°C is observed for each network composition, besides a broad transition around -25°C (Figure 4). As revealed by DSC-experiments (see next section), the transition at 60°C must be ascribed to the melting of crystalline polyODVE-domains. For increasing fractions of polyODVE in the networks, the absolute value for the tanδ-maximum at -25°C (T_g of the amorphous parts) decreases continuously because the higher degrees of crystallinity of the polyODVE-segments lower the contribution to this T_g-transition. Further details about this crystallization and its applicability to shape memory properties will be given in the next two sections.

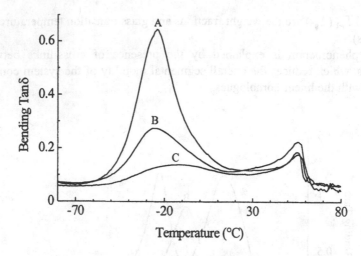

Figure 4. Tanδ versus temperature of *net*-poly(ODVE₇₀₀₀-*co*-BA) with the ratio
(w/w) of polyODVE/polyBA equal to 20/80 (A), 40/60 (B) and 70/30 (C).

3.1.3. *Side-chain Crystallization in PolyODVE-containing Networks*

The crystallization of the polyODVE-containing networks (Figure 4) is ascribed to the side chain crystallization of the octadecyl groups. The average number of CH_2-units participating in side-chain crystallization (side-chain crystallization number N_{cr}) can be obtained from the melting enthalpy (ΔH_m). N_{cr} is calculated by dividing ΔH_m by the number of repeating units in the polymer chain and by 3.08 kJ/mol (value for the melting enthalpy of one CH_2 unit in n-alkanes [27]). DSC results reveal that, for polyODVE fractions ranging from 20 to 40%, N_{cr} and the melting point (T_m) of the crystalline fraction of the polyODVE-segments remain almost unchanged compared to the corresponding values for pure polyODVE (Table 4). This confirms the almost quantitative phase separation as observed by DMA.

TABLE 4. Side-chain crystallization of polyODVE in segmented networks

Sample	ΔH^{*}_{m} (kJ/mol) [a]	T_m (°C)	N_{cr} [b]
PolyODVE bis-macromonomer (M_n =7000 g/mol)	32.2	47	10.4
net-poly(ODVE₇₀₀₀(20)-*co*-BA(80))	25.1	41	8.1
net-poly(ODVE₇₀₀₀(30)-*co*-BA(70))	26.6	41	8.6
net-poly(ODVE₇₀₀₀(40)-*co*-BA(60))	27.0	42	8.8

[a] ΔH^{*}_{m} = ΔH_m / (number of repeating units)
[b] N_{cr} = average amount of CH_2-units participating in side-chain crystallization (ΔH^{*}_{m} / 3.08 kJ/mol)

3.1.4. *Shape Memory Properties*

This side-chain crystallization, in combination with the chemical cross-links between both components, lead to shape memory properties for the *net*-poly(ODVE-*co*-BA) networks, even at low fractions of polyODVE (e.g. 20 wt%). Above the melting point (50°C) of the crystalline fraction of the polyODVE-segments, the networks become soft elastomeric materials. After deformation of the heated materials (200%), cooling to room temperature can freeze in the secondary shape. After the cooling process, elastic recovery to the primary shape is limited to an average value of 17%. The primary form can be recovered by heating above 50°C. Even after five consecutive shape memory tests, there is no detectable irreversible loss of the primary form (Figure 5).

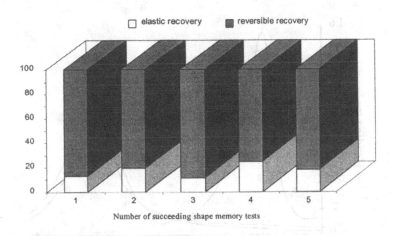

Number of succeeding shape memory tests

Figure 5. Elastic and reversible recovery of the primary form for five successive shape memory tests on *net*-poly(ODVE$_{7000}$(20)-*co*-BA(80))

3.2. POLY(2-ALKYL-2-OXAZOLINE)-CONTAINING SEGMENTED NETWORKS

3.2.1. *Synthesis of Hydrophilic and Amphiphilic Networks*

Both hydrophilic and amphiphilic segmented networks based on bis-macromonomers of polyMeOx or polyEtOx were synthesized. Hydrophilic copolymer networks were obtained by using the hydrophilic acrylate monomer 2-hydroxyethyl acrylate (HEA) as comonomer. The free radical copolymerization of HEA with the acrylate end groups of the bis-macromonomers was initiated with photo-initiators (365 nm) at room temperature. Series of segmented networks with differing type and length of the bis-macromonomer and with changing ratios of both components were prepared.

For the synthesis of the amphiphilic copolymer networks, MMA was chosen as hydrophobic comonomer. Materials with low soluble fractions could only be obtained when the copolymerization was performed with a thermal initiator at elevated temperatures.

3.2.2. *Phase Morphology*

DSC and DMA analysis showed in most cases single transitions for both the hydrophilic and amphiphilic segmented networks, independently from the difference in T_g's between the homopolymers (T_g(polyHEA) $\approx$ 10°C, T_g(polyMMA) $\approx$ 105°C, T_g(polyMeOx) $\approx$ 70°C, T_g (polyEtOx) $\approx$ 30°C [22]). This indicates the high degree of compatibility of the applied segments in these materials. As can be observed in Figure 6, the position of the transition depends on the nature of both constituents. The T_g can further be fine-tuned between 20 and 130°C by variation of the composition of both components in the networks. More phase-separated morphologies were only detected for some networks in which polyMeOx- or polyEtOx-bis-macromonomers with higher molecular weights were used.

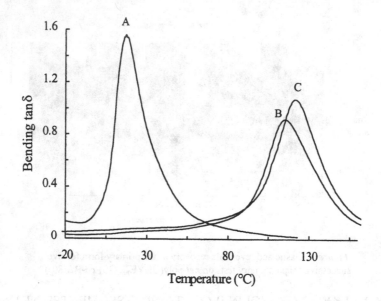

Figure 6. Tanδ versus temperature of polyMeOx- and polyEtOx- containing hydrophilic (A) and amphiphilic (B and C) segmented networks: *net*-poly(MeOx$_{2000}$(20)-*co*-HEA(80)) (A); *net*-poly(MeOx$_{2000}$ (20)-*co*-MMA(80)) (B) and *net*-poly(EtOx$_{2500}$(20)-*co*-MMA(80)) (C).

3.2.3. *Swelling Properties*

As expected, the hydrophilic networks show high degrees of swelling in water. The amount of swelling of the networks can be controlled by variation of the chain length of the bis-macromonomer (Figure 7). Higher molecular weights between the cross-links result in increasing degrees of swelling. Between 20 and 45 wt% of the bis-macromonomer, the composition of the network has no significant influence on the degree of swelling. The use of these materials as hydrogels will be further investigated.

The last two decades, hydrogels have been extensively studied for a variety of biomedical applications, controlled drug release, blood-contacting applications, etc. A more detailed description of the phase morphology and swelling behavior of the amphiphilic segmented networks *net*-poly(MeOx-*co*-MMA) and *net*-poly(EtOx-*co*-MMA) will be given in a forthcoming publication [28].

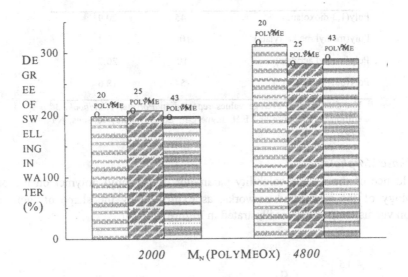

Figure 7. Equilibrium degree of swelling in water of *net*-poly(MeOx$_{2000}$-*co*-HEA) and *net*-poly(MeOx$_{4800}$-*co*-HEA) for different ratios of polyMeOx/polyHEA

3.3. POLY(1,3-DIOXOLANE)-CONTAINING SEGMENTED NETWORKS

PolyDXL is a crystallizable (T_m= 50-60°C), moderately hydrophilic polymer with a low T_g (-45°C). We have selected this polymer as building block for a variety of macromolecular segmented substances, because of its ability to degrade under controlled conditions. The methods for chain scission of the polyacetal segments are based on the hydrolytic instability of the acetal function in acid environment or on the low ceiling temperature of polyDXL [10,24,29]. The ceiling temperature is the temperature at which the rate of polymerization is equal to the rate of depolymerization.

For the construction of the polyDXL-containing amphiphilic segmented networks, a variety of hydrophobic acrylate monomers have been selected as comonomer. From the physical data of the corresponding homopolymers, presented in Table 5, it can be seen that poly(meth)acrylates with both high and low T_g's have been used. The corresponding copolymer networks possess a wide array of physical properties depending on the nature of vinyl monomer, the ratio of the two constituents and the molecular weight of polyDXL. Both transparent, strong materials as well as tough, opaque elastomers can be made by proper selection of these parameters.

TABLE 5. Glass transition temperatures and solubility parameters (δ) of polyDXL and the homopolymers of the selected vinyl monomers.

Polymer	T_g (°C) [a]	δ (MPa$^{1/2}$) [a]
Poly(1,3-dioxolane)	-45	20.9 [30]
Poly(methyl methacrylate)	105	19.4
Poly(methyl acrylate)	10	20.7
Poly(butyl acrylate)	-54	18.6

[a] The data are the average values reported in *Polymer Handbook*, 3rd edition, J. Brandrup and E.H. Immergut, Eds., J. Wiley & Sons, NY 1989.

3.3.1. *Phase Morphology*

The influence of the T_g and solubility parameter of the vinyl polymer on the phase morphology of the segmented networks, as expressed by the shape of their tanδ-transition versus temperature, is illustrated in Figure 8.

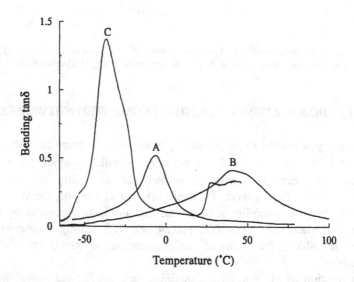

Figure 8. Tanδ versus temperature for three polyDXL-containing segmented networks with different nature of the vinyl polymer: *net*-poly(DXL$_{4000}$-*co*-MA) (A); *net*-poly(DXL$_{4000}$-*co*-MMA) (B); *net*-poly(DXL$_{4000}$-*co*-BA) (C) (w/w 50/50).

The segmented networks *net*-poly(DXL$_{4000}$(50)-*co*-MA(50)) (curve A) and *net*-poly(DXL$_{4000}$(50) -*co*-MMA(50)) (curve B), for which the difference in solubility parameters between both constituents is relatively small (resp. 0.2 and 1.5 MPa$^{1/2}$), respectively have a compatible (single transition) and semi-compatible (broad transition) morphology [24]. In order to evaluate the influence of the cross-links on the physical properties, a series of polymer blends of polyDXL and polyMMA was studied. This investigation showed that the blends undergo phase separation as soon as the polyDXL-content exceeds 10% [9]. Therefore, it was concluded that the incorporation of the two polymer segments in a single network causes a forced compatibility, resulting in more miscible macromolecular architectures.

The DMA-curve of the corresponding network with polyBA (curve C), on the contrary, shows a superposition of two transitions at low temperatures corresponding to the T$_g$'s of both homopolymers, together with a melting transition of crystalline polyDXL segments around 25°C. The larger difference between the solubility parameters (2.3 MPa$^{1/2}$) and the low T$_g$ of polyBA could explain the ability of the network constituents to undergo phase separation and of the polyDXL-rich domains to crystallize.

For a given vinyl component, the phase morphology depends to a great extent on the ratio of both constituents in the networks. In Figure 9, the DMA-curves of segmented networks with different fractions of polyMMA are presented.

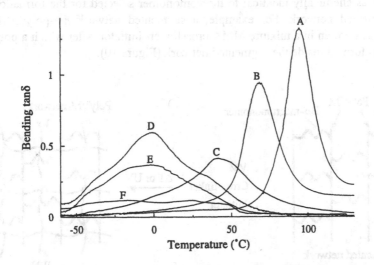

Figure 9. DMA-curves of *net*-poly(DXL$_{4000}$-*co*-MMA) with different ratios of polyDXL/polyMMA (w/w): 20/80 (A), 40/60 (B), 50/50 (C), 65/35 (D), 70/30 (E) and 85/15 (F) (Freq. 1 Hz, heating rate: 2°C/min.).

For increasing fraction of the polyDXL-segments, the maximum of the tanδ transition shifts to lower temperatures, which demonstrates the plasticizing effect of the bis-macromonomer. Simultaneously, the transition broadens, resulting in a plateau-like

92

transition over a wide temperature range for a network containing 85% of polyDXL (curve F). For networks with more than 50% polyDXL (curve D-F), a second transition at 25°C appears as a shoulder. This shoulder formation reveals the presence of a crystalline fraction of the polyDXL-segments, as confirmed by DSC-analysis ($T_m \approx 20$-30°C). For these networks, the absolute values of the tanδ maximum decrease continuously with increasing polyDXL-content (D $\Rightarrow$ F) because the corresponding increasing degrees of crystallinity lower the contribution of the polyDXL-segments to the T_g-transition.

Until now, materials with a variety of morphologies were obtained by changing the nature or the composition of the vinyl component in the networks. A novel approach was developed to control the morphology without changing these parameters [24]. The aim of this investigation was to define reaction conditions that allow the materials, with a certain overall composition of the components, to be engineered in order to create a microheterogeneous morphology (domains 100-200 Å). It is known that such a morphology may lead to materials with damping abilities over broad temperature or frequency ranges [31].

For this purpose, the segmented networks were applied as starting materials for the synthesis of sequential interpenetrating polymer networks (IPN's). Sequential IPN's are generally obtained in a two-step process in which the components for the formation of the second network are swollen and polymerized inside a previously cross-linked polymer. In our work, the vinyl monomer (VM) used for the construction of the second network was chemically identical to the comonomer selected for the formation of the first segmented network. For example, a segmented network *net*-poly(DXL$_{4000}$-*co*-MMA) was swollen in a mixture MMA/crosslinker/ initiator, after which a polyMMA-network is formed inside the segmented network (Figure 10).

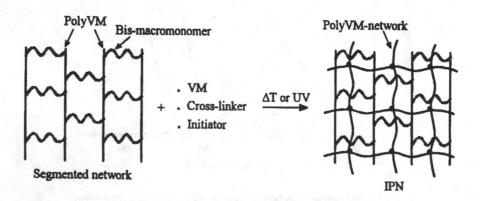

Figure 10. Reaction scheme for the synthesis of sequential IPN's based on segmented networks.

The weight ratio of the first to the second network depends on the time that the first network is swollen in MMA before the formation of the IPN. By a proper choice of the segmented networks and the time of swelling, IPN's could be obtained in which the overall ratio of both components always remains the same but in which the division of the vinyl polymer segments over both networks differs considerably.

Figure 11 shows the DMA-curves of three IPN's with the same overall ratio of polyDXL/polyMMA and the same cross-link degree in the second polyMMA-network (3 wt% cross-linker) but with different polyMMA-fractions in the first segmented network. It is clearly seen that an increasing amount of polyMMA in the first network determines whether two transitions, one broad transition or only one single transition appear in the DMA-curves. More in general, it could be concluded that the vinyl polymer segments in the first network promote the miscibility of both networks in the IPN's, which allows the control of the phase morphology within a broad range of domain characteristics.

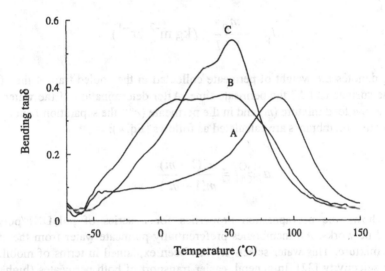

Figure 11. DMA-curves of three IPN's with the overall ratio (w/w) polyDXL/polyMMA equal to 55/45 but with the fraction of polyMMA in the segmented network equal to resp. 0% (A), 30% (B) and 40% (C), (Redrawn from [24]).

3.3.2. *Pervaporation Properties*

The application of the polyDXL-containing segmented networks, as membranes for dehydration of water/ethanol mixtures by the pervaporation technique, was investigated [32]. Pervaporation is a membrane technique that has been developed on an industrial level for applications where separation by conventional methods is difficult or impossible, e.g. for azeotropic mixtures, thermosensitive products, close-boiling mixtures, etc. [33,34,35]. One of the main industrial interests in this field is the removal of water from ethanol-rich mixtures, which originates from a possible substitution of fossil fuels by the use of ethanol as a biomass energy source.

In the continuous search for materials with better selectivity and permeability, different polymers were frequently combined in the form of polymer blends and block- or graft-copolymers. Recently, two cross-linked polymer structures, copolymer networks and

IPN's, have been investigated for the same purpose [36,37]. These network structures have the additional advantage that the chemical cross-links may be used to control the degree of swelling and to improve the dimensional stability of the membrane.

The aim of our work was to use the polyDXL-based amphiphilic segmented networks as model membranes to investigate the influence of different parameters such as the degree of swelling, cross-link density, hydrophilic/hydrophobic balance and the temperature on the dehydration process of ethanol. Some results of this investigation are briefly reviewed (see also [32]).

During a pervaporation experiment, the feed mixture remains in contact with one side of a dense membrane, while the permeate is removed in the vapor state at the opposite side by vacuum or gas sweeping. The permeation rate (J_p) is calculated by weighing the trapped permeate (Eq. 3):

$$J_p = \frac{m_p}{A . \Delta T} \quad (\text{kg m}^{-2} \text{ hr}^{-1})$$
(3)

where m_p denotes the weight of permeate collected in the cooled trap, A the effective membrane surface and ΔT the permeate time. After determination of the water weight fractions in the feed mixture (m) and in the permeate (m'), the separation factors for the water selective membranes are calculated as follows (Eq. 4):

$$a_{\text{EtOH}}^{\text{H}_2\text{O}} = \frac{m'(1-m)}{m(1-m')}$$
(4)

Table 6 shows the pervaporation properties for a series of polyDXL/polyMMA segmented networks. All membranes preferentially permeate water from the ethanol-rich feed mixture. This water selectivity has been explained in terms of mobility and solubility selectivity [32]. In general, easier transport of both permeates (higher flux) results in a decrease of the selectivity. While the permeation through a homopolymer network of polyMMA is negligible, the presence of 10% of the hydrophilic polyDXL is already sufficient to influence the flux and the selectivity considerably. This and other observations were indirect evidence for the miscibility on a molecular level of the two polymer segments in the investigated networks and are in agreement with the results obtained by DMA.

TABLE 6. Pervaporation properties for the separation of the azeotropic mixture water/ethanol (4.4/95.6 w/w) with membranes consisting of different polyDXL/polyMMA ratio (T=25°C).

Membrane	$J_{10\mu m}$[a] $(g.m^{-2}.hr^{-1})$	$\alpha_{Eth.}^{H_2O}$
PolyMMA-network (3% cross-linker)	± 0	?
net-poly(DXL$_{4000}$(10)-co-MMA(90))	300	16
net-poly(DXL$_{4000}$(20)-co-MMA(80))	460	14
net-poly(DXL$_{4000}$(30)-co-MMA(70))	530	11
net-poly(DXL$_{4000}$(50)-co-MMA(50))	1170	6

[a] The permeation rates are normalized to a membrane thickness of 10 μm.

The membrane net-poly(DXL$_{4000}$(20)-co-MMA(80)) was submitted to a detailed investigation of its swelling and pervaporation characteristics. Swelling and membrane experiments were done over the whole range of water/ethanol mixtures (Figs. 12 and 13).

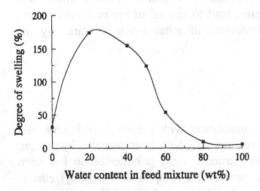

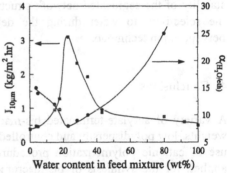

Figure 12. Degree of swelling of the membrane net-poly(DXL$_{4000}$(20)-co-MMA (80)) versus the water content in the water/ethanol binary mixture (T=25°C), (Redrawn from [32]).

Figure 13. The permeation rate $J_{10\mu m}$ (■) and selectivity α (●) of the membrane net-poly(DXL$_{4000}$(20)-co-MMA(80)) versus the composition of the feed mixture (T = 25°C). (Redrawn from [32])

It was observed that interaction effects of polyDXL with the solvents ("coupling effect") provokes extensive plasticizing effects at certain compositions of the binary mixture. As expected, there is a direct correlation between the swelling and the permeation behavior. A more swollen membrane results in a higher flexibility of the polymer chains, and thus in a decrease in the energy required for transport through the membrane.

For a certain composition of the feed mixture, the selectivity could be improved by increasing of the degree of cross-linking of the membrane material. An additional low molecular weight cross-linking agent tetraethylene glycol diacrylate (TEGDA) was introduced during the synthesis procedure. It can be concluded from Table 7, which shows the pervaporation results of the membrane net-poly(DXL$_{4000}$(20)-co-MMA(80)) with different amounts of TEGDA, that the flux decreases and α increases, as the degree of cross-linking increases. The higher degrees of cross-linking limit the swelling, thus provoking mobility selectivity.

TABLE 7. Dependence of the pervaporation characteristics of the membrane net-poly(DXL$_{4000}$(20)-co-MMA(80)) on the degree of cross-linking (feed: azeotropic mixture, T = 25°C).

TEGDA (wt% relative to polyMMA)	$J_{10\mu m}$ (g.m^{-2}.hr^{-1})	$\alpha_{Eth.}^{H_2O}$
0	460	14
5	320	16
10	220	21

In general, it was concluded that the compatibility, together with the mechanical stability of the segmented network structures, lead to control of the permeability and the selectivity to water during the dehydration of ethanol-rich mixtures by the pervaporation technique.

4. Conclusions

A variety of acrylate-terminated bis-macromonomers with predetermined molecular weights, low polydispersity and controlled functionality have been prepared by making use of cationic polymerization procedures. Cationic vinyl polymerization has been applied for the synthesis of bis-macromonomers of hydrophobic poly(vinyl ether)s while cationic ring opening polymerization has been applied for the synthesis of hydrophilic bis-macromonomers of polyDXL and poly(alkyl oxazoline)s.

The bis-macromonomers have been radically copolymerized with vinyl monomers such as methyl methacrylate, butyl acrylate and hydroxyethyl acrylate for the synthesis of hydrophobic, amphiphilic or hydrophilic segmented copolymer networks. It was shown that the morphology, hydrophilic/hydrophobic balance and swelling properties of the network structures are controlled by the nature and the ratio of the two components and by the molecular weight of the bis-macromonomers.

Besides the possible use of the hydrophilic and amphiphilic segmented networks as hydrogels with controllable swelling properties, some other interesting properties were investigated. The hydrophobic segmented networks based on poly(octadecyl vinyl ether) showed shape memory properties while the polyDXL-containing networks were applied as pervaporation membranes for the dehydratation of water/ethanol mixtures and as building blocks for IPN's with morphologies that may lead to materials with damping abilities over broad temperature or frequency ranges.

97

Acknowledgement. F. Du Prez thanks the F.W.O. (Fund for Scientific Research - Flanders) for financial support of postdoctoral research.

5. References

1. Paul, D.R. and Sperling, L.H. (1986) *Adv. in Chem. Series* **211**, ACS, Washington DC.
2. Klempner, D. and Frisch, K.C. (1980) *Polymer Alloys II: Blends, Blocks, Grafts, and Interpenetrating Networks*, Plenum Press, New York.
3. Velichkova, R.S. and Christova, D.C. (1995) *Prog. Polym. Sci.* **20**, 819-887.
4. Chang, P.S. and Buese, M.A. (1993) *J. Am. Chem. Soc.* **115**, 11475-11484.
5. Tezuka, Y., Murakami, Y., and Shiomi, T. (1998) *Polymer* **39**, 2973-2976.
6. Iván, B., Kennedy, J.P., and Mackey, P.W. (1991) in Dunn, R.L. and Ottenbrite, R.M. (eds), *Polymer Drugs and Delivery Systems*, ACS Symp. Book Series **469**, Washington DC, 194-212.
7. Kennedy, J.P. (1995) *Trends Polym. Sci.* **3**, 386-395.
8. Kennedy, J.P. (1994) *Macromol.Symp.* **85**, 79-96.
9. Tan, P., Walraedt, S.R., Geeraert, J.M., and Goethals, E.J. (1995) in Mishra, M.K. and Nuyken, O. (eds.), *Macromolecular Engineering: Recent Advances*, Plenum Press, New York, 163-169.
10. De Clercq, R.R. and Goethals, E.J. (1992) *Macromolecules* **25**, 1109-1113.
11. Matyjaszewski, K. (1996) *Cationic Polymerizations: Mechanisms, Synthesis, and Applications*, Marcel Dekker, New York.
12. Nuycken, O. and Pask, S. (1989) in Mark, H., Bikales, N., Overberger, C., and Menges, S. (eds.), *Encyclopedia of Polymer Science and Engineering*, Wiley & Sons, **16**, 494-532.
13. Miyamoto, M., Sawamoto, M., and Higashimura, T. (1984) *Macromolecules* **17**, 2228-2230.
14. Sawamoto, M. (1993) *Trends Polym. Sci.* **1**, 111-115.
15. Bennevault, V., Peruch, F., and Deffieux, A. (1995) *Macromol. Chem. Phys.* **196**, 3075-3090.
16. Goethals, E.J., Haucourt, N., and Peng, L. (1994) *Macromol. Symp.* **85**, 97-113.
17. Haucourt, N., Goethals, E.J., Schappacher, M., and Deffieux, A. (1992) *Makromol. Chem., Rapid Commun.* **13**, 329-336.
18. Lievens, S.S. and Goethals, E.J. (1996) *Polym. Int.* **41**, 277-282.
19. Goethals, E.J., De Clercq, R.R., De Clercq H.C., and Hartmann, P.J. (1991) *Makromol. Chem. Macromol. Symp.* **47**, 151-162.
20. Aoi, K. and Okada, M. (1996) *Prog. Polym. Sci.* **21**, 151-208.
21. Kobayashi, S. (1990) *Prog. Polym. Sci.* **15**, 751-823.
22. Christova, D., Velichkova, R., and Goethals, E.J. (1997) *Macromol. Rapid Commun.* **18**, 1067-1073.
23. Kobayashi, S., Uyama, H., Narita, Y. and Ishiyama, J. (1992) *Macromolecules* **25**, 3232-3236.
24. Du Prez, F.E. and Goethals, E.J. (1995) *Macromol. Chem. Phys.* **196**, 903-914.
25. Ito, K. (1994) in Mishra, M.K. (ed.), *Macromolecular Design: Concept and Practice*, Polymer Frontiers International, New York, 129-160.
26. Lipatov, Y.S. (1994), in Klempner, D., Sperling, L.H., and Utracki, L.A. (eds.), *Interpenetrating Polymer Networks, Adv. in Chem. Series* **239**, ACS, Washington DC, 125-139.
27. Platé, N.A. and Shibaev, V.P. (1987) *Comb Shaped Polymers and Liquid Crystals*, Plenum, NY, chs 1-3.
28. Christova, D., Velichkova, R., Du Prez, F.E., and Goethals, E.J. (1998) *to be published*.
29. Goethals, E.J., De Clercq, R.R., and Walraedt, S.R. (1993) *J. Macromol. Sci., Pure Appl. Chem.* **A30**, 679-688.
30. Alamo, R., Fatou, J.G., and Bello, A. (1983) *Polym. J.* **15**, 491-498.
31. Corsaro, R.D. and Sperling, L.H. (1990) *Sound and vibration damping with polymers*, ACS Symp. Book Series **424**, Washington DC.
32. Du Prez, F.E., Goethals, E.J., Schué, R., Qariouh, H., and Schué, F. (1998) *Polym. Int.* **46**, 117-125.
33. Huang, R.Y.M. (1991) *Pervaporation Membrane Separation Processes*, Membrane Science and Technology, Series 1, Elsevier, Amsterdam.
34. Zhang, S. and Drioli, E. (1995) *Sep. Sci. Technol.* **30**, 1-31.

98

35. Aminabhavi, T.M., Khinnavar, R.S., Harogoppad, S.B., and Aithal, U.S (1994) *J. Macromol. Sci.-Rev. Macromol. Chem. Phys.* **34**, 139-204.
36. Kerres, J.A. and Strathmann, H. (1993) *J. Appl. Polym. Sci.* **50**, 1405-1421.
37. Lee, Y.K., Tak, T.M., Lee, D.S., and Kim, S.C. (1990) *J. Memb. Sci.* **52**, 157-172.

RATE CONSTANTS AND REACTIVITY RATIOS IN CARBOCATIONIC POLYMERIZATIONS

H. MAYR

Institut für Organische Chemie der Ludwig-Maximilians-Universität München, Karlstrasse 23, D-80333 München, Germany

Abstract. Rate constants for the reactions of carbocations with substituted ethylenes can be determined when the reactions terminate at the [1:1]-product stage. Correlation equations are presented which allow one to derive rate constants for these reactions from the electrophilicity parameters (E) of the carbocations and the nucleophilicity (N) and slope parameters (s) of the π-systems. Since the slope s is usually close to 1, initiation and propagation rate constants of carbocationic polymerizations can be estimated from the data presented in Schemes 7–11. The special features of fast reactions (e.g., propagations), which do not have enthalpic barriers, are discussed.

1. Introduction

The determination of rate constants for the reactions of carbocations with olefins, the key-step of carbocationic polymerizations of alkenes remains to be a challenge [1]. Two facts hamper the derivation of such rate constants from the kinetics of carbocationic polymerizations (Scheme 1).

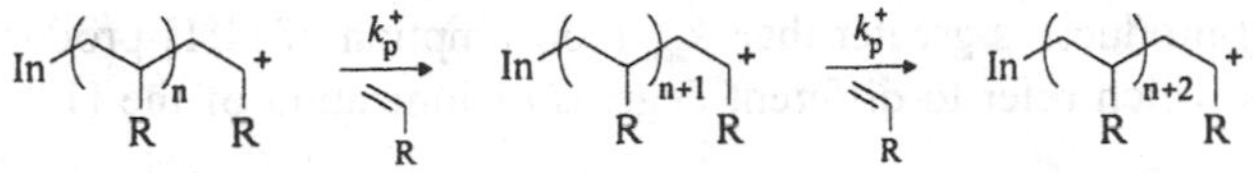

Scheme 1. Propagation steps in cationic polymerizations of olefins (In: Initiator).

(1) Though π-delocalized propagating carbocationic entities can directly be observed by UV-Vis spectroscopy, propagating cations with different chain length are almost indistinguishable by spectroscopic methods [2, 3]. For that reason it is difficult to derive rate constants from the absorbance of the propagating cations.

(2) The rates of consumption of monomers provide apparent propagation rate constants, important quantities for reactor design. However, only with the additional knowledge of the concentration of the propagating species, one can derive rate constants for the attack of carbocations at olefins from such data [4]. Since this information is difficult to obtain, the magnitude of k_p^+ is still controversially discussed for many polymerizations [1, 5, 6].

J.E. Puskas et al. (eds.), Ionic Polymerizations and Related Processes, 99–115.

2. Kinetic Methods

Rate constants for the reactions of carbocations with olefins can more easily be determined, if the reacting cation and the carbocation produced by the attack at the olefin are of different structural type. In such cases, the two types of cations posses different spectroscopic properties, and the kinetics can be derived by detection of one of them.

Almost two decades ago, Dorfman was the first one to directly measure rate constants for the reactions of phenyl substituted carbenium ions with alkenes [7]. The benzhydryl cation was produced from benzhydryl bromide by an 80 ns electron pulse in 1,2-dichloroethane and monitored at the maximum of its optical absorption band at 449 nm [7, 8]. By plotting the pseudo-first-order rate constants against the concentration of the alkenes in the solution, straight lines were obtained. Their slopes yielded the second-order rate constants listed in Scheme 2. Only alkenes, which are sufficiently nucleophilic to successfully compete with the combination of the carbocations with halide anions, can be studied in this way, and it was not possible to determine rate constants for the reactions of the benzhydryl cation with propene or 1,3-butadiene.

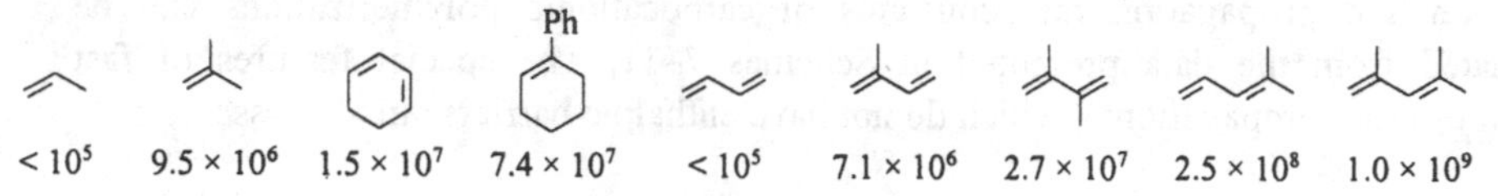

$$< 10^5 \qquad 9.5 \times 10^6 \qquad 1.5 \times 10^7 \qquad 7.4 \times 10^7 \qquad < 10^5 \qquad 7.1 \times 10^6 \qquad 2.7 \times 10^7 \qquad 2.5 \times 10^8 \qquad 1.0 \times 10^9$$

Scheme 2. Rate constants (L mol^{-1} s^{-1}) for the reactions of the benzhydryl cation with alkenes in 1,2-dichloroethane at 24 °C (from ref. [7]).

A different, more general approach, which allows the determination of smaller rate constants by conventional kinetic methods has been developed and used by us during the past twelve years [9–11]. It is based on the fact that under certain conditions, carbocations react with olefins to yield [1:1]-products exclusively because the consecutive reactions with olefins are slower. Such situations will be encountered if k_a (formation of [1:1]-product) is greater than k_{app} (consumption of [1:1]-product). Let us consider two cases which refer to different degrees of ionization of the [1:1]-products (Scheme 3):

Case I describes the situation that the [1:1]-product is predominantly ionic ($K_I \gg 1$). In this case, $k_{app} = k_b$, and termination at the [1:1]-product stage can be achieved if $k_a > k_b$. This situation is usually encountered if the new carbocation is better stabilized [12, 13] than the initial one. Faust's capping reactions of isobutylene polymers with 1,1-diarylethylenes [14] or reactions of carbocations with enamines to give iminium ions [15] fall into this category.

The problem: Since such reactions simultaneously profit from the increase in carbocation stabilization [12] and from the conversion of a π into a σ-bond, they are usually very fast [11], and only in rare cases can "Case I" type reactions be monitored with conventional techniques of chemical kinetics.

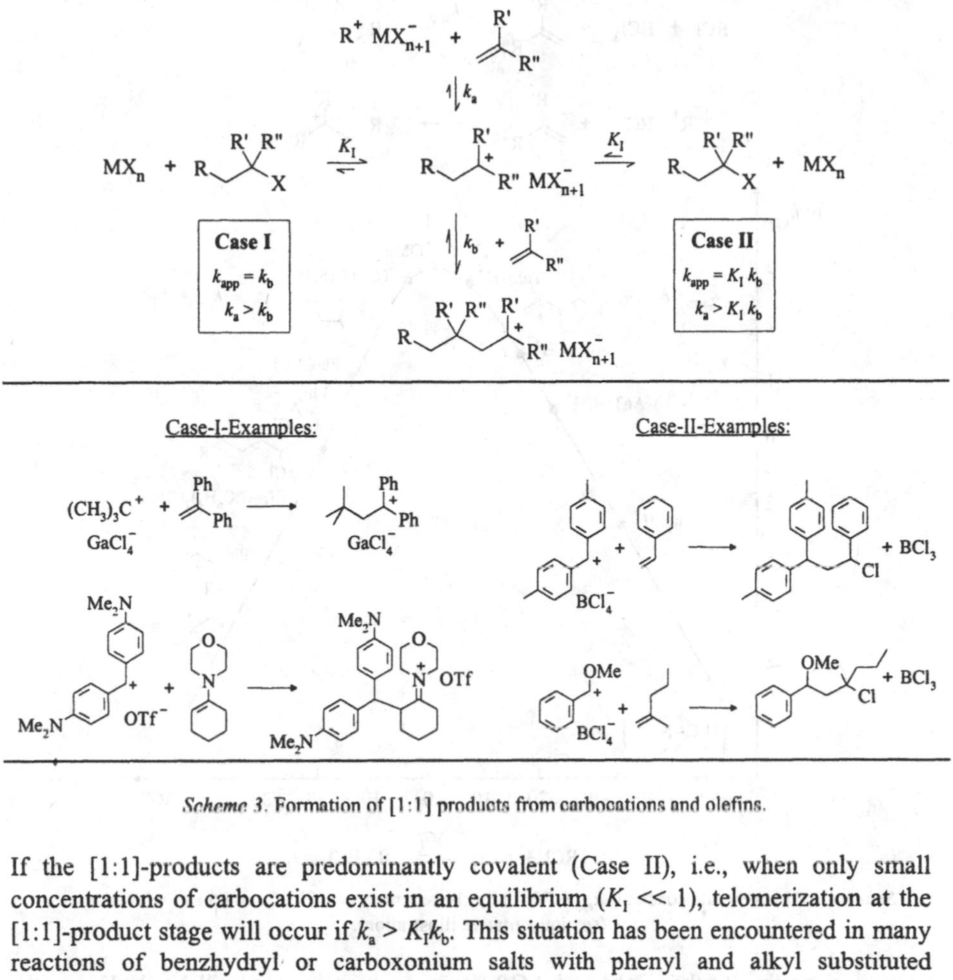

Scheme 3. Formation of [1:1] products from carbocations and olefins.

If the [1:1]-products are predominantly covalent (Case II), i.e., when only small concentrations of carbocations exist in an equilibrium ($K_I \ll 1$), telomerization at the [1:1]-product stage will occur if $k_a > K_I k_b$. This situation has been encountered in many reactions of benzhydryl or carboxonium salts with phenyl and alkyl substituted ethylenes [11, 16, 17]. Since neither k_a nor k_b are affected by the nature of the complex counterions $[MX_{n+1}]^-$, lowering of K_I by reducing the strength of the Lewis acid MX_n generally increases the selectivity for the formation of [1:1]-products. A lower limit for the strength of the Lewis acid is set by the requirement that the Lewis acid MX_n must be strong enough to fully ionize RX in order to have clearly defined conditions for the kinetic experiments [18, 19].

We have shown that for a given Lewis acid (used in excess in a certain solvent) a V-type curve defines the relationship between the apparent addition rate constants and the chloride affinity of the carbocations [18, 19]. Moving from left to right in Scheme 4, one first observes an increase of k_{app} as the chloride affinity is increased. Since $\Delta G_I° < 0$ in this range, one is working with predominantly ionized material, and a typical rate-equilibrium relationship [20] (Brønsted relationship [21]) holds: The reactivity increases with the Lewis acidity of the carbocations.

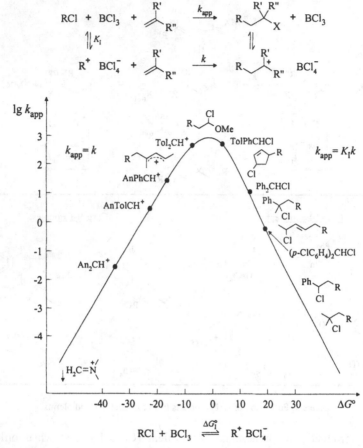

Scheme 4. Apparent rate constants k_{app} ($-70°C$) for the reactions of RCl/BCl_3-mixtures with alkenes (semiquantitative illustration).

As one is moving further right and $\Delta G_1°$ becomes positive, one will be dealing with predominantly covalent material. Now the apparent rate constants will decrease with increasing Lewis acidity of the cations, since structural variation affects the equilibrium constants to a greater degree than the rate constants [22].

Scheme 4, which has experimentally been verified for benzhydryl systems [18, 19], can be used to design [1:1]-product formation by Lewis acid initiated reactions of alkyl chlorides with alkenes.

The low apparent reactivities of tertiary alkyl chlorides and of 1-phenylalkyl chlorides, for example, explain why all the benzhydryl systems shown in Scheme 4 give high yields of [1:1]-products with styrene or isobutylene [16]. The high reactivities of α-chloroethers or 3-chlorocyclopentenes are in accord with the observation that [1:1]-products cannot be obtained by BCl_3 induced additions of RCl to methyl enol ethers or to cyclopentadiene. For the reasons discussed above, their production was achieved, however, from bis(p-methoxyphenyl)methyl chloride and vinyl ethers or cyclopentadiene in presence of the weaker Lewis acid $ZnCl_2$ [16, 23].

Case II-type reactions are usually slower than case I-type reactions, since the driving force of the CC-bond-forming step, which comes from $\pi \rightarrow \sigma$ conversion, is reduced by converting a stabilized carbocation into a less stabilized one [11]. If the carbocation produced in the CC-bond forming step is much less stabilized than the attacking carbocation, this step may become reversible, as recognized by the dependence of the reaction rates on nature and concentration of the negative counterions. Since variation of $[MX_{n+1}]^-$ will affect the rate of halide transfer, independence of the observed rate constant of the nature of $[MX_{n+1}]^-$ indicates that the first step of this reaction sequence is rate-determining. This situation has been observed for the reactions of several bis(p-anisyl)carbenium salts with 2-methyl-1-pentene as shown in Scheme 5 [10]. The analogous reaction of $An_2CH^+BCl_4^-$ with 2-methyl-2-butene has been found to slow down at low anion concentrations, however, as expected for a reaction in which the initial ionic adduct partially yields products and partially reverts to the reactants [10].

$[MX_{n+1}]^-$ / mmol L^{-1}	k_2 / L mol^{-1} s^{-1}
BCl_4^- / 4.60×10^{-4}	2.64×10^{-2}
$BClBr_3^-$ / 8.70×10^{-4}	2.70×10^{-2}
$B(OMe)Cl_3^-$ / 7.01×10^{-4}	2.81×10^{-2}
$SnCl_5^-$ / 4.22×10^{-5}	2.48×10^{-2}

Scheme 5. Evidence for the irreversibility of the CC-bond forming step (–70°C, CH_2Cl_2).

3. Reactivity Scales

By this method, it has become possible to determine several hundred rate constants for reactions of carbocations and related electrophiles with nucleophiles [24, 25]. These data, supplemented by rate constants obtained by Laser-flash-photolytic methods [26], have been subjected to correlation analysis. It was found that the rate constants can be expressed by three parameters, one for the electrophiles (electrophilicity E) and two for the nucleophiles (slope s and nucleophilicity parameter Nu or N) [24, 25]. As discussed elsewhere [24], $E(An_2CH^+)$ was set 0 and s(2-methyl-1-pentene) was set 1.0. The nucleophilicity may then be expressed by lg $k = Nu + s\ E$ [eq. (1)] in which the parameter Nu corresponds to the intercept of the correlation line with the ordinate at $E = 0$ (eq. 1). Equation (2), i.e., lg $k = s\ (N + E)$ uses the nucleophilicity parameter N which corresponds to the intercept of the correlation line with the abscissa (lg $k = 0$). Both equations are equivalent mathematically, but have different merits in their practical use. It is an advantage of equation (1) that it corresponds to the well-established representations of LFERs, and Nu formally corresponds to the rate constant of the reaction of the corresponding nucleophile with An_2CH^+. It does not have practical meaning, however, for $Nu > 9$ (diffusion controlled reaction with An_2CH^+) or for $Nu < -6$ (the reaction with An_2CH^+ is too slow to be observable). The most serious disadvantage of Nu is, that the intersections of the correlation lines with the ordinate ($E = 0$) are often far outside of the range accessible with our experimental methods. As a consequence, the determination of Nu requires far-reaching extrapolations, and the magnitude of Nu will strongly depend on the slope s which is often not known precisely.

For that reason, we prefer eq. (2), an unconventional form of a linear free enthalpy relationship where N is given by the intersection of the correlation lines with the abscissa ($\lg k = 0$). Since these intersections are within or close to the experimental range, the value of N depends little on s, and the nucleophilicity parameters N provide a reliable semiquantitative measure of nucleophilic reactivities. It should once more be stressed that the *calculation* of $\lg k$ by either equation (1) or (2) will give identical results.

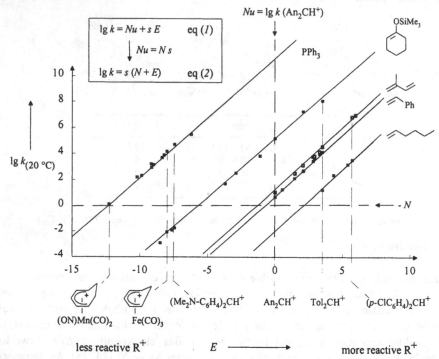

Scheme 6. Linear free enthalpy relationships for the reactions of carbenium ions with nucleophiles.

Scheme 7 provides a rough information on the relative reactivities of most classes of monomers utilized for carbocationic polymerization [23, 27–30]. The scale covers 14 orders of magnitude from monoalkylated ethylenes at the bottom to ketene acetals on top, the most nucleophilic π-systems in this list. One can recognize that besides many alkyl and aryl substituted ethylenes, this listing includes many silylated enol ethers but only few alkyl enol ethers. Why?

In the discussion of Scheme 4 it has been shown, that α-chloroethers, the [1:1] addition products from alkyl chlorides RCl and enol ethers, are highly reactive, and for that reason it is problematic to terminate reactions with alkyl enol ethers at the [1:1]-product stage, a prerequisite for applying our kinetic methods. Even by using weaker Lewis acids for the ionization of RCl, we often have trouble to avoid polymerization of the alkyl enol ethers [16]. Electrophilic attack at silyl enol ethers, on the other hand, yields α-siloxycarbenium ions, which are efficiently desilylated to give carbonyl compounds, and the [1:1]-products are usually obtained in high yield [27].

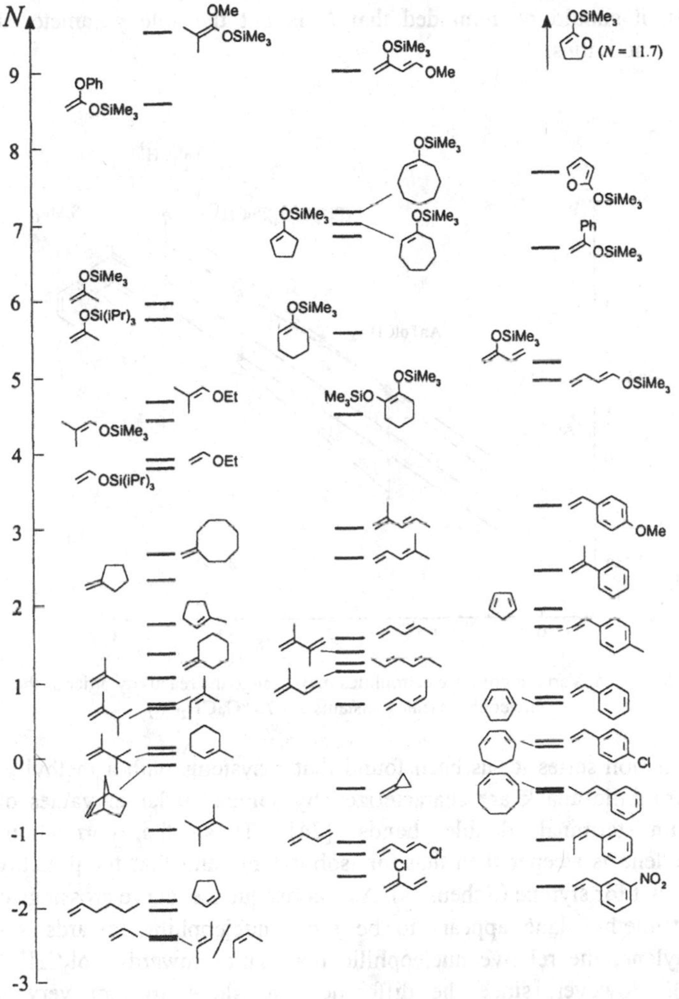

Scheme 7. Nucleophilicity parameters *N* of olefinic monomers.

Several comparisons show that alkyl enol ethers and structurally analogous silyl enol ethers are of similar nucleophilicity [26, 27]. For that reason one can assume that the *N* parameters of silyl enol ethers can usually be considered to be a good approximation for *N* of the corresponding alkyl enol ethers.

It is worth mentioning that the least reactive enol ethers are only slightly more nucleophilic than the most reactive alkyl substituted ethylenes and dienes shown in

Scheme 7. Enol ethers with acceptor substituents can, therefore, be expected to have similar reactivities as unsaturated hydrocarbons, and copolymerizations between such types of compounds become conceivable.

In accord with the results of copolymerization experiments [31], 1,3-dienes and alkyl substituted ethylenes differ only little in nucleophilicities [23, 30], and the reactivities of the styrenes investigated cover more than five orders of magnitude [29].

Before turning to more detailed discussions of structure-nucleophilicity relationships, it should be reminded that N is not the sole parameter determining nucleophilic reactivities.

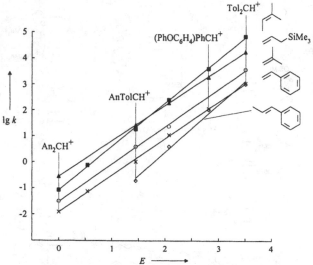

Scheme 8. Variation of the electrophiles may change the reactivity order of the nucleophiles (rate constants at –70 °C, CH_2Cl_2).

In several reaction series it has been found that π-systems with a methyl group at the site of electrophilic attack are characterized by somewhat larger values of s than π-systems with terminal double bonds [24]. Thus, the correlation line for trimethylethylene is steeper than that for isobutylene, and that for β-methylstyrene is steeper than that for styrene (Scheme 8). As a consequence, curve crossings occur: While allyltrimethylsilane appears to be more nucleophilic towards An_2CH^+ than trimethylethylene, the relative nucleophilic reactivities towards Tol_2CH^+ are reverse (Scheme 8). However, since the differences in slope are not very large, only nucleophiles with similar values of N may reverse the relative order of reactivity upon variation of the electrophile.

Scheme 9 compares nucleophiles with terminal double bonds, i.e., systems with closely similar s-values. In each of the three columns one substituent (H, CH_3, or $OSiMe_3$) is kept constant while the other is varied. As expected, in all three columns,

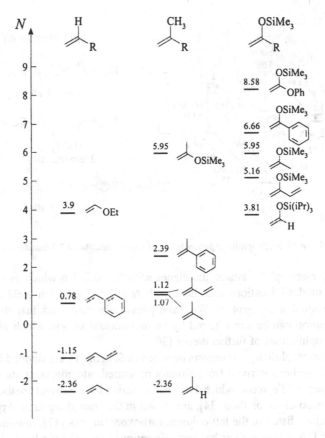

Scheme 9. Nucleophilicity parameters *N* of mono- and 1,1-disubstituted ethylenes.

the compound with R = OX is most reactive, followed by R = Ph. An interesting reversal of the relative activating effect of methyl *vs.* vinyl is observed as one moves from the left to the right column. While 1,3-butadiene is about one order of magnitude more reactive than propene, isoprene and isobutylene are similar in reactivity, and 2-(trimethylsiloxy)-1,3-butadiene is noticeably less reactive than 2-(trimethylsiloxy)propene. Since the electron demand of the carbocations generated by electrophilic attack at these double bonds decreases from the left to the right, this reactivity order reflects the frequently observed trend that vinyl is more efficient in stabilizing systems with high electron demand while alkyl is more efficient for systems with low electron demand [32, 33].

Scheme 10 summarizes the reactivities of methyl substituted 1,3-butadienes with at least one terminal double bond [30]. It can be seen that the seven dienes can be attributed to three categories: 1,3-butadiene, which yields a terminally monoalkylated

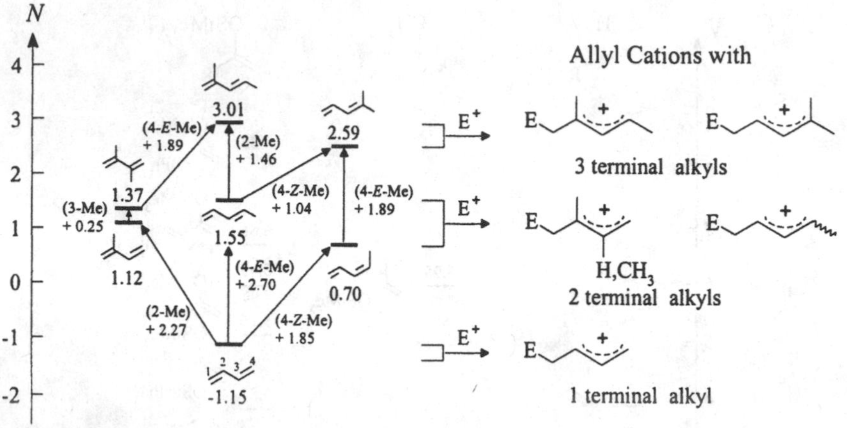

Scheme 10. Nucleophilicity parameters *N* of methyl substituted 1,3-butadienes.

allyl cation by electrophilic attack, the dienes with $N = 0.7$–1.6 which yield terminally dialkyl substituted allyl cations and those with $N = 2.6$–3.0 which yield allyl cations with three terminal alkyl groups. We have previously discussed that the effects of methyl substitution can be summarized by an incremental system which allows one to estimate nucleophilicities of further dienes [30].

The electrophilicity parameters depicted in Scheme 11 are divided in two parts: Those which have been derived from directly measured rate constants are summarized in the white part while those which have been derived by indirect methods, e.g., by using the diffusion-clock method [34], are shown in the shaded top area. Apart from the benzhydryl cations listed in the left column, carboxonium ions [17], iminium ions [35], tropylium ions [36, 37] as well as heterocyclic cations [24, 38] have been characterized. The large variety of the metal-coordinated carbocations which have been investigated kinetically, are not included in this listing [39]. The dashed lines between the two columns relate α-methoxy substituted benzyl cations with their phenylogous counterparts (*p*-methoxy substituted benzhydryl cations) and show that insertion of a *p*-phenylene group between MeO and the carbenium center causes a slight reduction in electrophilicity.

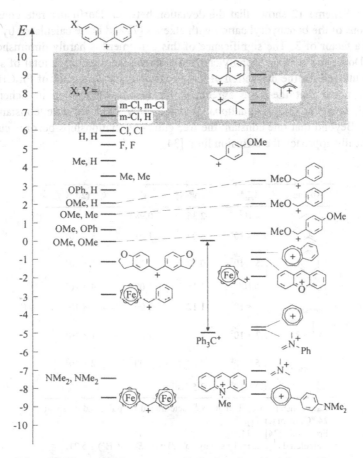

Scheme 11. Electrophilicity parameters E of carbocations (shaded part refers to indirectly determined E-values).

Because of the wide range of reactivity covered by eqs. (1) and (2), it is possible to neglect the differences in steric strain generated when combining different electrophiles in Scheme 11 with different nucleophiles in Scheme 7. Since steric effects cannot be ignored if several kcal/mol of strain are created in the transition states, bulky reagents, e.g. trityl cations, are not included in this listing. Depending on the front strain of the reaction partner, one can expect an effective E-parameter between 0 and –5 for the parent trityl cation.

The p-methoxyphenylethyl cation is the only cation in the white part of the list which can be produced by electrophilic attack at an olefin. One can combine its E-value with the N-value of p-methoxystyrene to derive a rate constant of $k_p^+ = 10^8$ L mol^{-1} s^{-1} for the propagation step of cationic p-methoxystyrene polymerization on the basis of eq. (2) (assuming $s = 1$). Since it is not yet clear, whether the propagation rate constants change with growing chain length, this value can only be predicted for the early stages of the polymerization.

How reliable are the rate constants derived from this correlation? The most objective test is to compare calculated rate constants with those determined in a foreign laboratory. Scheme 12 shows that the deviation between Dorfman's rate constants for the reactions of the benzhydryl cation with alkenes [7] and those calculated by eq. (2) is less than a factor of 3. The significance of this agreement is hardly diminished by the fact that Dorfman's data for isobutylene and isoprene were among a total of seven rate constants utilized for the derivation of the electrophilicity parameter of Ph_2CH^+ [24]. It should be noted that the good agreement for the lowest system in Scheme 12 is accidental, however, since equations 1 and 2 only hold for rate constants $< 10^8$ L $mol^{-1}s^{-1}$. Beyond that rate constant the free enthalpy relationships become curved and asymptotically approach the diffusion limit [34].

Alkene	k_{exp} [a]	N [b]	s [b]	k_{calc} (20 °C) [c]
	$< 10^5$	−2.36	0.98	2×10^3
	9.5×10^6	1.07	1.02	8×10^6
	1.5×10^7	0.72	1.10	1×10^7
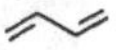	$< 10^5$	−1.15	1.00	4×10^4
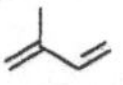	7.1×10^6	1.12	1.00	5×10^6
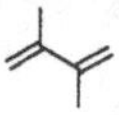	2.7×10^7	1.37	1.00	1×10^7
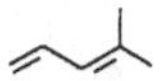	2.5×10^8	2.59	1.00	2×10^8
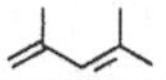	1.0×10^9	3.5 (estim.)		(2×10^9)

[a] Rate constants k_{exp} (L mol^{-1} s^{-1}) determined by pulse radiolysis at 24 °C, from ref. [7].
[b] From refs. [24] and [30].
[c] Calculated k_{calc} (L mol^{-1} s^{-1}) by eq. (2) with $E(Ph_2CH^+) = 5.71$.

Scheme 12. Rate constants (L mol^{-1} s^{-1}) for the reactions of the benzhydryl cation with alkenes.

Scheme 13 shows that rate constants up to 10^7 L mol^{-1} s^{-1} have directly been measured for the reactions of benzhydryl cations with isobutylene and styrene. Since second-order rate constants around 10^7 show little temperature dependence (see below) the data for Ph_2CH^+ and $(p\text{-}ClC_6H_4)_2CH^+$ were plotted as measured at ambient temperature. In order to predict the propagation rate constants k_p^+ for the cationic polymerization of isobutylene and styrene, one would need the electrophilicity parameters E for the 2,4,4-trimethyl-2-pentyl cation and the 1,3-diphenyl-1-butyl cation, which are shown on the right of Scheme 13. Since their electrophilicities E are greater than that of the parent benzhydryl cation, the correlation in Scheme 13 requires that the corresponding k_p^+ values are considerably greater than 10^7 L mol^{-1} s^{-1}, in agreement with our previous work [5].

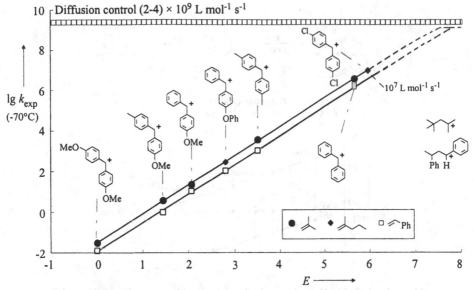

Scheme 13. Directly measured rate constants for the reactions of benzhydryl cations with isobutylene (or 2-methyl-1-pentene) and styrene.

Fast reactions like the propagation steps show some peculiarities as illustrated for the reactions of benzhydryl cations with 2-methyl-1-pentene and 2-methyl-2-butene in Schemes 14 and 15 [40]. In both reaction series the free enthalpy relationships (i.e., the correlations between $\Delta G^{\ddagger}$ and E) remain linear up to rate constants of 7×10^{7} L mol^{-1} s^{-1}. In both reaction series, the increase of reactivity from the dimethoxy compound to the dimethyl compound is entirely due to a decrease in the enthalpy of activation ($\Delta H^{\ddagger}$) while the entropy of activation ($\Delta S^{\ddagger}$) remains constant. Though the activation parameters for the reaction of the dichloro-substituted benzhydryl cation with 2-methyl-1-pentene have not been determined, it can be assumed that the further decrease of $\Delta G^{\ddagger}$ is caused by constant $\Delta S^{\ddagger}$ and a further decrease of $\Delta H^{\ddagger}$ (dashed line in Scheme 14).

The increase of reactivity from X = Y = Me to X = Y = Cl in Scheme 15 has a different origin, however. Since $\Delta H^{\ddagger}$ for the reaction of (p-MeC$_6$H$_4$)$_2$CH^{+} with 2-methyl-2-butene is zero, the further increase in reactivity for the dichloro substituted benzhydryl cation is achieved by reduction of the negative value of $\Delta S^{\ddagger}$.

We have to conclude from these data that variation of the substituents in the benzhydryl cations exclusively affects $\Delta H^{\ddagger}$ as long as there is an enthalpic barrier. Fast reactions, which proceed without enthalpic barrier, can further be accelerated by an increase of the entropy of activation (Scheme 16).

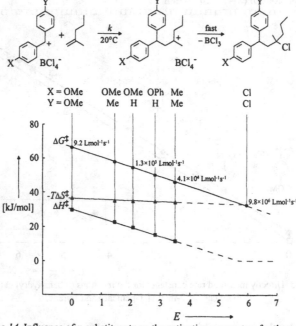

Scheme 14. Influence of p-substituents on the activation parameters for the reactions of benzhydryl cations with 2-methyl-1-pentene.

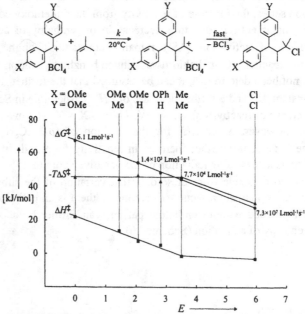

Scheme 15. Influence of p-substituents on the activation parameters for the reactions of benzhydryl cations with 2-methyl-2-butene.

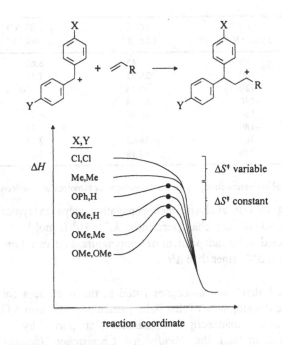

Scheme 16. Schematic enthalpy profiles for the reactions of benzhydryl cation with alkenes.

This behavior can be rationalized in the following way: As long as there is a barrier in ΔH, the different carbocations will approach the π-system along the same reaction path. They will travel along a limited number of similar trajectories, and constant $\Delta S^{\ddagger}$ will be observed.

When there is no enthalpic barrier from reactants to products, however, other ways of approach, all with a continuous decrease of ΔH become accessible, and it is easy to imagine that the width of the approach corridor will increase with increasing exothermicity of the reactions. The variety of trajectories will continuously grow and the same electronic effects which lead to a variation of $\Delta H^{\ddagger}$ at constant $\Delta S^{\ddagger}$ will lead to a variation of $\Delta S^{\ddagger}$ for $\Delta H^{\ddagger} \approx 0$.

The change from enthalpic to entropic effects is not restricted to reactions of carbocations with olefins but should be observable for any bimolecular reaction. The left column of Scheme 17 lists some activation entropies typical for bimolecular reactions. In combination with $\Delta H^{\ddagger} = 0$, one calculates the values of $\Delta G^{\ddagger}(20°C)$ given in the middle column and the corresponding rate constants in the right column. These are the maximum rate constants for bimolecular reactions with the corresponding values of $\Delta S^{\ddagger}$ if negative values of $\Delta H^{\ddagger}$ are excluded. As a consequence, isoentropic reaction series (with $\Delta S^{\ddagger}$ as listed in Scheme 17) will terminate at the rate constants given in the right column of Scheme 17; faster reactions will be controlled by the entropy of activation $\Delta S^{\ddagger}$.

$\Delta S^{\ddagger}$ / J mol^{-1} K^{-1}	$\Delta G^{\ddagger}$ / kJ mol^{-1}	lg k (20 °C) / L mol^{-1} s^{-1}
−80	23.45	8.61
−100	29.32	7.56
−120	35.18	6.52
−140	41.04	5.47
−160	46.90	4.43
−180	52.77	3.38
−200	58.63	2.34
−220	64.49	1.29

Scheme 17. Typical activation entropies and rate constants for bimolecular reactions with $\Delta H^{\ddagger} = 0$.

According to our analysis, the propagation steps in typical carbocationic polymerizations of alkenes are characterized by $k_p^+ > 10^8$ L mol^{-1} s^{-1}. Their rates can, therefore, be expected to be independent of temperature, and variations in the reaction conditions will affect $\Delta S^{\ddagger}$ rather than $\Delta H^{\ddagger}$.

Acknowledgment. I thank my associates listed in the references for their excellent contributions to the development of this field, particularly Dr. Armin Ofial for his help in preparing this manuscript. Financial support by the *Deutsche Forschungsgemeinschaft* and the *Fonds der Chemischen Industrie* is gratefully acknowledged.

4. References

1. Plesch, P.H. (1993) *Prog. Reaction Kinetics* **18**, 1–62.
2. Charleux, B., Rives, A., Vairon, J.-P., and Matyjaszewski, K. (1996) *Macromolecules* **29**, 5777–5783.
3. Vairon, J.-P., Charleux, B., and Rives, A. (1996) *Macromol. Symp.* **107**, 307–317.
4. Vairon, J.-P., Rives, A., and Bunel, C. (1992) *Makromol. Chem., Macromol. Symp.* **60**, 97–105.
5. Mayr, H., Roth, M., and Lang, G. (1997) *Cationic Polymerization, Fundamentals and Applications*, ACS Symposium Series, Vol. 665, pp. 25–40.
6. Roth, M. and Mayr, H. (1996) *Macromolecules* **29**, 6104–6109.
7. Wang, Y. and Dorfman, L.M. (1980) *Macromolecules* **13**, 63–65.
8. Dorfman, L.M., Sujdak, R.J., and Bockrath, B. (1976) *Acc. Chem. Res.* **9**, 352–357.
9. Schneider, R., Grabis, U., and Mayr, H. (1986) *Angew. Chem.* **98**, 94–95; *Angew. Chem. Int. Ed. Engl.* **25**, 89–90.
10. Mayr, H., Schneider, R., Schade, C., Bartl, J., and Bederke, R. (1990) *J. Am. Chem. Soc.* **112**, 4446–4454.
11. Mayr, H. (1990) *Angew. Chem.* **102**, 1415–1428; *Angew. Chem. Int. Ed. Engl.* **29**, 1371–1384.
12. Problems to define "stabilities" of carbocations have been discussed. See ref. [13], and refs. cited therein.
13. Mayr, H. (1996) Fundamentals of the reactions of carbocations with nucleophiles, in Matyjaszewski, K. (ed.), *Cationic Polymerization: Mechanisms, Synthesis and Applications*, Marcel Dekker, New York, pp. 51–136.
14. a) Hadjikyriacou, S., Fodor, Z., and Faust, R. (1995) Synthetic applications of nonpolymerizable monomers in living cationic polymerization: Functional polyisobutylenes by end-quenching, *J. Macromol. Sci., Pure Appl. Chem.* **A32**, 1137–1153. b) Roth, M., Mayr, H., and Faust, R. (1996) Examination of models for carbocationic polymerization: The influence of chain length on carbocation reactivities, *Macromolecules* **29**, 6110–6113. c) Fodor, Z. and Faust, R. (1994) Living carbocationic polymerization of p-methylstyrene and sequential block copolymerization of isobutylene with p-methylstyrene, *J. Macromol. Sci., Pure Appl. Chem.* **A31**, 1985–2000. d) Li, D. and Faust, R. (1995)

Polyisobutylene-based thermoplastic elastomers. 3. Synthesis, characterization, and properties of poly(α-methylstyrene-b-isobutylene-b-α-methylstyrene) triblock copolymers, *Macromolecules* **28**, 4893–4898. e) Hadjikyriacou, S. and Faust, R. (1995) Living carbocationic homopolymerization of isobutyl vinyl ether and sequential block copolymerization of isobutylene with isobutyl vinyl ether. Synthesis and mechanistic studies, *Macromolecules* **28**, 7893–7900. f) Hadjikyriacou, S. and Faust, R. (1996) Amphiphilic block copolymers by sequential living cationic polymerization: Synthesis and characterization of poly(isobutylene-b-methyl vinyl ether), *Macromolecules* **29**, 5261–5267. g) Schlaad, H., Erentova, K., Faust, R., Charleux, B., Moreau, M., Vairon, J.-P., and Mayr, H. (1998) Kinetic study on the capping reaction of living polyisobutylene with 1,1-diphenylethylene. 1. Effect of temperature and comparison to the model compound 2-chloro-2,4,4-trimethylpentane, *Macromolecules*, in print.

15. Hering, N. (1996) *Quantifizierung der Nucleophilie von Enaminen*, Diploma thesis, Technische Hochschule Darmstadt.
16. Mayr, H. and Striepe, W. (1983) *J. Org. Chem.* **48**, 1159–1165.
17. Mayr, H. and Gorath, G. (1995) *J. Am. Chem. Soc.* **117**, 7862–7868.
18. Mayr, H. (1989) in Schinzer, D. (ed.), *Selectivities in Lewis Acid-Promoted Reactions*, NATO ASI Series C, Vol. 289, Kluwer Academic Publishers, Dordrecht, pp. 21–36.
19. Mayr, H., Schade, C., Rubow, M., and Schneider, R. (1987) *Angew. Chem.* **99**, 1059–1060; *Angew. Chem. Int. Ed. Engl.* **26**, 1029–1030.
20. Pross, A. and Shaik, S.S. (1982) *J. Am. Chem. Soc.* **104**, 1129–1130.
21. Brönsted, J.N. (1923) *Recl. Trav. Chim. Pays-Bas* **42**, 718–728.
22. Dau-Schmidt, J.-P. and Mayr, H. (1994) *Chem. Ber.* **127**, 205–212.
23. Mayr, H., Schneider, R., Irrgang, B., and Schade, C. (1990) *J. Am. Chem. Soc.* **112**, 4454–4459.
24. Mayr, H. and Patz, M. (1994) *Angew. Chem.* **106**, 990–1010; *Angew. Chem. Int. Ed. Engl.* **33**, 938–957.
25. Mayr, H., Kuhn, O., Gotta, M.F., and Patz, M. (1998) *J. Phys. Org. Chem.* **11**, 642–654.
26. Bartl, J., Steenken, S., and Mayr, H. (1991) *J. Am. Chem. Soc.* **113**, 7710–7716.
27. Burfeindt, J., Patz, M., Müller, M., and Mayr, H. (1998) *J. Am. Chem. Soc.* **120**, 3629–3634.
28. Roth, M., Schade, C., and Mayr, H. (1994) *J. Org. Chem.* **59**, 169–172.
29. Pock, R. and Mayr, H. (1986) *Chem. Ber.* **119**, 2497–2509.
30. Mayr, H. and Hartnagel, M. (1996) *Liebigs Ann.*, 2015–2018.
31. Kennedy, J.P. and Marechal, E. (1982) *Carbocationic Polymerization*, Wiley, New York.
32. Olah, G.A. and Westerman, P.W. (1973) *J. Am. Chem. Soc.* **95**, 7530–7531.
33. Larsen, J.W., Bouis, P.A., and Riddle, C.A. (1980) *J. Org. Chem.* **45**, 4969–4973.
34. Roth, M. and Mayr, H. (1995) *Angew. Chem.* **107**, 2428–2430; *Angew. Chem. Int. Ed. Engl.* **34**, 2250–2252.
35. Mayr, H. and Ofial, A.R. (1997) *Tetrahedron Lett.* **38**, 3503–3506.
36. Henninger, J., Mayr, H., Patz, M., and Stanescu, M.D. (1995) *Liebigs Ann.*, 2005–2009.
37. Mayr, H., Müller, K.-H., Ofial, A.R., and Bühl, M. (1998) submitted.
38. Mayr, H., Henninger, J., and Siegmund, T. (1996) *Res. Chem. Intermed.* **22**, 821–838.
39. Mayr, H., Patz, M., Gotta, M.F., and Ofial, A.R. (1998) *Pure Appl. Chem.*, in press.
40. Patz, M., Mayr, H., Bartl, J., and Steenken, S. (1995) *Angew. Chem.* **107**, 519–521; *Angew. Chem. Int. Ed. Engl.* **34**, 490–492.

PHOTOCHEMICALLY INDUCED CATIONIC PHOTOPOLYMERIZATION OF VINYL ETHERS AND OXETANES

O. NUYKEN, M. RUILE
Technische Universität München,
Lehrstuhl für Makromolekulare Stoffe
Lichtenbergstr. 4, D-85747 Garching, Germany

Abstract. 2,3-dihydrofuran can be polymerized by means of cationic photoinitiators yielding 100 % polymer in less than 120 seconds at room temperature. The product is colorless, has good film forming properties and does not show any side reaction such as aldehyde formation. Several mono-, bi- and trifunctional oxetanes were synthesized in bulk and in solution cationically. Selected photoinitiators have been applied. It was found that sulfonium salts are very efficient due to good solubility, almost no discoloration of the product and storage stability in the monomer in the absence of light. The conversion was determined by quantitative IR-spectroscopy. Conversion between 75 % and 85 % was found in all cases. The shrinkage during polymerization was much lower than for vinyl monomers. No inhibition by oxygen was observed. Monomer layers thicker than 5.5 mm could be polymerized. The products are transparent and almost colorless. The glass transition temperature of the crosslinked polymers was above 37 °C (temperature of the human body).

1. Introduction

Photoinduced polymerizations are of considerable interest due to energy, economic and environmental issues. These polymerizations are applied in coatings [1], adhesives, printing inks [2] and they may become of some importance in the field of dental applications too. Their main advantage over the common cationic polymerization is the reaction in bulk and the crosslinking, which is essential for many of the applications mentioned above. Among the most promising monomers for this type of polymerization are vinylethers and cyclic ethers. Both types have been studied in some detail. Therefore, this presentation is subdivided into two chapters: vinylethers and oxetanes.

117

J.E. Puskas et al. (eds.), Ionic Polymerizations and Related Processes, 117–142.

2. Photoinduced Cationic Polymerization of Vinyl Ethers

2.1. MOTIVATION

Most of the commercial coatings are based on acrylates cured with photoinitiators such as *Lucirin TPO* (trade name: product from BASF AG, Ludwigshafen/Rhein-Germany) producing radicals if it is irradiated with light of a high pressure mercury lamp, which is able to initiate a radical polymerization [3]. Due to toxic problems and expected stronger legal requirements in the near future one has to search for alternatives early enough. Among the possible substances vinylether seemed to be the most promising class of compounds, moreover it was commercially available in large quantities.

What are the demands to an alternative system?

1. The monomer has to be converted 100 % into a (crosslinked) polymer in a very short time.
2. The polymer should be colorless.
3. The polymerization must be possible on open air and it should take place at room temperature.

2.2. PRELIMINARY RESULTS OF THE PHOTOINDUCED CATIONIC POLYMERIZATION OF ALKYLVINYLETHER

Indeed the polymerization of alkylvinylether, initiated by CRIVELLO salts [4,5,6] (see section 2.4) was very fast. Furthermore, the resulting polymer had excellent adhesion properties towards metal, paper and wood. Therefore it seemed to be the ideal substance for the basic material of the new generation of coatings. However, "analysis by nose" and later by more sophisticated instrumentation indicated a problem, namely the formation of acetaldehyde during the polymerization. Since acetaldehyde is on the list of unacceptable compounds, the application of vinylethers in the coatings would be rather limited if it is not possible to avoid this aldehyde formation. Although the detectable amount of acetaldehyde was very small we could not find a suitable way to avoid its formation. This is probably due to the fact that traces of water can add onto vinylether resulting a semiacetale. The following ways of decomposition of this semiacatale are imaginable [7,8]:

(1)

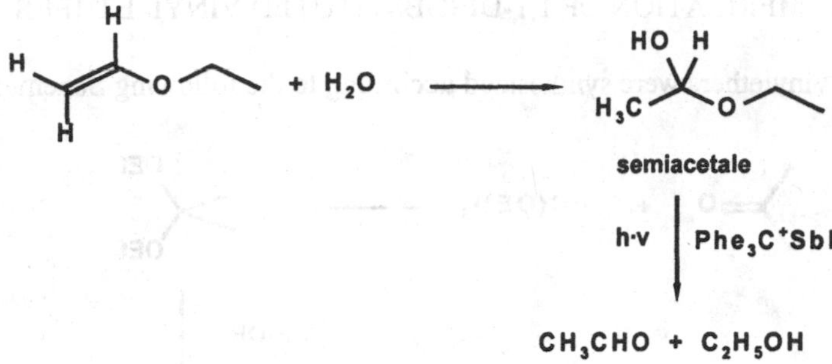

$$(2)$$

We favor the second route [8] sience we did not find any $H_2C=CH_2$ in the gas analysis. The second route runs via a semi-acetal which hydrolyses as soon as the proton production starts.

2.3. ALTERNATIVES TO VINYL ETHER

If one believes in the mechanism given above, small structure variations should solve the problem:

first variation:

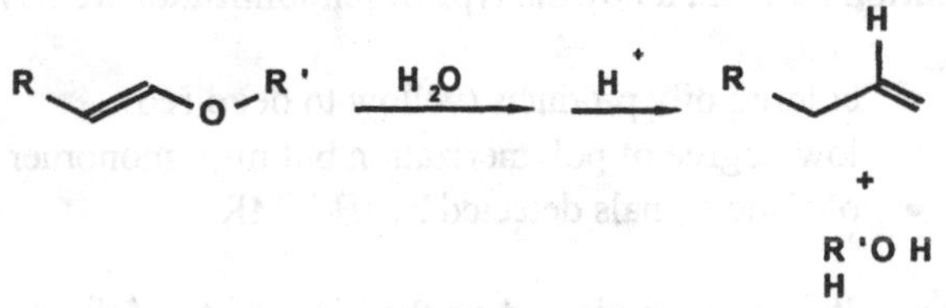

Hydrolysis would result in an extended aldehyde, which would be less problematic then acetaldehyde.

second variation:

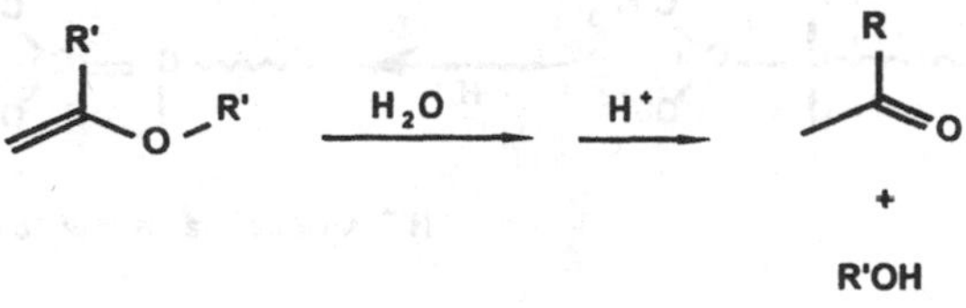

Addition of water onto a 1,1-disubstituted vinyl ether would yield a semiketal. Its proton induced hydrolysis would result in a ketone instead of an aldehyde.

2.4. PHOTOPOLYMERIZATION OF 1,1-DISUBSTITUTED VINYL ETHERS

1,1-disubstituted vinylethers were synthesized according to the following Scheme:

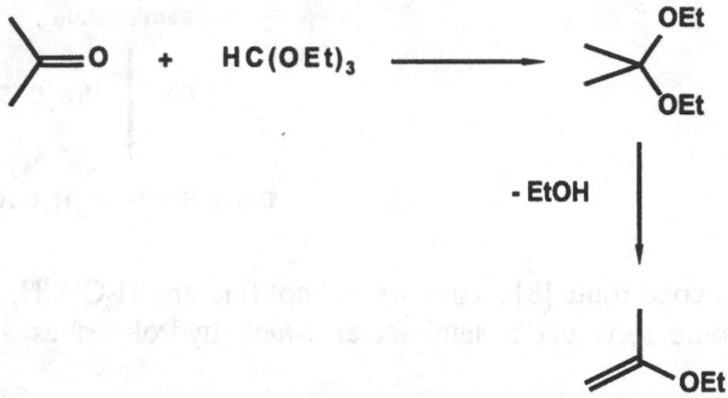

These compounds can be polymerized with high rate at room temperature initiated by photoinduced protons. However, the results were rather disappointing. Although we have made many attempts in order to improve the situation by variation of the substitution pattern and by the type of photoinitiator we have always observed:

- colored oily products (yellow to deep red)
- low degree of polymerization but high monomer conversion
- olefinic signals detected by ^{1}H NMR

These results can be explained on the basis of the following Scheme, in which transfer reactions play a key role [9,10].

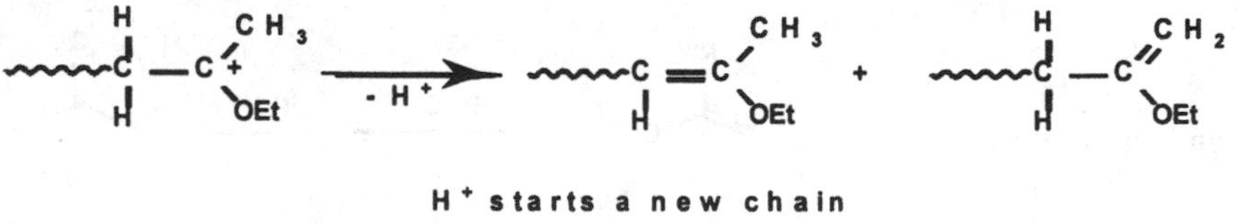

H$^+$ starts a new chain

The following photoinitiators were tested:

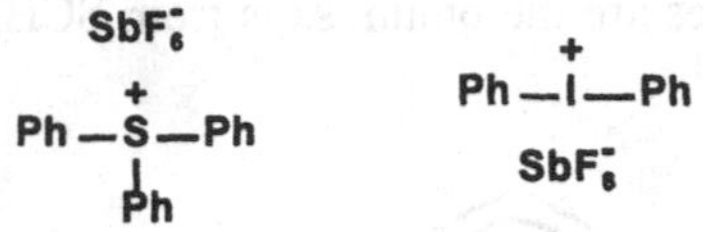

$$Ph-\overset{+}{\underset{Ph}{S}}-Ph \quad SbF_6^-$$

$$Ph-\overset{+}{I}-Ph \quad SbF_6^-$$

(CRIVELLO - type)

Pyridinium OEt PF_6^-

Biphenyl pyridinium N—OEt PF_6^-

Isoquinolinium N—OEt PF_6^-

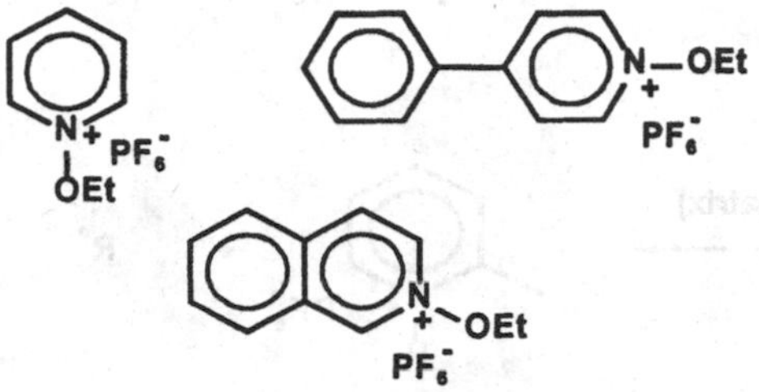

(SCHNABEL - YAGCI - type)

and

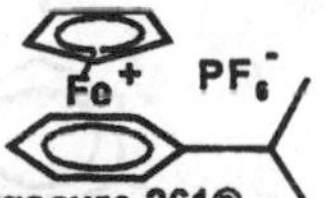

$Fe^+ \quad PF_6^-$

rgacure 261®

The formation of a superacid by photoirradiation of a "CRIVELLO salt" is shown in the following Scheme:

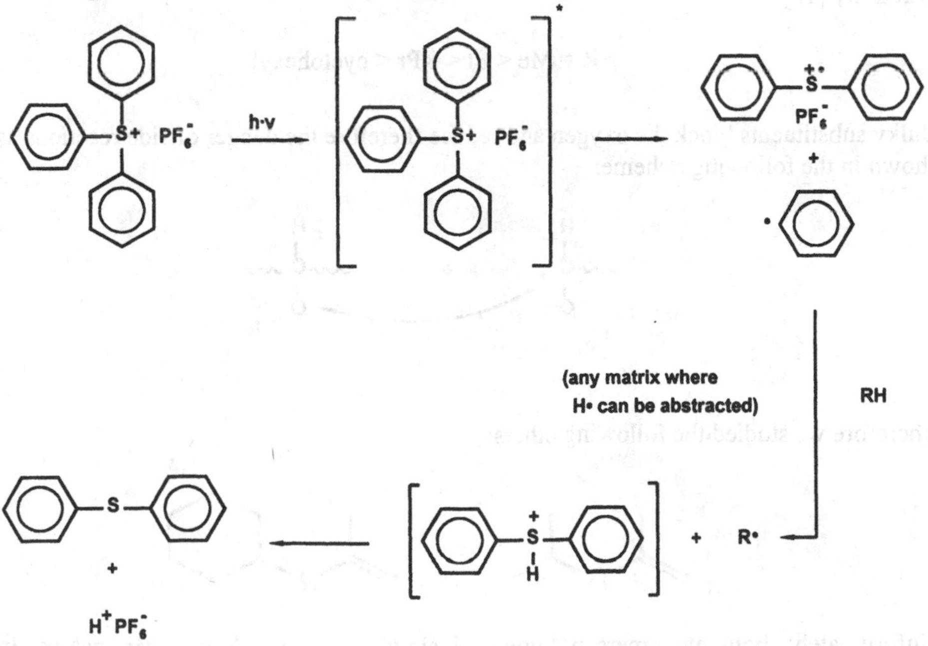

(any matrix where
H• can be abstracted)

RH

$H^+ PF_6^-$

Although the formation of a superacid is the key process one should not ignore the formation of radicals and their possible reactions.

Other possible substances for proton sources are the onium salts from SCHNABEL and YAGCI [11]:

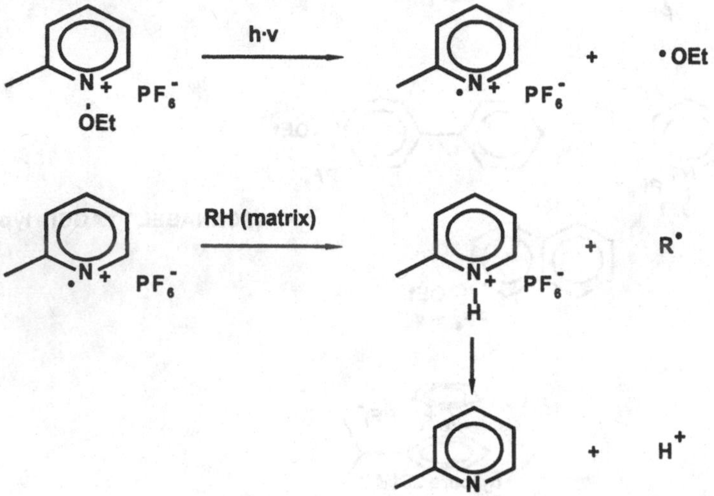

For the polymerization of unsubstituted vinylether one has found the following order in reactivity [12]:

$$R = Me < Et < i\text{-}Pr < cyclohexyl$$

Bulky substituents block the oxygen and reduce therefore the danger of side reactions as shown in the following Scheme:

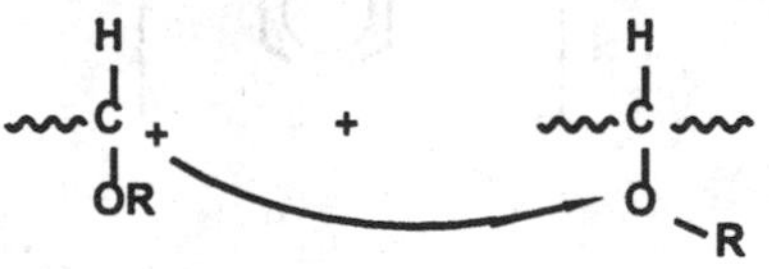

Therefore we studied the following ethers:

Unfortunately, both monomers polymerized slowly, resulting low molar masses. In addition, the products were colored.

Another variation did not result in the desired properties as shown in the following list:

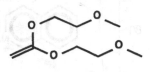

α-ethoxystyrene   **1,1-bis(methoxyethoxy)ethene**

- low rate (40 % conversion in 2 h)
- low DP
- oily product
- side reactions

- high rate
- solid, but colored
- decomposition during purification

2.4.1. *Variation of the temperature*

Polymerization should work at room temperature. However, it seemed worthwhile to study the polymerization at lower temperatures. Reducing the reaction temperature from 25 °C to - - 10 °C had a remarkable effect on the degree of polymerization which increased from 6 to 71. Certainly, it is of academic interest to study the temperature effect. However, the commercial view is strict; it is not possible to apply systems which do not work at room temperature.

2.4.2. *Polymerization of 1-Ethoxy-propene*

Although, we have made several attempts to improve the polymerization by variation of the substitution pattern of the monomers and by changing the type of photoinitiator, we did not find a way to overcome the problem that the polymerization rates were too slow.

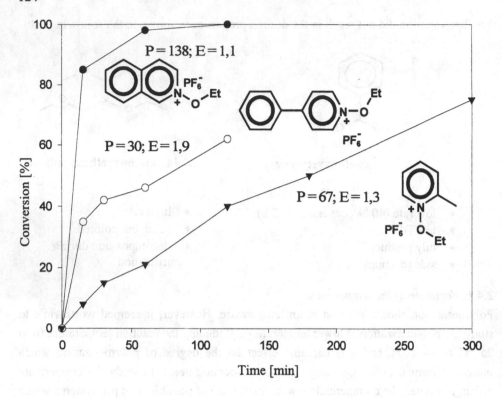

Figure 1: Polymerization of 1-Ethoxy-propene at room temperature- effect of different initiators ($E = M_w/M_n$)

2.5. CYCLIC VINYL ETHERS

All attempts to polymerize 1-ethoxy-cyclohex-1-ene and 1-ethoxy-cyclopent-1-ene were unsatisfactory.

We have observed:
- low rate of polymerization
- low degree of polymerization
- side reactions

2.5.1. *3,4-Dihydro-2H-pyrane, 2,3-Dihydrofurane*

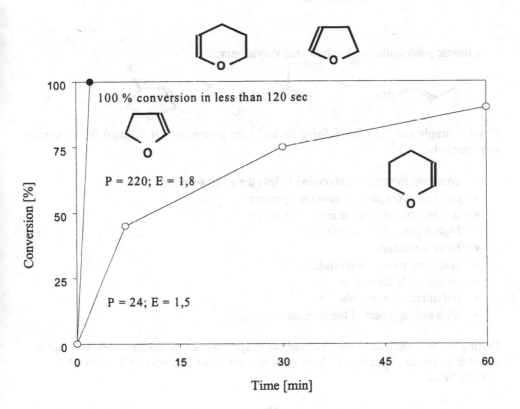

100 % conversion in less than 120 sec

P = 220; E = 1,8

P = 24; E = 1,5

Figure 2. Photopolymerization of 3,4-dihydro-2H-pyrane and 2,3-dihydrofuran initiated by Crivello salt

Typical competing systems for coatings are:

acrylates:

Lucirin TPO + (+ diacrylates)

cationic photoinitiators + substituted vinylethers:

From a simple test 2,3-dihydrofuran looked very promising - it fulfilled the following requirements:

- complete monomer conversion in less than 120 seconds
- reaction takes place at room temperature
- it is an easy obtainable monomer
- high degree of polymerization
- no side reactions
- colorless, transparent product
- no aldehyde formation
- polymerization in bulk
- high boiling point of the monomer.

During the polymerization we observed the evaporation of monomer traces. In order to avoid this undesired process we have studied a substituted monomer: 3-phenyl-2,3-dihydrofuran.

boiling point: 140 °C, 12 Torr

This monomer is an alternative to the unsubstituted 2,3-dihydrofuran. The cationic photopolymerization of this monomer yielded 100 % conversation in less than 120 sec. (exact kinetics are not yet available).

However, the synthesis of the monomer [13] is not simple [14] and needs improvement:

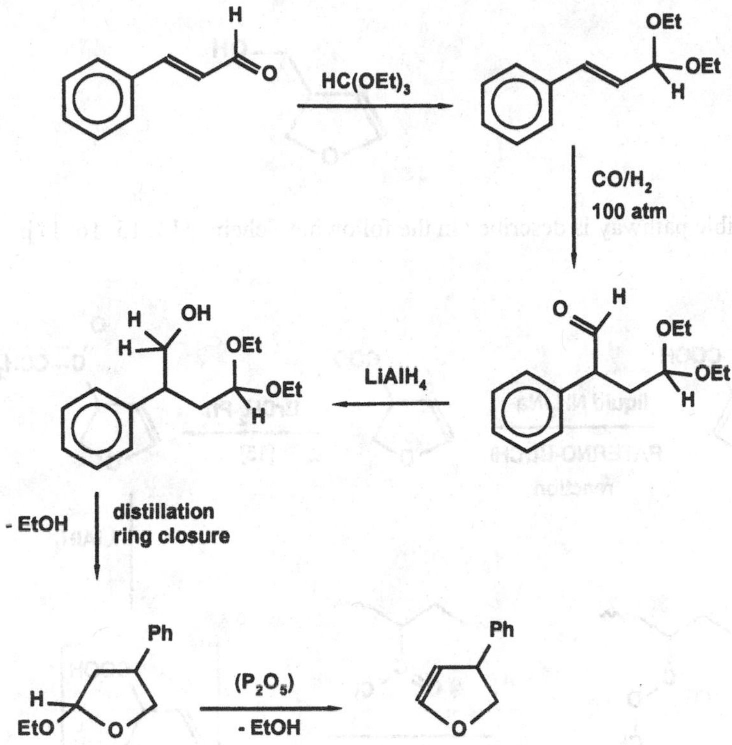

A proposal for the mechanism of the ring closure is shown below:

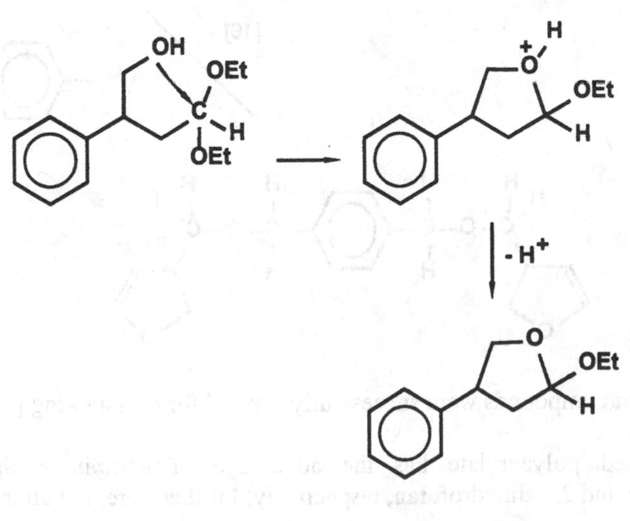

128

If one has to think about a bifunctional monomer - an intermediate monomer could be the following compound:

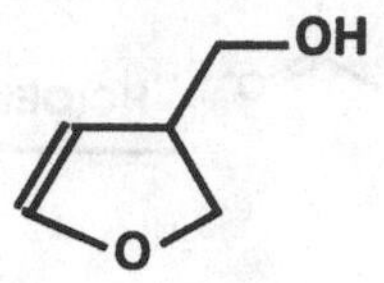

A possible pathway is described in the following Scheme [14, 15, 16, 17]:

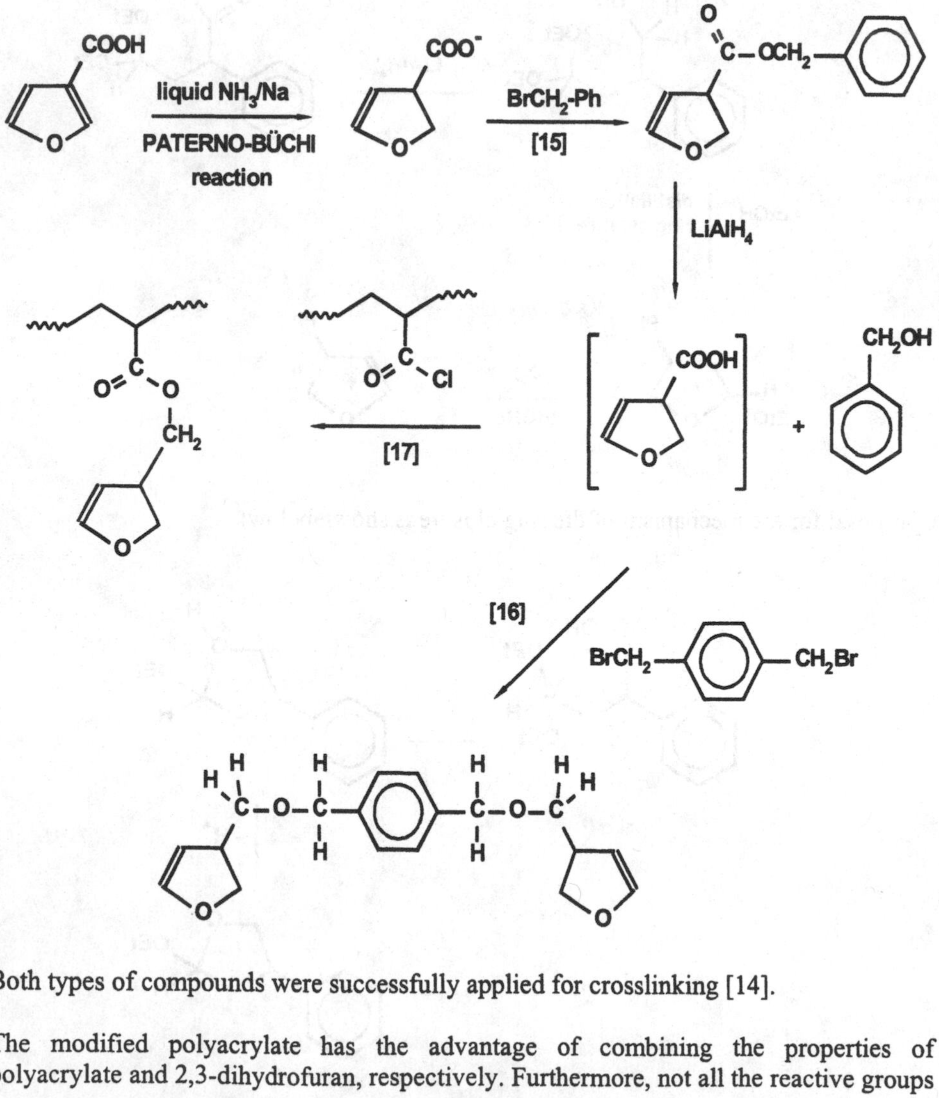

Both types of compounds were successfully applied for crosslinking [14].

The modified polyacrylate has the advantage of combining the properties of polyacrylate and 2,3-dihydrofuran, respectively. Furthermore, not all the reactive groups must be converted in order to get insoluble (crosslinked) material. In addition, the danger of migration of unreacted monomer is minimized.

Since the synthesis of 4-hydroxymethyl-2,3-dihydrofuran we tried to extend the hydroxyalkyl substituent by one CH_2 group, because of the easy synthesis.

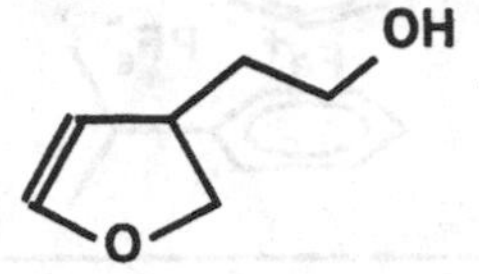

According to the following Scheme the synthesis [18,19,20,21,22] looked simple and the starting compounds are cheap and commercially available. The synthesis of the monomer leads to a bicyclic compound.

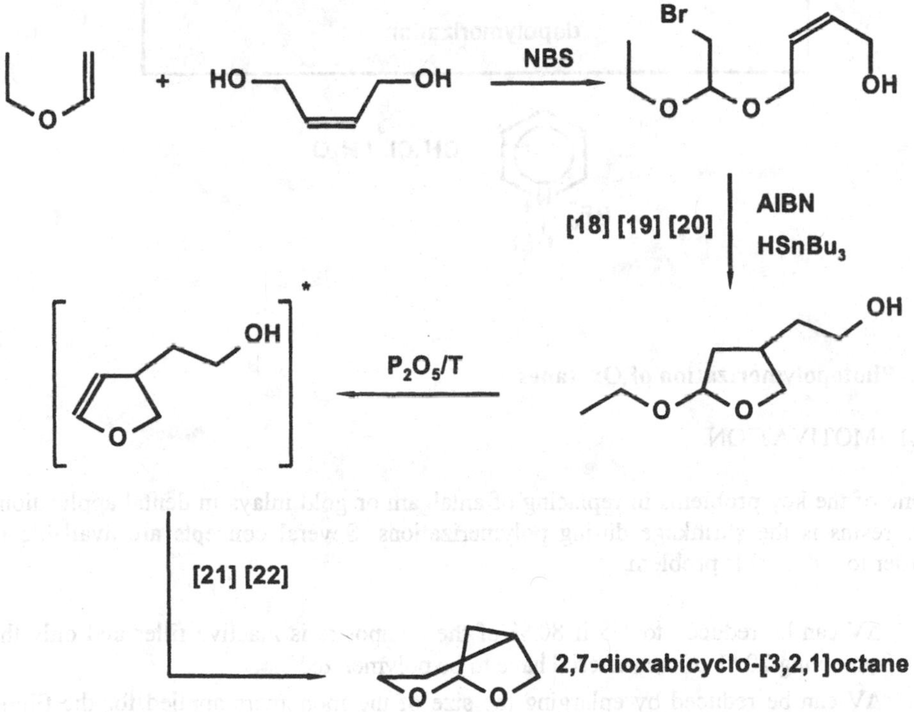

This bicyclic compound can be polymerized cationically via ring-opening of the 5-membered ring by photoinitiators. In addition, it is possible to depolymerize the obtained polymer by use of a photo-Brönsted-acid [14]:

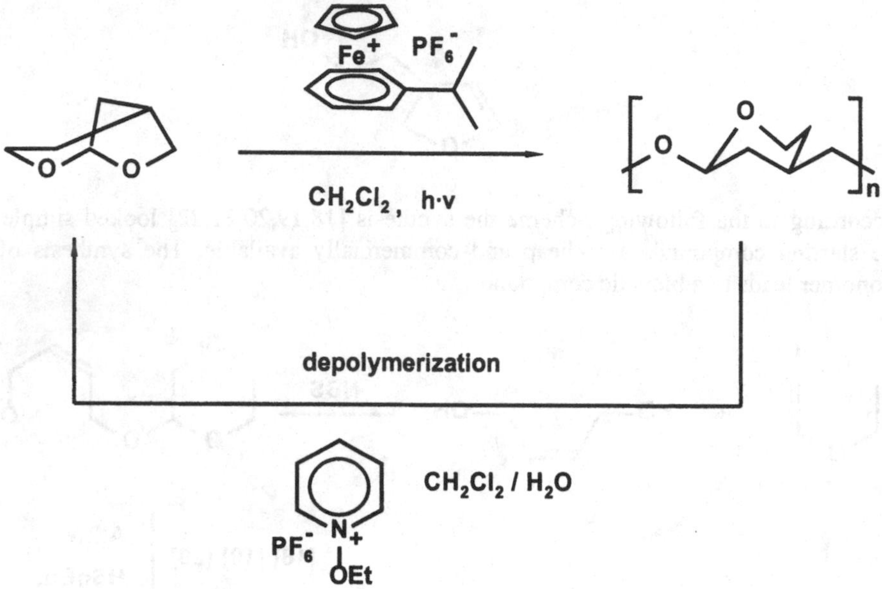

3. Photopolymerization of Oxetanes

3.1. MOTIVATION

One of the key problems in replacing of amalgam or gold inlays in dental applications by resins is the shrinkage during polymerizations. Several concepts are available in order to reduce this problem.

1. ΔV can be reduced to 1/5 if 80 % of the composite is inactive filler and only the remaining 20 % of the material have to be polymerized.
2. ΔV can be reduced by enlarging the size of the monomers applied for the filling material.

ethylene $\rightarrow$ PE 66 % (M : 28)
MMA $\rightarrow$ PMMA 21.2 % (M : 100)
N-VC $\rightarrow$ PNVC 7.4 % (M : 193)

The favored monomers for dental applications are *BIS-GMA* (2,2-bis-4(2-hydroxy-3-methacryloxypropyloxy)phenylpropane):

TEDMA (Triethylenglycoldimethylacrylate)

What are the requirements for a suitable synthetic dental filling material?

- low shrinkage of the volume during polymerization
- easily induced polymerizations at room temperature
- no inhibition of the polymerization with oxygen
- low price
- simple handling
- no toxicity
- monomer must be bifunctional
- low water adsorption of the polymers
- product must be colorless, it should not change with applications

3.2. OXETANES, OXIRANES, THF

Cyclic ethers have been considered to be suitable substances.
Selected properties of cyclic ethers are listed in the Table below. Oxetanes are easy to synthesize [24], their cationic polymerization can be initiated by photoinitiators like sulfonium and iodonium salts undisturbed by oxygen and little affected by moisture [25,26,27]. Most important factors which control the reactivity of the rings are basicity of the ring oxygen, ring stress and sterical hindrance. Since ring stress is similar for oxiranes and oxetanes but basicity is higher for oxetanes we consider oxetanes an interesting family of compounds for applications in which reduced shrinkage in volume is important [28].

TABLE 1

molecular weight	44.1	58.1	72.1
$\dfrac{\text{density}}{\text{g} \cdot \text{mL}^{-1}}$	0.882	0.893	0.889
$\dfrac{\text{decrease}}{\text{vol.} - \%}$	23	17 [23]	10
pK_s	-3.7	- 2.02	- 2.1
$\dfrac{\text{ring strain}}{\text{kJ} \cdot \text{mol}^{-1}}$	114	107	23

([23] estimated in relationship of oxirane and tetrahydrofuran)

3.3. CATIONIC POLYMERIZATION OF OXETANES

3.3.1. *Classical methods*

Ring-opening polymerization of oxetanes can be initiated by means of protonic acids [29], oxonium salts [30], hexafluorophosphates [31] and Lewis acids in the presence of coinitiators [32,33,34]. Side reactions such as ring formation and backbiting can be reduced by lowering the reaction temperature [35,36]. These reactions can be more or less ignored in case of dental composites if bi- and higher functional monomers are applied.

Linear polyoxetanes were synthesized as early as 1954 [37,38] and became commercial products (3,3-bis(chloromethyl)oxetane: *Penton, Pentaplast*) which show mechanical properties comparable with nylon-6 and a remarkable chemical resistance [39].

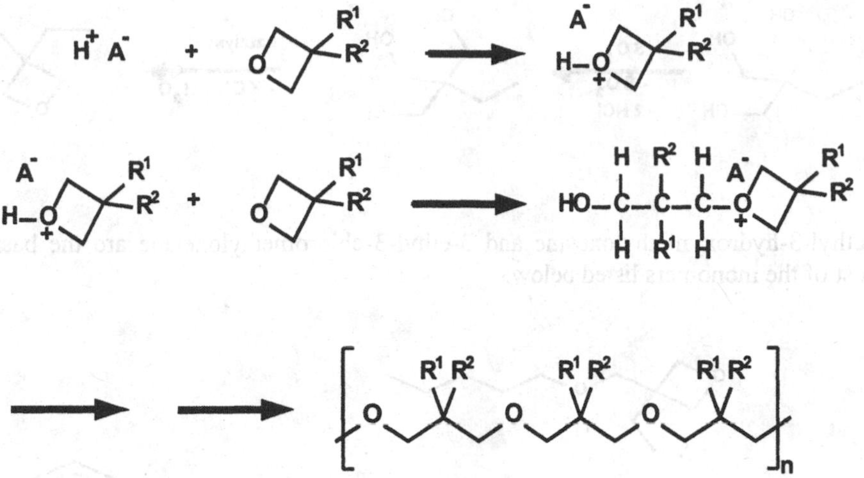

3.3.2. *Monofunctional monomers*

The simplest and probably cheapest method for the synthesis of monofunctional oxetane is the transesterification of diethyl carbonate with 1,1,1-trimethylolpropane yielding a cyclic carbonate as intermediate which split off CO_2 to form 3-ethyl-3-hydroxymethyloxetane [27].

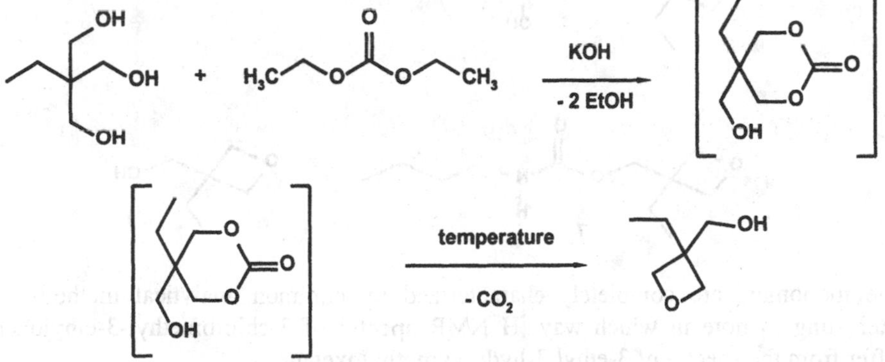

Another interesting module for the synthesis of oxetane derivatives can be synthesized according to the following Scheme:

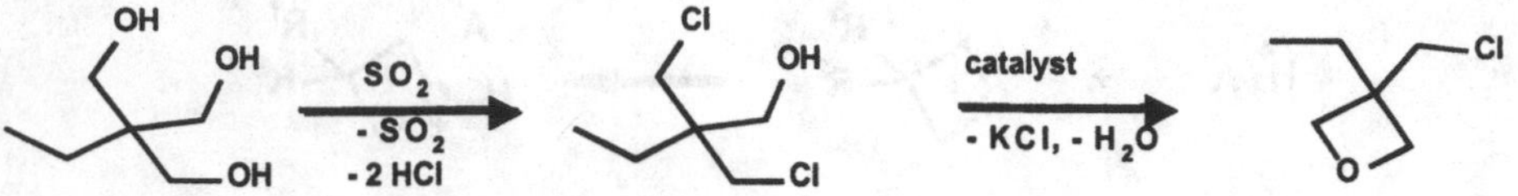

3-ethyl-3-hydroxymethyloxetane and 3-ethyl-3-chloromethyloxetane are the basis for most of the monomers listed below:

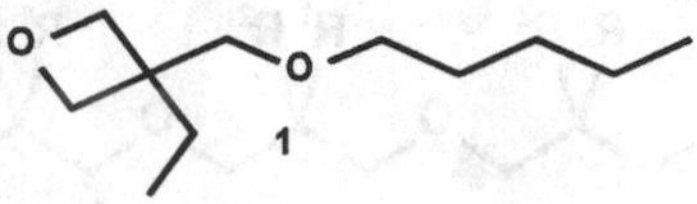

1

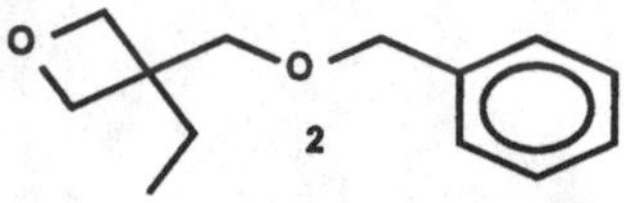

2

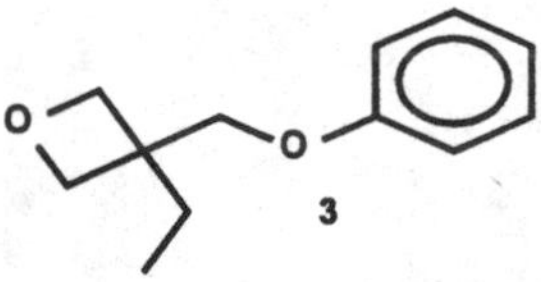

3

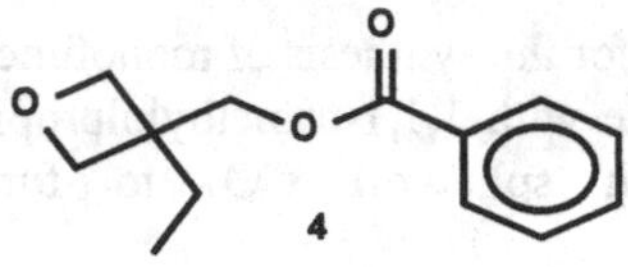

4

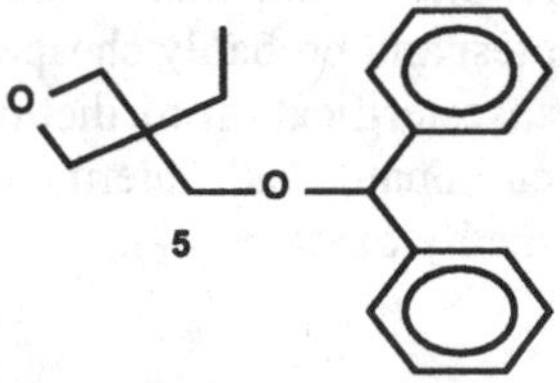

5

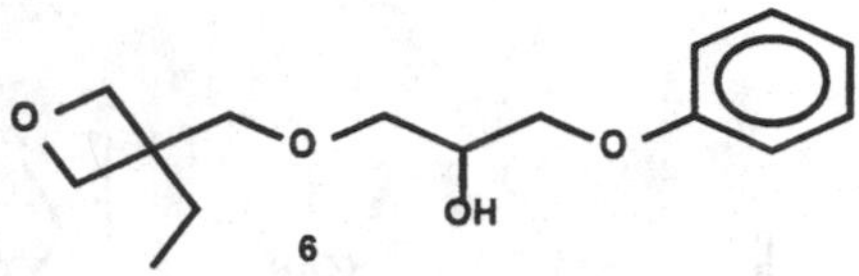

6

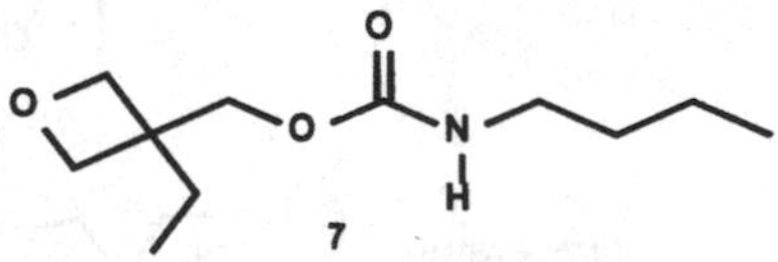

7

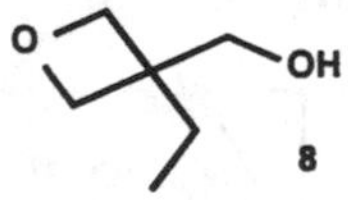

8

The monomers are completely characterized by common analytical methods. It is interesting to note in which way ^{1}H NMR spectra of 3-chloromethyl-3-ethyloxetane differ from the spectra of 3-ethyl-3-hydroxymethyloxetane:

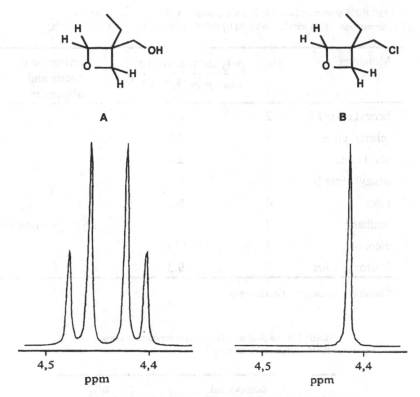

Figure 3. Section of the ¹H NMR spectra of 3-ethyl-3-hydroxymethyloxetane (A) and 3-chloromethyl-3-ethyloxetane (B).

Since the oxetane ring is not completely planar, one can distinguish between H_a and H_b for 3-ethyl-3-hydroxymethyloxetane due to the stabilization of the ring conformers by hydrogen bridging which is not possible for this 3-cloromethyl derivative. This is the reason for the observation of a singulett pattern in the second case.

Monomers 1 - 9 were polymerized by means of $RF_3 \cdot OEt_2$. The quantitative data are summarized in Table 2.

It is interesting to note that the highest rates of polymerization were observed for monomers which contain hydroxyl groups or labile hydrogen (benzyl ether II). Esters and urethane are either not polymerizable or react with very low rates only. This can easily explained on the basis of their pK_a-values:

TABLE 2. Polymerization of different oxetanes in $CDCl_3$ solution, monomer concentration = 0.41 mol·L^{-1}, initiated by BF$_3$·OEt$_2$ (c = 0.041 mol·L^{-1}), T = 27 °C

Monomer		polymerization rate [40] mol·min^{-1}L^{-1}·10^{-3}	formation of cycles and oligomers
benzyl ether I	2	2.5	-
phenyl ether	3	3.0	-
alkyl ether	1	2.8	-
benzyl ether II	5	6.1	-
ester	4	6.3	+
urethane	7		no polymerization
alcohol	8	11.6	-
hydroxyl ether	9	9.3	-

([40] linear regression (20% conversion))

TABLE 3. pK$_a$-values of selected compounds having functional groups similar to 1 – 8

compound	pK$_s$
ethyl phenyl ether	-6.44
diethyl ether	-3.59
methanol	-2.2
isopropanol	-3.2
ethyl benzoate	-7.78
N-methylacetamide	-0.46
alcohol 8	-2.08

Functional groups should be less suitable for cationic polymerizations which have pK$_a$-values higher (less negative) than that of oxetane. This is in agreement with our experimental results.

The relatively low molar masses for the polymers from hydroxyl moiety containing monomers indicate that OH causes chain transfer by oxonium groups, which stabilize via proton transfer.

TABLE 4. Characterization of the polyoxetanes

Monomer	Nr.	T_g °C	M_n RI detection	$\dfrac{M_w}{M_n}$
alkylether	1	- 93	12400	1.8
benzylether I	4	- 53	33100	2.9
phenylether	3	- 5	8000	1.7
benzylether II	5	- 1	1600	1.5
ester	4	- 39	oligomers	-
alcohol	8	13	1200	1.7
hydroxyether	9	- 25	3200	1.4

3.3.3. *Bi- and trifunctional oxetanes*

The route for the preparation of the bi- and trifunctional monomers listed below is adapted from that of the monofunctional monomers.

All these monomers are solids. However, they can be converted into liquids by addition of traces of monofunctional oxetanes.

Different photoinitiators have been tested for the cationic polymerization of selected oxetanes: Sulfonium salts [41,42], iodonium salts [43], N-oxide [44], metal complex [45] and sulfonic acid ester [46].Without going into too many details one can summarize the tests of initiators in the following way: All tested sulfonium and iodonium salts were initiators for oxetanes. However, only *Degacure KI 85B* fulfilled all the requirements: fast reaction, suitable solubility, almost no discoloration of the product, storage stability dissolved in the monomer (avoiding light).

Degacure KI 85B

The light induced polymerization of the bifunctional oxetanes was studied in bulk with a commercial UNILUX AC (KULZER). The monomer/initiator was filled in a specially designed cell, shown in Figure 4:

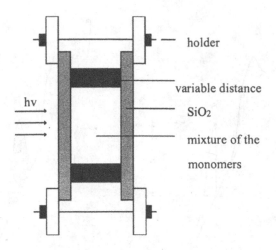

Figure 4. Cell for photopolymerization of oxetanes

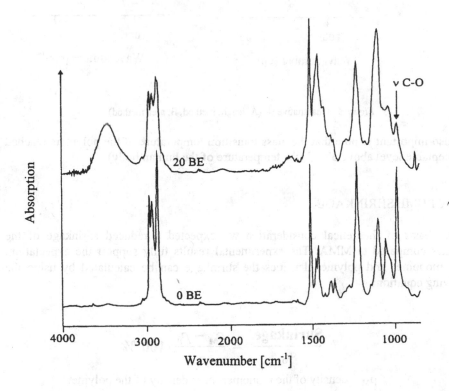

Figure 5. IR-spectra of an oxetane (dibenzyl ether **9** before (0 BE) and after (20 BE) irradiation, BE: irradiation units (Belichtungseinheiten)

140

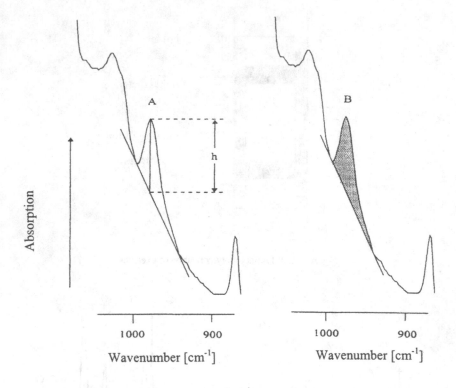

Figure 6. Quantitative IR (A: height method, B: area method)

It is also important to note that the glass transition temperature of the polymers reached an acceptable level above 37 °C (the temperature of the human body).

3.4. VOLUME SHRINKAGE

On the basis of theoretical consideration we expected a reduced shrinkage of the oxetanes compared to MMA. The experimental results fully support the expectation. From monomer and polymer densities the shrinkage can be calculated by using the following equation:

$$\frac{\textbf{Shrinkage}}{\%} = \frac{\rho_M - \rho_P}{\rho_P} \cdot \textbf{100 \%}$$

ρ_M = density of the monomer, ρ_P = density of the polymer

Some of the results are given in the following Table.

TABLE 5. Shrinkage in volume of selected oxetanes

Monomer	$\rho_{monomer}$ g·cm^{-1}	$\rho_{polymer}$ g·cm^{-1}	shrinkage %	conversion %
Alcohol 8	1.0209	1.1	7.8	100
Benzyl ether 2	1.0236	1.07	4.9	100
Dibenzyl ether 9	1.0657	1.11	3.9	70 - 80

TABLE 6. Glass transition temperature, conversion of selected difunctional oxetanes after certain exposure time

Monomer	exposure time BE	conversion A[47]	B[48]	T_g °C
Benzyldiether[49]	5	84	78	43
	10	79	72	50
	20	80	76	57
	5	76	66	50
	10	78	67	49
	20	82	72	61
Phenyldiether[50]	20	84	75	76

([47]peak height method, [48]peak area method, [49]The pure monomers are solids, they are converted into liquids by addition of 1 % monofunctional oxetane, [50]as previous note)

Table 6 documents quite clearly that oxetane fulfill many of the requirements listed at the beginning of this chapter.

142

4. References

1. Garrat, P.G. (1981) *Ind. Lachier. Betr.* **49**, 41.
2. Hashimoto, K. and Saraiya, S. (1981) *J. Radiat. Curing* **8**, 4.
3. Pappas, S.P. (Ed.) (1978) *UV-Curing: Science and Technology*, Technology Marketing Corp., Norwalk, CT.
4. Crivello, J.V. (1984) *Adv. Polym. Sci.* **62**, 1.
5. Crivello, J.V. and Jo, K.D. (1993) *J. Polym. Sci.: Part A: Polym. Chem.* **31**, 1473.
6. Crivello, J.V. and Jo, K.D. (1993) *J. Polym. Sci.: Part A: Polym. Chem.* **31**, 1483.
7. Murad, E. (1961) *J. Am. Chem. Soc.* **83**, 1328.
8. Schwertlick, K. (1986) (Hrsg.), *Organikum* **16**. Auflage, p. 269, VEB Dt. Verlag d. Wissenschaften, Berlin.
9. Eley, D.D. and Pepper, D.C. (1947) *Trans. Faraday Soc.* **43**, 112.
10. Eley, D.D. and Richards, A.W. (1949) *Trans. Faraday Soc.* **45**, 425.
11. Yagci, Y., Kornowski, A., and Schnabel, W. (1992) *J. Polym. Sci.: Part A: Polym. Chem.* **30**, 1987.
12. Jakobi, M. and Henne, A. (1983) *J. Radiat. Curing* **10**(4), 16.
13. Botteghi, C. (1975) *Gazz. Chim. Ital.* **105**, 233.
14. Raether, B. (1996) *Dissertation*, TU München.
15. Lowe, G. and Swain, St. (1985) *J. Chem. Soc. Perkin Trans.* **I**, 393.
16. Böhner, R., *Dissertation*, München 1995.
17. Voit, B., *Dissertation*, Bayreuth, 1990.
18. Dulcere, J.-P., Crandall, J., Faure, R., Santelli, M., Agati, V., and Mikoubi, M.N. (1993) *J. Org. Chem.* **58**, 5702.
19. Ueno, Y., Chino, K., Watanabe, M., Moriya, O., and Okawara, M. (1982) *J. Am. Chem. Soc.* **104**, 5564.
20. Stork, G. and Mook, R., Jr. (1983) *J. Am. Chem. Soc.* **105**, 3720.
21. Miyashita, M., Yoshikowski, A., and Grieco, P.A. (1977) *J. Org. Chem.* **42**, 3772.
22. Kigasawa, K., Hiiragi, M., Wakisaka, K., Kusama, O., Kawasaki, K., and Sugi, H. (1977) *J. Heterocycl. Chem.* **14**, 473.
24. Pattison, D.B. (1957) *J. Am. Chem. Soc.* **79**, 3455.
25. Sasaki, H. and Crivello, J.V. (1992) *J. Macromol. Sci., Chem.* **A29**, 915.
26. Crivello, J.V. and Sasaki, H. (1993) *J. Macromol. Sci., Chem.* **A30**, 173.
27. Crivello, J.V. and Sasaki, H. (1993) *J. Macromol. Sci., Chem.* **A30**, 189.
28. Arnett, E.M. and Wu, C.Y. (1962) *J. Am. Chem. Soc.* **84**, 1684.
29. Liu, Y.-L., Hsiue, G.-H., and Chiu, Y.-S. (1994) *J. Polym. Sci., Part A, Polym. Chem.* **32**, 2543.
30. Kops, J. and Spanggaard, H. (1981) *Polym. Bull. (Berlin)* **4**, 505.
31. Black, P.E. and Worsfold, D.J. (1976) *Can. J. Chem.* **54**, 3325.
32. Rose, J. B.. (1956) *J. Am. Chem. Soc.* **78**, 542.
33. Rose, J.B. (1956) *J. Am. Chem. Soc.* **78**, 546.
34. Penczek, I. and Penczek, S. (1963) *Makromol. Chem.* **67**, 203.
35. Bucquoye, M. and Goethals, E.J. (1954) *Macromol. Chem.* **179**, 1681.
36. Goethals, E.J. (1977) *Adv. Polym. Sci.* **23**, 103.
37. Farthing, A.C. and Reynolds, R.J.W. (1954) *J. Polym. Sci.* **12**, 503.
38. Farthing, A.C. (1955) *J. Chem. Soc.*, 3648.
39. Penczek, S. and Kubisa, P. (1989) in Eastmond G.C., Ledwith A., Russo S. and Sigwalt P. (eds.), *Comprehensive Polymer Science*, Pergamon Press, Oxford, , Vol. 3, p. 751f
40. Shankar, S.I. and Hakettstown, N.J. (1983) US 4.400.541, *Inv. Chem. Abstr.* **99**, 175.369.
41. Dektar, J.L. and Hacker, N.P. (1991) *J. Am. Chem. Soc.* **112**, 6004.
42. Crivello, J.V. and Lam, J.H.W. (1977) *Macromolecules* **10**, 1307.
43. Johnen, N., Kobayashi, S., Yagci, Y., and Schnabel, W. (1993) *Polym. Bull. (Berlin)* **30**, 279.
44. Crivello, J.V. and Lee, J.L. (1983) *Makromol. Chem.* **184**, 463.
45. Gaur, H.A., Gronenboom, C.J., Hageman, H.J., Hakvoort, G.T.M., Oosterhof, P., Overeen, T., Polam, R.J., and Van der Werf, S. (1984) *Makromol. Chem.* **185**, 1795.

KINETICS OF THE LIVING POLYMERIZATION OF ISOBUTYLENE

J. E. PUSKAS*, M. G. LANZENDÖRFER, H. PENG,
A. J. MICHEL, L. B. BRISTER, C. PAULO
Department of Chemical and Biochemical Engineering
The University of Western Ontario, London, ON, Canada N6A 5B9

Abstract. This paper briefly reviews present understanding of the kinetics of living isobutylene polymerizations, and discusses the effects of $[I]_0/[TiCl_4]_0$ ratio, solvents and initiator type. It is shown that the reaction order with respect to $[TiCl_4]$ is closer to one at $[I]_0/[TiCl_4]_0 \geq 1$, and closer to two at $[I]_0/[TiCl_4]_0 >> 1$. Considerable kinetic differences were found between conditions using methylcyclohexane or hexane as the nonpolar component of the solvent mixture. The reason for this is not clear at this time. The type of initiator was also found to have an effect on the $[TiCl_4]$ reaction order; this is proposed to be due to the complexation equilibrium between initiator and Lewis acid. This paper also presents a new fiberoptic mid-IR real-time monitoring system for more convenient kinetic investigations.

1. Introduction

Spearheaded by the Kennedy school, there has been a great deal of interests in the living carbocationic polymerization of olefins such as isobutylene (IB). This technique can be used for the effective synthesis of various polymeric structures such as blocks, grafts, stars and hyperbranched polymers [1-7]. In order to have control over the polymerization, the kinetics and mechanism of the process have to be understood so they have been studied extensively [7-21]. Kaszas *et al.* [17] first suggested that in living IB polymerizations coinitiated by Lewis acids such as $TiCl_4$, the chlorine-terminated inactive chain ends are in dynamic equilibrium with a minute concentration of active ionic species. The suggestion that the active species are ion pairs [22] was corroborated by Storey *et al.* [13]. While various systems differ in the nature of initiators (I), additives (such as electron pair donor (ED) or proton trap) and solvents, the kinetic order with respect to initiator has been reported to be invariably 1 [7-21]. The reaction order in monomer was found to be 1 in most cases [7-14,17,21], but apparent zero order monomer dependence has also been reported [15,18-20]. However, the reaction order with respect to the Lewis acid $TiCl_4$ has been unclear until recently. Most research groups have reported close to second order dependency on $TiCl_4$ concentration, but Kaszas *et al.* [11] and Kennedy *et al.* [21] had claimed first order

143

J.E. Puskas et al. (eds.), Ionic Polymerizations and Related Processes, 143-159.
© 1999 *Kluwer Academic Publishers. Printed in the Netherlands.*

dependence under certain conditions. Kaszas *et al.* [11] obtained first order dependence in MeCHx/MeCl cosolvents under conditions where $[I]_0/[TiCl_4]_0 \geq 1$, while close to second order dependency was reported in Hx/MeCl or Hx/MeCl$_2$ cosolvents under conditions where $[I]_0/[TiCl_4]_0 < 1$ (with one exception [21]). The published fractional order values (1.7 – 2.2, [7,12,13] signalled kinetic complexity of the system. Our recent investigations [23] targeted the TiCl$_4$ order, and found n = 1; 1.38 and 1.76, depending on the reaction conditions.

This paper will discuss the effects of reaction conditions on the kinetics of living IB polymerization, and compare selected experimental data with kinetic simulation (the details of kinetic simulation will be discussed elsewhere) [24]. Further, a new fiberoptic mid-IR monitoring method for more convenient kinetic investigations will be demonstrated.

2. Experimental

2.1. MATERIALS

Titanium tetrachloride (TiCl$_4$), and 2,6-di-*t*-butylpyridine (D*t*BP) were used as received (Aldrich). Hexane (Hx) and methylcyclohexane (MeCHx) (Caledon) were freshly distilled from CaH$_2$ prior to use. 2,4,4-Trimethyl-2-chloropentane (TMPCl), α-methylstyrene epoxide (MSE) and 2,4,4-trimethyl-pentyl-epoxide-1,2 (TMPO-1) were synthesized as described in [25], and purified by vacuum distillation. Purity was checked by ^{1}H NMR. Methyl chloride (MeCl) and isobutylene (IB) were dried by passing the gases through drying columns packed with CaCl$_2$/BaO.

2.2. PROCEDURES

2.2.1. *Polymerizations*
Polymerizations were carried out in a Mbraun LabMaster 130 glove box under dry nitrogen at -80°C. The moisture (< 1 ppm) and oxygen (5 ppm) content were monitored. The hexane bath was cooled with an FTS Flexi Cool Immersion Cooler. A 500 ml round bottom flask equipped with overhead stirrer was charged with MeCl and Hx (40/60 v/v) or MeCl and MeCHx (40/60 v/v). TMPCl was added to the mixture, followed by the addition of D*t*BP and IB (concentrations are specified in the text or figure captions). Appropriate amounts of IB and MeCl were condensed into prechilled graduated cylinders before addition to the reactor. The reaction was started by the addition of a chilled stock solution of TiCl$_4$ in Hx or MeCHx. Samples were taken at specified times into chilled culture tubes containing methanol for gravimetric conversion analysis. A reaction was also followed by the simultaneous direct monitoring of IB consumption and PIB formation by a fiberoptic immersion Transmission (TR) IR probe (Remspec) attached to a BioRad FTIR instrument, using BioRad WinIR software. Reactions were terminated at specified times by addition of prechilled methanol to the charge. Solvents were evaporated and the polymers were

purified by redissolving them in Hx, washing with distilled water, drying over MgSO$_4$, filtering and precipitating from methanol and drying in a vacuum oven. Conversions were determined gravimetrically or by the ratio of the relevant IR band intensities at time zero and the sampling time.

2.2.2. *Polymer Analysis*

Polymer molecular weights (MW) and molecular weight distributions (MWD) were determined by size exclusion chromatography (SEC) using a Waters system equipped with six Styragel-HR columns (100, 500, 10^3, 10^4, 10^5 and 10^6 Å pore sizes), thermostated to 35 °C, a Waters 410 DRI detector thermostated to 40 °C and a Waters 996 PDA detector set at 254 nm. THF, freshly distilled from CaH$_2$, was employed as the mobile phase and was delivered at 1 ml/min. MWs were calculated by using the Universal Calibration Principle which was shown to be valid for linear PIBs as cited in [26].

3. Results and Discussion

3.1. EFFECTS OF THE REACTION CONDITIONS IN IB POLYMERIZATIONS

3.1.1. *Effect of the [I]$_0$/[TiCl$_4$]$_0$ Ratio*

3.1.1.1. [I]$_0$/[LA]$_0$ ≥ 1. Kaszas *et al.* [11] investigated this range in order to demonstrate that no excess TiCl$_4$ is necessary to initiate IB polymerization. Under the reported conditions initiation and propagation proceeded simultaneously, and the reaction was described using a first order kinetic/mechanistic scheme as shown below:

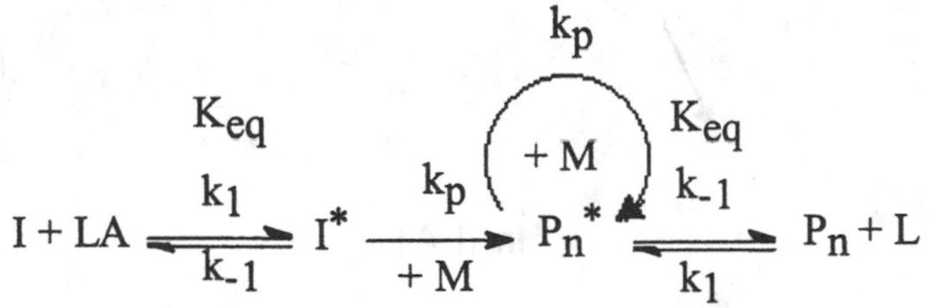

Scheme 1. Mechanism for first order LA dependence in living IB polymerizations.

$$\ln \frac{[I]_0}{[I]} = k_1 [LA]_0 \, t \qquad (1)$$

$$\ln \frac{[M]_0}{[M]} = k_p' \, [I]_0 \, [LA]_0 \, t \tag{2}$$

where $k_p' = k_p K_{eq}$. Kaszas et al. [11] found that this scheme described living IB polymerizations at $[I]_0/[LA]_0 \geq 1$. Kamigaito et al. [27] used a similar first order kinetic scheme to describe living vinyl ether polymerizations.

The TiCl$_4$ order can be investigated by varying the TiCl$_4$, while keeping all other variables constant. The rate equation then can be written as follows:

$$d[M]/dt = k_{app} \, [TiCl_4]_0^n \tag{3}$$

and

$$\ln(d[M]/dt) = \ln k_{app} + n\ln[TiCl_4]_0 \tag{4}$$

The slope of the $\ln(d[M]/dt)$ - $\ln[TiCl_4]_0$ will give the reaction order in TiCl$_4$. We recently reported n = 1 for $[I]_0/[TiCl_4]_0$ = 1 and 4 in MeCHx/MeCl 60/40 at –80°C, and n = 1.38 for $[I]_0/[TiCl_4]_0$ = 1, 2 and 4 in Hx/MeCl 60/40 at –80°C [23]. Figure 1 shows new experimental data for $[I]_0/[TiCl_4]_0$ = 1; 1.35; 2.7 and 5.4 in Hx/MeCl 60/40 at –80°C.

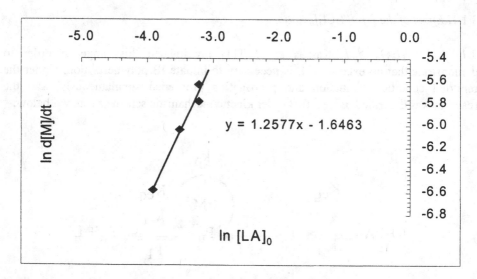

Figure 1. $\ln(d[M]/dt) - \ln[LA]_0$. $[TMPCl]_0$ = 0.054 mol/l, $[IB]_0$ = 2 mol/l, $[DtBP]_0$ = 0.008 mol/l, Hx/MeCl = 60/40, T = -80 °C.

From Figure 1 the TiCl$_4$ reaction order is n = 1.26. This result further reinforces our finding that the reaction order is closer to 1 at $[I]_0/[TiCl_4]_0 \geq 1$. Figures 2 and 3 shows comparison of experimental initiator and monomer consumption plots for $[I]_0/[LA]_0$ = 2. The polymerizations were simulated with the PREDICI software package [28] using the first order scheme.

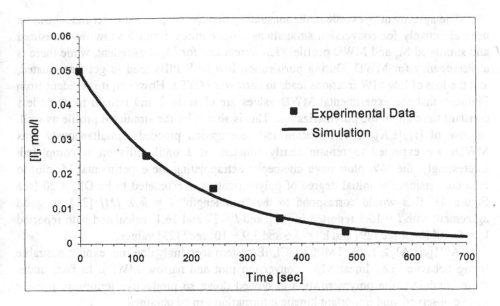

Figure 2. Comparison of experimental and simulated initiator consumption plots.
[TMPCl]$_0$ = 0.050 mol/l, [TiCl$_4$]$_0$ = 0.025 mol/l, [IB]$_0$ = 2 mol/l,
[DtBP]$_0$ = 0.007 mol/l, Hx/MeCl = 60/40, T = -80 °C.

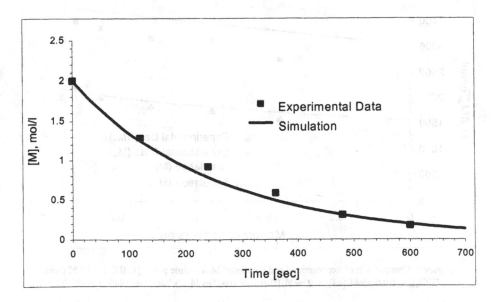

Figure 3. Comparison of experimental and simulated monomer consumption plots.
[TMPCl]$_0$ = 0.050 mol/l, [TiCl$_4$]$_0$ = 0.025 mol/l, [IB]$_0$ = 2 mol/l,
[DtBP]$_0$ = 0.007 mol/l, Hx/MeCl = 60/40, T = -80 °C.

The agreement is excellent, demonstrating that the simple first order model can be used effectively for conversion simulations [24]. Figures 4 and 5 show the measured and simulated M_n and MWD profiles. The agreement for M_n is excellent, while there is a discrepancy for MWD. During purification low MW PIBs tend to get fractionated, and the loss of low MW fractions leads to narrower MWDs. However, it is evident from Figure 5 that the experimental MWD values are close to 2 and remain more or less constant throughout the polymerization. This is shown by the simulated profile as well. In case of $[I]_0/[LA]_0 \geq 1$, initiation and propagation proceeds simultaneously thus MWDs are expected to remain nearly constant at 2 until initiation is completed. Interestingly, the MW plots have intercepts; extrapolating the experimental M_n plot to zero conversion, the initial degree of polymerization is estimated to be $DP_n \approx 20$ (see Figure 4). This would correspond to the "run length" $l = k_p/k_{-1}[M]$ [22], in good agreement with $l = 16.5$ reported in [22], and $l = 17$ and 15.1, calculated with reported $k_p = 6 \times 10^8$ l/mol.sec [29] and $k_{-1} = 3.6$ and 3.9×10^7 sec^{-1} [23] values.

At $[I]_0/[LA]_0 \geq 1$, the TMPCl/TiCl$_4$/IB system seemingly does not exhibit desirable living behavior (i.e., linear M_n – conversion plot and narrow MWD). In fact, under these conditions the polymerization is slowed down so productive ionization periods can be observed, and important kinetic information can be obtained.

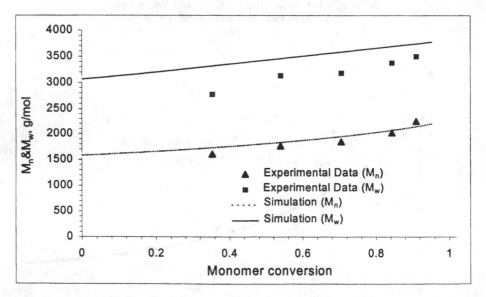

Figure 4. Comparison of experimental and simulated MW – time plots. [TMPCl]$_0$ = 0.050 mol/l, [TiCl$_4$]$_0$ = 0.025 mol/l, [IB]$_0$ = 2 mol/l, [DtBP]$_0$ = 0.007 mol/l, Hx/MeCl = 60/40, T = -80 °C.

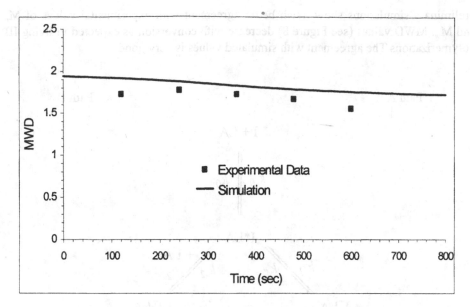

Figure 5. Comparison of experimental and simulated MWD – time plots. $[TMPCl]_0 = 0.050$ mol/l, $TiCl_4]_0 = 0.025$ mol/l, $[IB]_0 = 2$ mol/l, $[DtBP]_0 = 0.007$ mol/l, Hx/MeCl = 60/40, T = -80 °C.

3.1.1.2. $[I]_0/[LA]_0 < 1$. Under these conditions the order of the reaction relative to $[TiCl_4]$ has previously been reported to be n = 1.7 – 2.2 [7,12,13]. Our investigations revealed a reaction order of n = 1.76 [23]. Based on our results and the studies discussed above, we proposed the complex mechanism shown in Scheme 2, comprising a series of consecutive and competitive reactions.

This mechanism accounts for the experimental fact that the reaction order in $TiCl_4$ varies, depending on reaction conditions. While both Paths A and B may proceed simultaneously, at $[I]_0/[TiCl_4]_0 < 1$ Path B dominates and the $TiCl_4$ order is closer to 2. This comprehensive scheme can be simplified to the format of Scheme 1 as described [24], with the understanding that at $[I]_0/[TiCl_4]_0 < 1$, K_{eq} and k_p' will be dependent on $[TiCl_4]_0$ and has to be obtained experimentally for each $[TiCl_4]_0$. Comparison of experimental and simulated monomer consumption profiles for $[I]_0/[LA]_0 = 1/16$ is shown in Figure 6. The agreement is excellent. Since initiation was instantaneous, initiator consumption could not be monitored.

Figures 7 and 8 show experimental and simulated MW and MWD profiles. Experimental M_n values increase linearly with conversion as expected for living polymerizations. There is good agreement with simulated values, except at low conversion where the simulation shows an apparent intercept. This intercept is governed by the relative rate of initiation and propagation, that is, by k_1. In the simplified model k_1 is a composite rate constant so its components cannot be adjusted individually. The complex model uses individual rate constants and will address this problem;

150

preliminary simulations show much better agreement with experimental values of M_n and M_w. MWD values (see Figure 8) decrease with conversion as expected in living IB polymerizations The agreement with simulated values is very good.

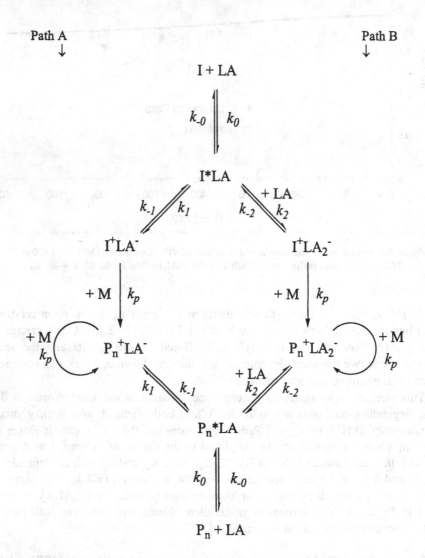

Scheme 2. Comprehensive mechanism for living IB polymerizations [23].

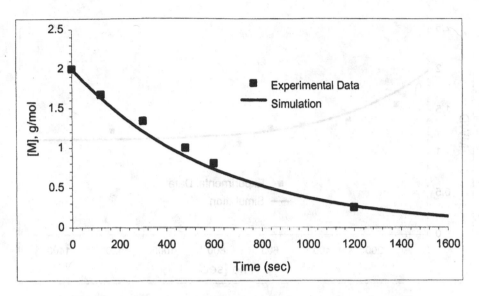

Figure 6. Comparison of experimental and simulated monomer conversion profiles.
[TMPCl]$_0$ = 0.004 mol/l, [TiCl$_4$]$_0$ = 0.064 mol/l, [IB]$_0$ = 2 mol/l, [DtBP]$_0$ = 0.007 mol/l.
Hx/MeCl = 60/40, T = -80 °C.

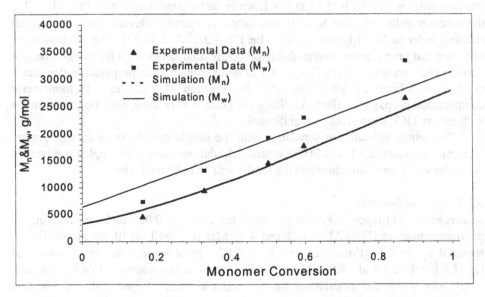

Figure 7. Comparison of experimental and simulated MW profiles. [TMPCl]$_0$ = 0.004 mol/l,
[TiCl$_4$]$_0$ = 0.064 mol/l, [IB]$_0$ = 2 mol/l, [DtBP]$_0$ = 0.007 mol/l, Hx/MeCl = 60/40, T = -80 °C.

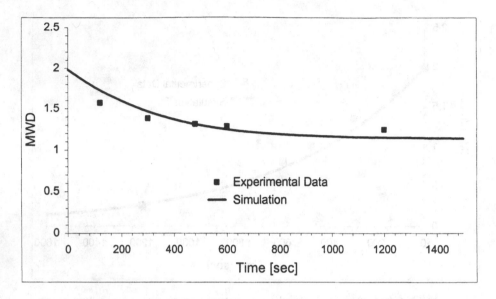

Figure 8. Comparison of experimental and simulated MWD profiles. $[TMPCl]_0 = 0.004$ mol/l, $[TiCl_4]_0 = 0.064$ mol/l, $[IB]_0 = 2$ mol/l, $[DtBP]_0 = 0.007$ mol/l, Hx/MeCl = 60/40, T = -80 °C.

In sum, in living IB polymerizations under conditions where $[TiCl_4]_0 \leq [I]_0$, the reaction order in $[TiCl_4]$ is close to n = 1, while under conditions where $[TiCl_4]_0 > [I]_0$, the reaction order is close to 2. In this latter, practically relevant range the overall reaction order in $[TiCl_4]_0$ was found to be 1.7 – 2.2 [7,12,13,23]. The fraction order indicates that propagation most probably proceeds simultaneously with monomeric and dimeric counteranions; at $[I]_0/[LA]_0 \ll 1$ dimeric counteranion propagation dominates (Path B in Scheme 2), while "starving" the reaction of $TiCl_4$ leads to monomeric counteranion propagation (Path A). Since the latter is very slow, only polymerizations with excess $TiCl_4$ have practical significance.

Our results indicate that simulation with the simple model can be used to predict monomer conversion, M_n and MWD values. Simulations using the complex mechanism are underway in our laboratory and the results will be published later.

3.1.2. *Effect of Solvents*

Kaszas *et al.* [11] reported $k_p' = 0.33$ $l^2/mol^2.sec$ and $k_l = 0.02$ l/mol.sec for living IB polymerizations at $[I]_0/[LA]_0 = 1$, 2 and 4 in MeCHx/MeCl 60/40 v/v at –90°C. We reported $k_p' = 0.54$ $l^2/mol^2.sec$ and $k_l = 0.024$ l/mol.sec for the same system at $[I]_0/[LA]_0 = 1$ and 4 at –80°C. The fact that k_p' was independent of $[TiCl_4]_0$ for both cases, was interpreted as evidence for first order kinetics. Figure 9 shows new data measured in our lab, together with data points calculated from Kaszas' paper [11]. The reaction order is close to one in both cases (n = 1.06 and 0.87, respectively).

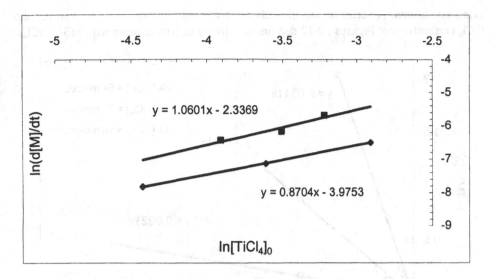

Figure 9. ln(d[M]/dt) – ln[LA]$_0$. [TMPCl]$_0$ = 0.054 mol/l, [IB]$_0$ = 2 mol/l, [DtBP]$_0$ = 0.008 mol/l, McCHx/McCl = 60/40, T = -90 °C.

Comparison of these data with those reported for Hx/MeCl 40/60 (n = 1.26, see 3.1.1.1., k_p' ≈ 3.4 l^2/mol^2.sec and k_1 = 0.22 l/mol.sec, [23]) reveals striking differences between the two solvent mixtures at [TiCl$_4$]$_0$ ≤ [I]$_0$. With k_p = 6.0 × 10^8 l/mol.sec, from Eq. (2) we get K_1 = 5.5 × 10^{-10} l/mol and k_{-1} = 3.6 × 10^7 sec^{-1} for MeCHx, and K_1 = 5.6 × 10^{-9} l/mol and k_{-1} = 3.9× 10^7 sec^{-1}. It is important to note that the rate constant of the deactivation step is identical for the two nonpolar solvents, while the composite equilibrium constant K_1 is quite different. The reasons for this discrepancy are not clear: the dielectric constants of the two non-polar solvents are nearly the same. Also, in the case of excess TiCl$_4$, the reaction rate is not affected by the solvents [23]. This implies that the solvent has a profound effect on Path A, while little or no effect on Path B. Thus the solvent seems to have a large effect on the equilibrium of the monomeric counterion formation, which underlines the importance of solvent selection for effective initiation. Further investigation of this solvent effect phenomenon is in progress in our laboratories.

3.1.3. The Effect of Initiator Type

It was recently reported that a novel initiator, α-methylstyrene epoxide (MSE) can be

effectively used to initiate the living polymerization of IB [25]. Linear first order monomer consumption plots together with linear M$_n$ – conversion plots demonstrated

154

living conditions. A series of experiments were carried out with MSE by varying the $TiCl_4$ concentration. Figures 10-12 demonstrate living polymerization with $MSE/TiCl_4$.

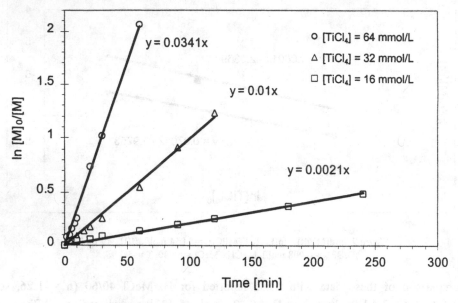

Figure 10. First order monomer consumption plots in MSE-initiated IB polymerizations.
$[MSE]_0 = 4.0 \times 10^{-3}$ M, $[IB]_0 = 2.0$ M, $[DtBP] = 7.0 \times 10^{-3}$ M,
Hx/MeCl = 60/40 (v/v), T = -80 °C.

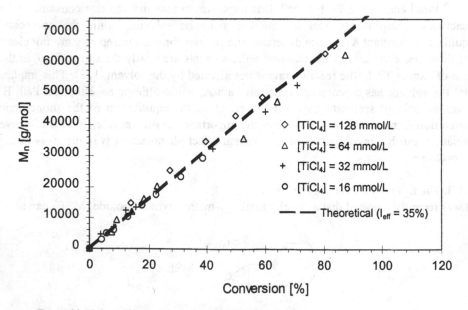

Figure 11. Mn – conversion plot in MSE-initiated polymerizations. $[MSE]_0 = 4.0 \times 10^{-3}$ M,
$[IB]_0 = 2.0$ M, $[DtBP] = 7.0 \times 10^{-3}$ M, Hx/MeCl = 60/40 (v/v), T = -80 °C.

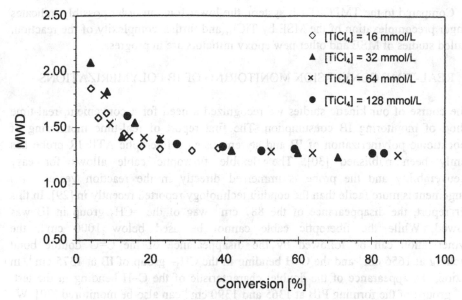

Figure 12. MWD – conversion plot in MSE-initiated polymerizations. $[MSE]_0 = 4.0 \times 10^{-3}$ M, $[IB]_0 = 2.0$ M, $[DtBP] = 7.0 \times 10^{-3}$ M, Hx/MeCl = 60/40 (v/v), T = -80 °C.

The [TiCl₄] order was found to be 1.56 with MSE, as shown in Figure 13.

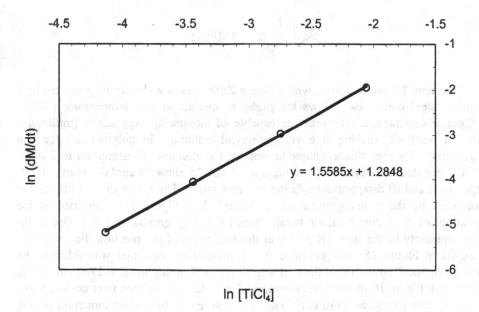

Figure 13. ln(d[M]/dt) – ln [TiCl₄]₀ plot for MSE-initiated living IB polymerizations. $[MSE]_0 = 4.0 \times 10^{-3}$ M, $[IB]_0 = 2.0$ M, $[DtBP] = 7.0 \times 10^{-3}$ M, Hx/MeCl = 60/40 (v/v), T = -80 °C.

Compared to the TMPCl/TiCl$_4$ system, the lower fraction order possibly indicates stronger precomplexation of the MSE by TiCl$_4$, and further complexity of the reaction. Detailed studies of MSE and other new epoxy initiators are in progress.

3.2. REAL-TIME CONVERSION MONITORING OF IB POLYMERIZATIONS

In the course of our kinetic studies we recognized a need for a convenient, real-time method of monitoring IB consumption. The first report of real-time monitoring of carbocationic polymerization of IB and styrene using a fiberoptic ATR IR probe has recently been published [30]. The flexible fiberoptic cable allows for easy maneuverability and the probe is immersed directly in the reaction vessel. This arrangement is more facile than the conduit technology reported recently in [29]. In this latter report, the disappearance of the 887 cm^{-1} wag of the =CH$_2$ group in IB was followed. While the fiberoptic cable cannot be used below 1000 cm^{-1}, the polymerization can be followed by the disappearance of the C=C double bond stretching at 1656 cm^{-1} and the C–H bending in the CH$_3$– group of IB at 1375 cm^{-1}. In addition, the appearance of the doublet characteristic of the C–H bending in the tert-butyl groups of the forming PIB at 1365 and 1390 cm^{-1} can also be monitored [30]. We present here the development of a new technique using a very sensitive fiber optic transmission (TR) infrared probe. This sensitive new technique allowed us to monitor IB polymerizations initiated by our newly discovered epoxi initiator TMPO-1.

TMPO-1

The new TR probe design, with a 3-inch ZnSe crystal with mirror, protected by a stainless steel outer case, allows the probe to operate at low temperature (-80°C) without freeze-fracture. The probe is capable of monitoring very dilute (millimol/l) solution reactions, making it a very powerful technique in polymer science and engineering. We used this technique to monitor the monomer consumption in TMPO-1/TiCl$_4$ initiated IB polymerization. Figures 14 and 15 show "waterfall" scans of PIB appearance and IB disappearance during polymerization. The appearance of PIB can be monitored by the growing signal at 1230 cm^{-1} (I in Figure 14), identified as the vibrations of the distorted carbon tetrahedrons (-C(CH$_3$)$_2$- groups) in PIB [. Due to the high sensitivity of the new TR probe, at the beginning of the reaction the 1656 cm^{-1} peak (II in Figure 15) was saturated thus monomer consumption was followed by monitoring the 2nd overtone of the C-H wag in the =CH$_2$ group in IB at 1780 cm^{-1} (III in Figure 15). Figure 16 shows a representative first order plot for monomer consumption, based on monitoring the 1780 cm^{-1} band. There seems to be a short induction period, after which the plot is linear, indicating living conditions. Figure 17 shows a representative SEC trace of a PIB obtained by TMPO-1/TiCl$_4$; the very narrow

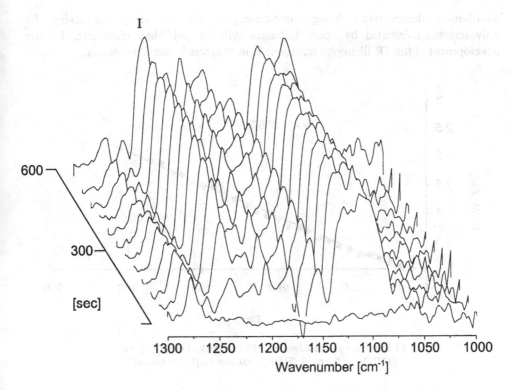

Figure 14. Real-time IR spectra of IB polymerization – PIB appearance.
[TMPO-1] = 0.1 mol/l, [TiCl$_4$] = 0.3 mol/l, [IB]$_0$ = 1 mol/l.

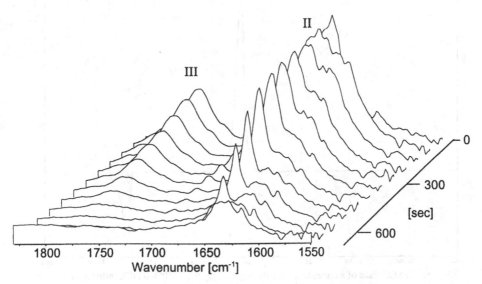

Figure 15. Real-time IR spectra of IB polymerization – IB disappearance.
[TMPO-1] = 0.1 mol/l, [TiCl$_4$] = 0.3 mol/l, [IB]$_0$ = 1 mol/l.

158

distribution demonstrates living conditions. A detailed study on living IB polymerization initiated by epoxi initiators will be published elsewhere. Further development of the TR fiberoptic technique is in progress in our laboratories.

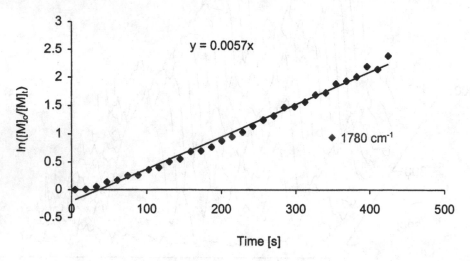

Figure 16. First order plot based on the 1780 cm^{-1} peak of IB. [TiCl$_4$] = 0.3 mol/l, [TMPO-1] = 0.1 mol/l, [DtBP] = 7.0 mmol/l, [IB]$_0$ = 1.0 mol/l.

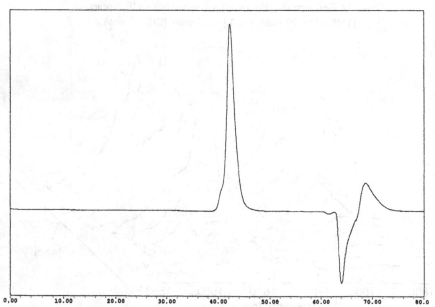

Figure 17. SEC trace of a representative PIB prepared by the TMPO-1/TiCl$_4$ initiator system. M$_n$ = 64,100; MWD = 1.17

Acknowledgement. The authors would like to thank S. Angepat, K. Diamond, D. Jamieson, and E. Tzaras for their contributions to the experimental work, Dr. P. Melling (Remspec) for consultation, and the Rubber Division of Bayer Inc., Canada, NSERC (Natural Science and Engineering Research Council) and MMO (Materials and Manufacturing Ontario, Canada) for financial support.

4. References

1. Kennedy, J.P., Ross, L.R., and Nuyken O. (1981) *Polym. Bulletin*, **5** (1), 5.
2. Keszler, B., Chang, V.S.C., and Kennedy, J.P. (1984) *J. Macromol. Sci.*, **A21** (3), pp. 307.
3. Puskas, J.E. and Wilds, C.J. (1998) *J. Polym. Sci.*, **36** (1), 85.
4. Wang, L., McKenna, S.T., and Faust, R. (1995) *Macromolecules*, **28**, 4681.
5. Nuyken, O., Gruber, F., Pask, S.D., Riederer, A., and Walter, M. (1993) *Makromol. Chem.* **194**, 3415.
6. Puskas, J.E., and Grassmüller, M. (1998) *Macromol. Symp*., **132**, 117.
7. Storey, R.F. and Shoemake, K.A. (1994) *Polymer Preprint* **35**(2), 578.
8. Gyor, M., Wang, H.C., and Faust, R. (1992) *J. Macromol. Sci.* **A29**, 639.
9. Storey, R.F., Chisolm, B.J., and Mowbray, E. (1993) *Polymer Preprint* **34**(1), 568.
10. Fodor, Z., Gyor, M., Wang, H.-C., and Faust, R. (1993) *J. Macromol. Sci.* **A30**, 349-363.
11. Kaszas, G. and Puskas, J.E. (1994) *Polymer Reaction Engineering* **2**(3), 251.
12. Storey, R.F., Chisolm, B.J., and Brister, L.B. (1995) *Macromolecules* **28**, 4055.
13. Storey, R.F. and Choate Jr., K.R. (1997) *Macromolecules* **30**, 4799.
14. Storey, R.F., Chisholm, B.J., and Lee, Y.J. (1992) *Polym. Preprint* **33**(2), 184.
15. Zsuga, M., Kelen, T., Balogh, L., and Majoros, I. (1992) *Polym. Bulletin* **29**, 127.
16. Kelen, T., Zsuga, M., Balogh, L., Majoros, I., and Deak, G. (1993) *Makromol. Chem., Macromol. Symp.* **67**, 325.
17. Kaszas, G., Puskas, J.E., and Kennedy, J.P. (1989) *J. Macromol. Sci.* **A26**, 1099.
18. Kaszas, G., Puskas, J.E., and Kennedy, J.P. (1988) *Makromol. Chem., Macromol. Symp.* **13-14**, 473.
19. Zsuga, M. and Kennedy, J.P. (1989) *Polym. Bulletin* **21**, 5.
20. Roth, M., Patz, M., Freter, H., and Mayr, H. (1997) *Macromolecules* **30**, 722.
21. Kennedy, J.P., Majoros, I., and Nagy, A. (1994) *Advances in Polymer Science* **112**, 1.
22. Puskas, J.E., Kaszas, G., and Litt, M. (1991) *Macromolecules* **24**, 5278.
23. Puskas, J.E. and Lanzendörfer, MG. (1998) *Macromolecules* **31**, 8684.
24. Puskas, J.E. and Peng, H. (1999) *Polymer Reaction Engineering*, accepted for publication.
25. Puskas, J.E., Pattern, W.E., Lanzendörfer, M.G., Jamieson, D., and Oraha Al-Rais, L. (1998) *Polymer Preprint* **39**(1), 325.
26. Puskas, J.E. and Hutchinson, R. (1993) *Rubber Chemistry & Technology* **66**(5), 742.
27. Kamigaito, M., Yamaoka, K., Sawamoto, M., and Higashimura, T. (1992) *Macromolecules*, **25**, 6400.
28. Wulkow, M. (1996) *Predici® Simulation Package for Polyreactions*.
29. Storey, R.F., Donnalley, A.B., and Maggio, T.L. (1998) *Macromolecules* **31**, 1523.
30. Puskas, J.E., Lanzendörfer, M.G, and Pattern, W.E. (1998) *Polym. Bulletin*, **40**(1), 55.
31. Thompson, H.W. and Torkington, P. (1945) *Trns. Faraday Soc.*, **41**, 246.

Acknowledgement. The authors would like to thank S. Angelat, K. Diamond, D. Jameson and S. Tzaras for their contributions to the experimental work, Dr. F. Meltzing (Leverkusen) for consultation, and the Rubber Division of Bayer Inc., Canada, NSERC (Natural Science and Engineering Research Council) and MMO (Materials and Manufacturing Ontario Canada) for financial support.

4. References

1. Kennedy, J.P., Ross, L.R. and Nuyken, O. (1981) *Polymer Bulletin* 5, 5 [illegible]
2. Kaszas, G., Gyor, M. [illegible] and Kennedy, J.P. [illegible] *J. Macromol. Sci.*, A25 [illegible]
3. [illegible] and [illegible] (1988) *J. Polym. Sci.* [illegible]
4. Wang, [illegible] and [illegible] (1991) *Macromolecules* 24, [illegible]
5. [illegible], Dubois, P., Park, [illegible] and Walter, [illegible] (1991) *Makromol. Chem.* [illegible]
6. Puskas, J.E. and Grassmuller, M. (1998) *Macromol. Symp.* 132, 117
7. Storey, R.F. and [illegible], K.A. (1998) *Polymer Preprint* [illegible]
8. Gyor, M., Wang, H.C. and [illegible] (19[illegible]) *J. Macromol. Sci.* A29, [illegible]
9. Storey, R.F., Chisholm, B.J. and Masse[illegible], E. (1997) *Polymer Preprint* 38(1), 502
10. Fodor, Z., Gyor, M., Wang, H.C. and Faust, R. (1993) *J. Macromol. Sci.* A30, 349-363
11. [illegible], G. and Puskas, J.E. (1994) *Polymer Reaction Engineering* 2(3), 251
12. Storey, R.F., Chisholm, B.J. and Brister, L.B. (1995) *Macromolecules* 28, 4055
13. Storey, R.F. and Choate, Jr., K.R. (1997) *Macromolecules* 30, 4799
14. [illegible], R.F., Chisholm, B.J. and Lee, Y. (1993) *Polymer* 34, [illegible]
15. [illegible], M., Keszler, T., Ivan, B. and Majoros, I. (1992) *Polymer Bulletin* 29, 17[illegible]
16. Keszler, T., Zsuga, M., Balogh, L., Majoros, I. and Deak, G. (199[illegible]) *Makromol. Chem.* [illegible]
17. Kaszas, G., Puskas, J.E. and Kennedy, J.P. (1988) *J. Macromol. Sci.* A25, 1[illegible]
18. Kaszas, G., Puskas, J.E. and Kennedy, J.P. (1988) *Makromol. Chem., Macromol. Symp.* [illegible]
19. Zsuga, M. and Kennedy, J.P. (1989) *Polymer Bulletin* 21, [illegible]
20. [illegible], Faust, R. and [illegible] (1997) *Macromolecules* 30, 72[illegible]
21. Kennedy, J.P., Majoros, I. and Nagy, A. (1994) *Advances in Polymer Science* 112, [illegible]
22. Puskas, J.E., Kaszas, G. and [illegible], M. (1991) *Macromolecules* 24, [illegible]
23. Puskas, J.E. and Lanzendorfer, M.G. (1998) *Macromolecules* 31, 8684
24. Puskas, J.E. and Peng, H. (1999) *Polymer Reaction Engineering*, accepted for publication
25. Puskas, J.E., Pattern, W.E., Lanzendorfer, M.G., Jamieson, D. and [illegible] (1998) *Polymer* [illegible] 18(1)
26. Puskas, J.E. and Hutchinson, R. (1993) *Rubber Chemistry & Technology* [illegible], 742
27. [illegible], M., [illegible] (1992) *Macromolecules* 25, 6400
28. [illegible], M. (1990) [illegible]
29. [illegible], D., [illegible] and [illegible] (1997) *Macromolecules* [illegible]
30. Puskas, J.E., Lanzendorfer, M.G. and Pattern, W.E. (1998) *Polym. Bulletin* 40(1) [illegible]
31. [illegible], R.W. and [illegible] (1945) *Trans. Faraday Soc.* 41, 2[illegible]

KINETICS AND MECHANISM OF LIVING CATIONIC POLYMERIZATION OF OLEFINS

R. F. STOREY, C. L. CURRY, AND T. L. MAGGIO
The University of Southern Mississippi
Department of Polymer Science
Hattiesburg, MS 39406-0076, USA

Abstract. Olefins such as isobutylene and styrene can be cationically polymerized in a controlled/living fashion under certain specific conditions. The best systems tend to employ either TiCl$_4$ or BCl$_3$ coinitiators. Lewis base additives are used to suppress both protic initiation and propagation by free ions. The growing chains are characterized by a reversible ionization equilibrium between a small concentration of active ion pairs and a much larger concentration of dormant chains. These systems deviate from ideal living behavior to a lesser or greater degree depending upon conditions. With select systems at low temperatures, the deviations are very small, and well defined polymers can be produced. Non-ideality results from unimolecular termination reactions, as opposed to bimolecular chain transfer to monomer. This makes clear the danger of excess aging of living chains, and it highlights the practical importance to synthetic procedures of a thorough understanding of the kinetics of propagation and chain breaking events.

Chain termination in TiCl$_4$-initiated isobutylene polymerizations, in the presence of a Lewis base at low temperature, involves principally carbocation rearrangement and secondarily β-proton elimination. Aliquots of polymerization reactions taken to very high monomer conversions (10^3 half-lives) were analyzed by ^{1}H NMR and GPC. Extent of rearrangement was measured by integration of appropriate end group resonances in the NMR spectrum; β-elimination was measured indirectly from the incidence of coupling of primary chains observed in GPC chromatograms. The relative importance of these two reactions was found to depend on temperature, nature and concentration of the Lewis base, and absolute concentration of TiCl$_4$. Non-complexing bases such as di-*tert*-butylpyridine, used in conjunction with low absolute concentrations of TiCl$_4$, produces the greatest relative proportion of β-elimination relative to rearrangement.

We recently developed a method for measurement of carbocationic polymerization kinetics in real time using ATR-FTIR spectroscopy. Kinetics of IB polymerization initiated using BCl$_3$/DMP in methylchloride were investigated with regard to the kinetic dependency on Lewis acid concentration.

161

J.E. Puskas et al. (eds.), Ionic Polymerizations and Related Processes, 161-175.
© *1999 Kluwer Academic Publishers. Printed in the Netherlands.*

1. Introduction

Controlled/living polymerization of isobutylene (IB) has been the subject of intense investigations by a number of research groups.[1-12] The vast majority of this work has been applied to systems using either TiCl$_4$ or BCl$_3$ coinitiators. For these particular Lewis acids, "livingness" is conferred by inherently low rates of chain transfer to monomer and β-proton elimination, and the use of Lewis base (LB) additives to suppress both protic initiation (TiCl$_4$ systems) and participation of free ions in propagation.

Controlled/living polymerization of IB are characterized by a reversible ionization equilibrium between a small concentration of active propagating carbenium ions and a much larger concentration of dormant (deactivated) chains, as shown in Scheme 1 for TiCl$_4$-coinitiated systems. Since this equilibrium is a central feature of the mechansim of controlled/living polymerization, its characterization has received a great deal of attention from researchers.

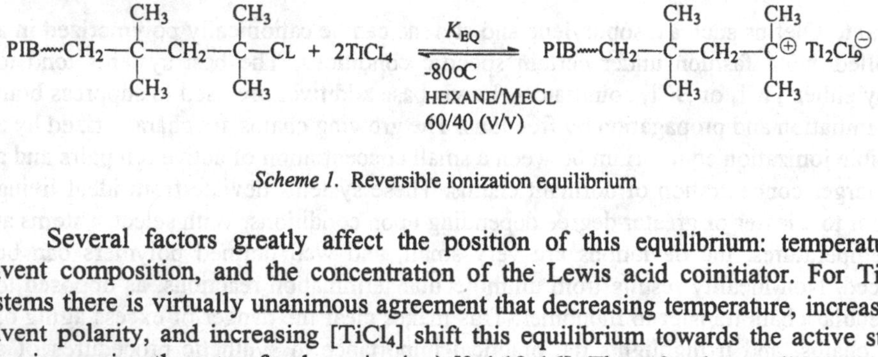

Scheme 1. Reversible ionization equilibrium

Several factors greatly affect the position of this equilibrium: temperature, solvent composition, and the concentration of the Lewis acid coinitiator. For TiCl$_4$ systems there is virtually unanimous agreement that decreasing temperature, increasing solvent polarity, and increasing [TiCl$_4$] shift this equilibrium towards the active state (thus increasing the rate of polymerization) [3,4,13-16]. Thus, for the conditions given in Scheme 1, the apparent activation energy for propagation has been found to be -8.5 kcal/mol by two different groups [3,17]. Scheme 1 also depicts dimeric counterions of the type Ti$_2$Cl$_9^-$. This reflects the fact that, when [TiCl$_4$] > the concentration of polymer chain ends (CE), a second-order dependence on [TiCl$_4$] has been consistently reported in a number of papers by at least two different groups [2-5]. It should be noted that second-order dependence can only result if ionization is first affected by monomeric TiCl$_4$ and the resulting TiCl$_5^-$ counterion rapidly reacts with additional TiCl$_4$ to form Ti$_2$Cl$_9^-$ [3]. There has been some controversy over TiCl$_4$ reaction order for polymerizations in which [TiCl$_4$] ≤ [CE]. Kaszas and Puskas [18] reported a first-order dependence on [TiCl$_4$] for IB polymerizations at -90°C; this work was recently revisited by Puskas and Lanzendörfer [19], and essentially the same conclusions were reached for polymerizations at -80°C. Storey and Donnalley have consistently found second-order dependence under these conditions at -80°C [6,20].

The successful development of controlled/living polymerization of IB has naturally prompted researchers to probe the limits of livingness of these system [17,21-23]. Living polymerization is an ideal condition that is approached more or less closely by real systems. It has been shown that under conditions of low [I] and high monomer conversions, the livingness of IB polymerizations coinitiated by TiCl$_4$ at –80°C is limited by a unimolecular termination process, namely carbenium ion rearrangement to form mixtures of isomerized chain end structures [21]. However, further polymerization experiments at –80°C have shown that carbenium ion rearrangement is also accompanied by a minor proportion of β-proton elimination, evidenced by the appearance of chain-chain coupling. Rearrangement and elimination occur concurrently, with the former being overwhelmingly more important at –80°C. Fodor *et al.* [17] have shown that at higher temperatures β-proton elimination is the predominant termination mechanism. Thus, non-ideality of TiCl$_4$-coinitiated IB polymerization (in the absence of protic initiation) results from unimolecular termination reactions, as opposed to bimolecular chain transfer to monomer. This makes clear the danger of excess aging of living chains, and it highlights the practical importance to synthetic procedures of a thorough understanding of the kinetics of propagation and chain breaking events. This paper will examine termination reactions in controlled/living IB polymerizations at very high monomer conversion, in the presence of various concentrations of active TiCl$_4$ and various types/concentrations of Lewis bases.

Compared to the large volume of work that has been performed with TiCl$_4$-coinitiated systems, few investigations have systematically probed the kinetics of IB polymerizations coinitiated with BCl$_3$. This paper will also address this topic using the relatively new technique of real-time ATR-FTIR monitoring of polymerization kinetics.

2. Experimental

2.1. MATERIALS

The source and preparation of 2-chloro-2,4,4-trimethylpentane (TMPCl) and 5-*tert*-butyl-1,3-(2-chloro-2-propyl)benzene (*t*-Bu-*m*-DCC) have been previously reported.[3,21] Hexane (Aldrich Chemical Co.) was dried prior to use by distillation from CaH$_2$. Isobutylene and CH$_3$Cl (AIRCO) were dried by passing the gases through columns packed with BaO and CaCl$_2$. TiCl$_4$ (99.9%, packaged under N$_2$ in SureSeal bottles), 2,4 lutidine (99%, 2,4-dimethylpyridine, DMP), 2,6-di-*tert*-butylpyridine (97%, DTBP), *N,N*-dimethylacetimide (99.8%, packaged under N$_2$ in SureSeal bottles, DMAc), pyridine (99.8%, packaged under N$_2$ in SureSeal bottles, PYR), and anhydrous methanol were used as received from Aldrich Chemical Co. BCl$_3$ (AIRCO) was obtained in a cylinder and sublimed immediately prior to use into a sealed tube chilled by liquid N$_2$.

2.2. INSTRUMENTATION

Molecular weights and molecular weight distributions, M$_w$/M$_n$, were determined using a gel permeation chromatography (GPC) system equipped with a Waters model 410 differential refractometer detector and an on-line three-angle laser light-scattering detector (MiniDawn, Wyatt Technology Corp.), as previously described [21].

^{1}H NMR spectra of the polymers were obtained on a Bruker AC-300 spectrometer using 5mm o.d. tubes. Sample concentrations were 15% (w/v) in CDCl$_3$

containing 0.03% TMS as an internal reference. The quantity $[P\text{-}tCl]_0/[P\text{-}tCl]$ was calculated according to

$$\frac{[P-tCl]_o}{[P-tCl]} = \frac{BA_o}{AB_o} \tag{1}$$

where $[P\text{-}tCl]$ is the concentration of tert-alkylchloride end groups, A represents the integrated peak area of the methylene protons (1.96 ppm) adjacent to the tert-alkylchloride end group, B represents the integrated peak area of the methylene protons (1.84 ppm) adjacent to the initiator residue, and the zero subscript refers to the quantities measured at ≈98% conversion.

A ReactIR™ 1000 reaction analysis system (light conduit type) (ASI Applied Systems, Millersville, MD), equipped with a DiComp™ (diamond-composite) insertion probe, a general purpose type PR-11 platinum resistance thermometer (RTD) and CN76000 series temperature controller (Omega Engineering, Stamford, CT) was used to collect infrared spectra of the polymerization components and monitor reactor temperature in real-time as previously described [24].

2.3. TICl₄-COINITIATED ISOBUTYLENE POLYMERIZATION

Isobutylene polymerizations were initiated using the tBu-m-DCC/TiCl₄/LB system ([IB] = 1.0 M) in 60/40 hexane/MeCl (v/v) where LB represents one of the following Lewis Bases: DMP, DTBP, DMAc, or Pyr. Reagent concentrations are listed in Table 1. All polymerizations were performed at –80°C under dry N_2 gas within an inert-atmosphere glovebox, and conditions were chosen to provide degrees of polymerization (DP) of 42 or 84, for monofunctional and difunctional initiators, respectively, with nearly complete IB conversion in 2.5 min. A typical procedure was as follows: to a 2L three-necked flask equipped with mechanical stirrer were added 530 mL of MeCl, 4.954g of tBu-m-DCC, 800 mL of hexane, 1.45 mol (114 mL) of IB, and 3.63 x 10^{-3} mol (0.42 mL) of DMP. The resulting masterbatch solution was stirred for 30 min, and then 50 mL aliquots were transferred into separate 25 x 200 mm culture tubes. Each tube was initiated by the addition of 5.91 x 10^{-3} mol (0.65 mL) of TiCl₄ (neat and at room temperature). Tubes were quenched at various reaction times up to 24 h with 7 mL of prechilled MeOH. After quenching, the contents of each tube were poured into a 16 oz jar containing 50 mL of MeOH at 23°C. After evaporation of most of the solvents, the remaining liquid was decanted from the precipitated polymer, and the latter was redissolved by addition of 10 mL of hexane and then reprecipitated by addition of 250 mL of MeOH. The samples were dried under vacuum overnight, then analyzed by GPC and ^{1}H NMR to investigate the structure and reactivity of the active centers.

TABLE 1. TiCl$_4$-Coinitiated Polymerization Systems

Target MW (g/mol)	Lewis Base	[Lewis Base] (M)	[TiCl$_4$] (M)
			0.12
		2.5 x 10^{-3}	0.23
	DMP		0.34
		5.0 x 10^{-3}	0.12
			0.34
5000 g/mol			0.12
[t-Bu-m-DCC] = 0.012 M		2.5 x 10^{-3}	0.23
	DTBP		0.34
		5.0 x 10^{-3}	0.12
			0.34
	Pyr	2.5 x 10^{-3}	0.12
	DMAc	2.5 x 10^{-3}	0.12
	No LB		0.12
2500 g/mol	DMP	2.5 x 10^{-3}	0.12
[TMPCl] = 0.024 M			

2.4. BCl$_3$-COINITIATED POLYMERIZATIONS AND ATR-FTIR DATA COLLECTION

All polymerizations were conducted under a dry nitrogen atmosphere in an M. Braun Labmaster 130 glove box equipped with an integral heptane bath maintained at -80°C. A typical experimental procedure was as follows: Background spectra for use in ATR-FTIR analysis were obtained by fitting a stainless steel sleeve to the end of the DiComp probe of the ReactIR 1000 and immersing it in the heptane bath at −80°C. After allowing sufficient time for thermal equilibration, a number of spectra were collected until a relatively constant peak profile was achieved. The sleeve was removed from the probe and replaced by a 250 mL four-neck round-bottom flask equipped with a mechanical stirrer and platinum resistance thermometer. The flask was then charged with 0.875 g (3.02 x 10^{-3} mol) t-Bu-m-DCC initiator, 175 mL MeCl (−80°C), 4.62 x 10^{-2} mL (4 x 10^{-4} mol) DMP, and 15.87 mL (0.2 mol) IB (−80°C). The solution was stirred until thermal equilibration was reached as indicated by the thermometer (~10-15 min), and polymerization was initiated by the rapid addition of the desired amount (3.45 mL (4.53 x 10^{-2} mol), 6.90 mL (9.06 x 10^{-2} mol), 13.7 mL (1.80 x 10^{-1} mol), or 27.4 mL (3.60 x 10^{-1} mol) of BCl$_3$ co-initiator. Polymer samples for GPC analysis were obtained by withdrawing a 5-10 mL aliquot from the reaction vessel at predetermined time intervals and adding it to a scintillation vial containing 10 mL anhydrous MeOH (-80°C). Temperature and ATR-FTIR data were collected with the ReactIR 1000 system during reagent addition and subsequent polymerization of the isobutylene. The ATR-FTIR data were comprised of spectra collected as the average of either 128 or 1024 scans, over the spectral ranges of 4000-2200 cm^{-1} and 1900-650 cm^{-1}, with 8 cm^{-1} resolution.

166

3. Results and Discussion

3.1. TICl₄-COINITIATED LIVING POLYMERIZATION OF ISOBUTYLENE

3.1.1. *Investigations of Proton Elimination with Variations in [TiCl₄], LB, and [LB]*

All of the TiCl₄-coinitiated IB polymerizations were designed to reach ≈98% conversion in 2.5 min at –80°C, at which time quenching with MeOH quantitatively preserves the normal *tert*-chloride chain end structure (Scheme 2).

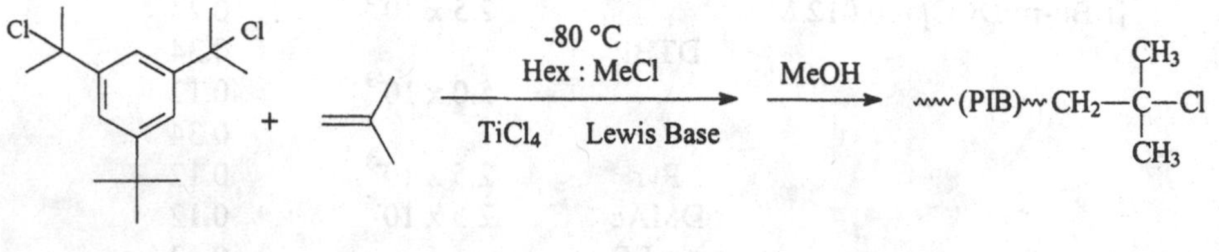

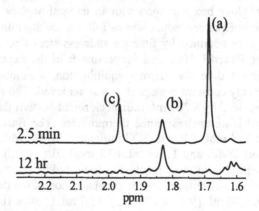

Scheme 2. Synthesis of *tert*-alkylchloride telechelic polyisobutylene

When the time of polymerization is allowed to reach on the order of 10^2 half-lives beyond ≈98% conversion (i.e., when the living chains are "aged"), ^{1}H NMR shows diminution of the *tert*-alkylchloride chain end resonances (1.67 and 1.96 ppm), along with the appearance of many new smaller resonances within the 1.5 - 2.3 ppm range (Figure 1).

Figure 1. ^{1}H NMR spectra of TiCl₄-coinitiated PIB quenched at 2.5 min and 12 h

In a recent paper, this chain-end "degradation" was attributed to carbenium ion rearrangements (e.g. hydride and methide shifts) which lead to isomerization of the PIB backbone and produce, upon deactivation, various isomeric *tert*-alkylchloride structures [21]. The depletion of *tert*-alkylchloride end groups was monitored via ^{1}H NMR, and the data were fitted to the first-order rate equation,

$$\ln \frac{[p-tCl]}{[p-tCl]_0} = k_r Kt \qquad (2)$$

where, k_r is first-order rate constant for rearrangement, and $K = K_{eq}[TiCl_4]^2$ (K_{eq} is defined in Scheme 1). The quantity $k_r K$, which is the apparent rate constant for end group depletion, was measured to be 8×10^{-5} s^{-1} for [CE] = 2.3×10^{-2} M, [TiCl$_4$] = 0.24 M, and [DMP] = 2.4×10^{-3} M at $-80°C$. Under these conditions the apparent rate constant for propagation was measured to be $k_{app} = 5.3 \times 10^{-2}$ s^{-1}. Since the latter was measured within the first minute or so of the polymerization reaction when rearrangement was still negligible, it is related to the second-order rate constant for propagation, k_p, as follows (for small K),

$$k_{app} = k_p[P^\oplus C^\bullet] = k_p K[P - tCl]_o \qquad (3)$$

where, $[P^\oplus C^\bullet]$ is the concentration of instantaneously active chain ends. From these two apparent rate constants can be obtained the ratio,

$$\frac{k_p}{k_r} = \frac{k_{app}}{k_{rapp}[P - tCl]_o} \qquad (4)$$

which, for this system under these conditions, assumes a value of 3×10^4 M^{-1}. Thus TiCl$_4$-coinitiated polymerization of IB at $-80°C$ is a Class 4-5 living system according to Matyjaszewski [25].

It was further concluded that β-proton elimination was not occurring at $-80°C$, based upon two findings with regard to active, aged monofunctional PIB chains: chain-chain coupling was not detected by GPC, and olefinic structures were not detected by ^{1}H NMR. However, upon change to a difunctional initiator and lower [TiCl$_4$], a discernable high molecular weight "shoulder" was observed in the GPC chromatograms of aged samples. Further experimentation utilizing a monofunctional initiator, TMPCl, showed that the high molecular weight peak corresponded to twice that of the targeted molecular weight. To verify our earlier studies, 2500 g/mol PIB samples initiated with a difunctional initiator were aged under various [TiCl$_4$] and [ED] conditions, with each showing no high molecular weight shoulder and a less than 3% increase in PDI upon aging. These results have led us to believe that β-proton elimination is also occurring upon aging, with the resulting terminal olefin leading to chain-chain coupling. However, at $-80°C$ it is slow relative to rearrangement and is difficult to detect. A difunctional initiator facilitates detection since the degree of coupling (or linking) is double that of a monofunctional initiator, at a given level of β-proton elimination.

Since it is assumed that chain-chain coupling is a result of β-proton elimination, studies were performed to determine the extent and mechanism of proton elimination (i.e., whether it is spontaneous or the result of direct nucleophilic attack by the Lewis base additive) as a function of polymerization conditions. Qualitative assessments of GPC chromatograms, of aged samples for which quantitative *tert*-alkylchloride end group decay was verified by ^{1}H NMR, were performed for systems utilizing several different LB, [LB], and [TiCl₄]. Of the various LBs used, DTBP showed the highest amount of coupling upon aging (Figure 2, left). Systems which contained no added LB also displayed coupling upon aging (Figure 2, right); these two facts suggest that the mode of elimination is mixed for the case of DTBP (i.e., partially spontaneous and partially by bimolecular abstraction by DTBP) and primarily spontaneous for complexed LBs, such as DMP. However, ^{1}H NMR analysis revealed no *endo*-olefins in any of the samples. If it is assumed that spontaneous elimination yields an isomeric mixture of *endo*- and *exo*-olefins, then absence of the former could be attributed to one of three things: either their concentration was below the limits of detection, *endo*-olefins take place in coupling events, or residual *endo*-olefins do not take place in coupling reactions and are rehydrochlorinated under living conditions.

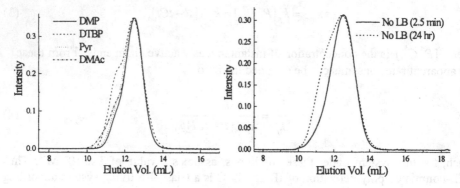

Figure 2. GPC traces for PIB samples aged for 24 h using (**left**) various LBs at concentrations of 2.5 x 10⁻³ M, and [TiCl₄] = 0.23 M, (**right**) no LB at [TiCl₄] = 0.12 M.

These tentative conclusions are supported by the fact that the degree of coupling increases drastically with increases in [DTBP], but exhibits only slight increases with increases in concentrations of the other Lewis bases. It was also noted that as [TiCl₄] is increased, the amount of chain-chain coupling decreases in all cases. Since low molecular weight recipes are generally formulated with high [TiCl₄], this accounts for the absence of chain-chain coupling in our previous studies, which were conducted on relatively low molecular weight polymers (2500 g/mol). Figure 3 gives a relative comparison of the extent of coupling between polymerization systems containing different [DTBP] and [DMP] at various [TiCl₄].

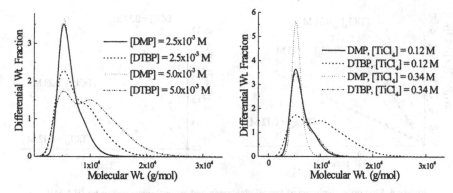

Figure 3. Differential molar mass plots for **(left)** varying [LB] at [TiCl₄] = 0.12 M, **(right)** varying [TiCl₄] at [LB] = 5.0 x10⁻³ M

3.1.2. *Kinetics of Termination Reactions in IB Polymerizations with Various [TiCl₄] and LB.*

The rate of *tert*-alkylchloride end group depletion was measured *via* ¹H NMR by integrating the area of peak (c) relative to peak (b) (in Figure 1) for a number of different aging times up to 24 h. The data were treated using the following modified expression for the first order consumption of *tert*-alkylchloride chain ends,

$$\ln \frac{[p-tCl]}{[p-tCl]_0} = k_r + k_e Kt \qquad (5)$$

where k_e is the rate constants for β-proton elimination. The combined rates of both chain termination reactions were measured for various [TiCl₄] in the presence of either DMP or DTBP.

Figure 4 shows that the depletion of *tert*-alkylchloride end groups follows a first-order kinetic dependency in all cases. It can also be noted that the apparent rate constant for chain end decay, $k_{dapp} = (k_r + k_e)K$, increases as [TiCl₄] increases, signifying that the combination of *tert*-alkylchloride rearrangement and β-proton elimination are accelerated with increases in chain end ionization. Furthermore, it is interesting to note that the rate of decay of PIB *tert*-alkylchloride chain ends is dependent upon the LB used within the polymerization system, with DTBP exhibiting k_{dapp} values 34.5% higher than DMP (Table 2).

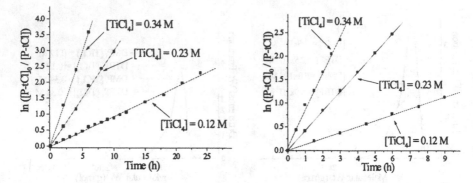

Figure 4. First order depletion of *tert*-alkylchloride end groups, measured using ^{1}H NMR with **(A)** [DMP] and **(B)** [DTBP] = 2.5 x 10^{-3} M at various [TiCl$_4$].

TABLE 2. Measured k_{dapp} Values for [LB] = 2.5x10^{-3} M at Various [TiCl$_4$]

[TiCl$_4$]	DMP (k_{dapp})	DTBP (k_{dapp})
0.12 M	0.593 hr^{-1}	0.784 hr^{-1}
0.23 M	0.299 hr^{-1}	0.412 hr^{-1}
0.34 M	0.093 hr^{-1}	.124 hr^{-1}

3.2. REAL-TIME ATR-FTIR MONITORING OF POLYMERIZATION KINETICS

This section describes investigations in which real-time *in situ* ATR-FTIR spectroscopic analysis has been applied to BCl$_3$-coinitiated IB polymerizations. We were specifically interested in examining the effect of [BCl$_3$] on rate of polymerization. The ReactIR 1000 reaction analysis system is used to monitor the intensity of characteristic peaks in the mid-IR spectrum, which allows the calculation of monomer concentration as a function of time. We have previously detailed the capabilities of this instrument and demonstrated its usefulness in generating accurate kinetic data on isobutylene polymerizations coinitiated by TiCl$_4$ [24]. Figure 5 illustrates the major absorbances of IB within the mid-infrared region. Our previous kinetic studies have focused on the primary aborbance located at 887 cm^{-1}.

Isobutylene polymerizations coinitiated with BCl$_3$ were carried out in methylchloride solvent at –80°C utilizing *t*-Bu-*m*-DCC as initiator and DMP as the LB. Figure 6 shows a profile of the relative absorbance (measured as peak height) of the 887 cm^{-1} peak as a function of time.

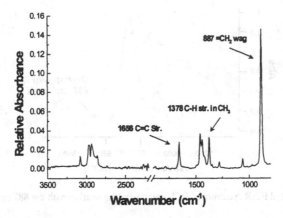

Figure 5. Partial FTIR spectrum of isobutylene illustrating major absorbances.

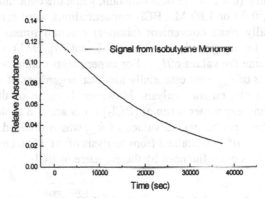

Figure 6. Real-time profile of isobutylene monomer depletion over time from analysis of 887 cm^{-1} peak with [BCl$_3$] = 0.22 mol/L.

During the course of the investigation it became evident that an overlap of the BCl$_3$ spectrum with the IB 887 cm^{-1} peak becomes significant at BCl$_3$ concentrations higher than about 0.9 mol/L, as evidenced by the partial FTIR spectrum of BCl$_3$ presented in Figure 7. Obviously, this overlap can introduce error into the profile of IB concentration over time. It has been demonstrated that there is another IB peak located at 1656 cm^{-1} that can be monitored to obtain kinetic data [26]. Although this peak is much less pronounced than the peak at 887 cm^{-1}, it does not appear to be influenced by the presence of BCl$_3$, and kinetic data was obtained by analyzing both spectral regions.

The data were interpreted by applying the kinetic rate law for a controlled/living polymerization with fast initiation,

$$r_p = -\frac{d[M]}{dt} = k_p[P^\oplus C^\bullet][M] = k_{app}[M] \qquad (6)$$

where r_p is the rate of polymerization [M] is the concentration of monomer.

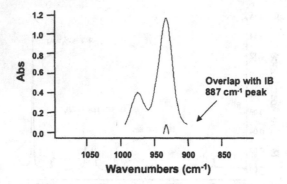

Figure 7. Partial FTIR spectrum of BCl₃ demonstrating overlap with the 887 cm⁻¹ IB peak.

Since we were interested in probing the effect of [BCl₃] on the system, the initial concentration of monomer (1 M), t-Bu-m-DCC initiator (1.51 x 10⁻² M), DMP (2 x 10⁻³ M) and reaction volume (0.2 L) were held constant, while the concentration of BCl₃ was varied as 0.22, 0.45, 0.90 or 1.80 M. BCl₃ concentrations were set relatively high to produce experimentally more convenient (shorter) reaction times. As presented in Figure 8, plots of the integrated form of eq 1, $\ln([M]_o/[M]_t)$ versus time, were constructed to determine the values of k_{app}. For experiments conducted at [BCl₃] = 0.22 and 0.45 M, the values of k_{app} were essentially identical regardless of whether the 887 or 1656 cm⁻¹ peak were chosen for analysis. However, because of the aforementioned spectral overlap, during experiments run at [BCl₃] = 0.9 and 1.8 mol/L, analysis of the 887 cm⁻¹ and 1656 cm⁻¹ peaks yielded values of k_{app} which differed by approximately 10%. We took the value of k_{app} obtained from analysis of the 1656 cm⁻¹ peak to be more accurate as this peak was not influenced by the presence of BCl₃.

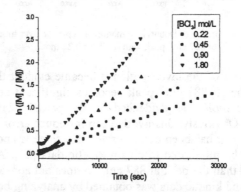

Figure 8. First-order kinetic plots for various [BCl₃]: [IB]₀ = 1.0 M; [DMP] = 2.0 x 10⁻³ M; [t-Bu-m-DCC] = 1.51 x 10⁻² M; MeCl solvent at –80°C.

The slopes obtained by performing linear regression on each set of data presented in Figure 8 are listed in Table 3. These values were taken as the apparent rate constant, k_{app}, for polymerization under those conditions. The kinetic order of the reaction with respect to [BCl₃] was determined by plotting $\ln(k_{app})$ versus $\ln[BCl_3]$, as shown in Figure 9. The slope obtained by performing linear regression on the data was found to be 0.55. This result was surprising as we had expected a first order dependence on the Lewis acid concentration. A dependence greater than unity, due to the formation of polymeric counterions, was considered unlikely for BCl_3, in contrast to Lewis acids such as titanium tetrachloride [3].

A probable explanation for the observed kinetic order with respect to $[BCl_3]$ is the decrease in polarity of the reaction medium as larger volumes of BCl_3 were added. At $[BCl_3] = 1.8$ M, BCl_3 made up approximately 14% of the reaction volume, whereas at $[BCl_3] = 0.22$ M it was only about 2% by volume. The reaction medium is expected to be less polar at higher $[BCl_3]$, since the dielectric constant of BCl_3 is ≈ 3 while that of MeCl is ≈ 19 [27]. Lower solvent polarity would act to suppress the rate of polymerization, and this would partially offset the increase in degree of ionization expected from increasing Lewis acid concentration.

TABLE 3: Apparent Rate
Constant at Varying $[BCl_3]$

$[BCL_3]$ M	$K_{app} \times 10^4$ s^{-1}
0.22	0.44
0.45	0.64
0.90	1.01
1.80	1.35

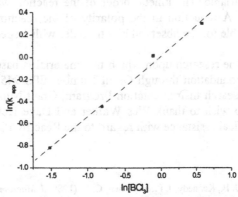

Figure 9. Kinetic order with respect to $[BCl_3]$ of IB polymerization rate.

4. Conclusions

Investigations were performed to evaluate the kinetics of *tert*-alkylchloride end group depletion in controlled/living PIB synthesis. The rate of carbenium ion rearrangement increased as the concentration of ionized (active) chains was increased, as would be expected. The extent of proton elimination was enhanced by an uncomplexed Lewis base (i.e., DTBP), which suggests that proton abstraction by DTBP increases the degree of olefin formation, and as a consequence, chain-chain coupling. Furthermore, this elimination process is impeded by increasing [TiCl$_4$]. Interestingly, coupling appears even in systems that lack an added Lewis Base, suggesting the occurrence of spontaneous β-proton elimination. Unpublished investigations within our laboratory have shown that olefin-terminated PIB rehydrochlorinates under controlled/living conditions. It can be reasoned from these results that chain end degradation is a complex combination of carbenium ion rearrangement and reversible proton elimination. It is assumed that when the PIB chain end eliminates to form a hindered olefin, (i.e., terminal *endo*-olefin or an olefin derived from a rearranged chain) these residual olefinic structures are eventually hydrochlorinated under the controlled/living conditions. Of course, *exo*-olefins undergo coupling, which is relatively more rapid than hydrochlorination, and thus no olefins are observed in the ^{1}H NMR spectrum. The fact that degradation occurs more rapidly with DTBP-containing systems implies that, along with spontaneous β-proton elimination, either an extremely low concentration of uncomplexed, yet nucleophilic impurity [23], or the DTBP itself, produces olefinic chain end structures that lead to chain-chain coupling.

The ReactIR 1000 reaction analysis system was employed to investigate the living carbocationic polymerization of isobutylene using the *t*-Bu-*m*-DCC/DMP/BCl$_3$ initiating system. By monitoring the polymerizations with real-time *in-situ* ATR-FTIR spectroscopy, profiles of the rate of monomer consumption over time were constructed. The kinetic data were interpreted by applying the first-order kinetic rate law for a controlled/living polymerization with fast initiation. During this investigation, the apparent rate constant, k_{app}, was determined for polymerizations conducted in which the [BCl$_3$] was the only variable. The kinetic order of the reaction with respect to [BCl$_3$] was found to be 0.55. A reduction in the polarity of the reaction medium at higher [BCl$_3$] may be responsible for the observed kinetic order with respect to [BCl$_3$].

Acknowledgements. The research upon which this material is based was supported by the National Science Foundation through Grant Number EPS-9452857, as well as the Defense University Research Instrumentation Program, Grant Number DAAG55-97-1-0067. The authors also wish to thank Wes Walker and Dave Hobart of ASI Applied Systems for their technical assistance with regard to the ReactIR 1000 system.

5. References

1. Kaszas, G., Puskas, J. E., Kennedy, J. P., and Chen, C. C. (1989) *J. Macromol. Sci.* **A26**, 1099.
2. Gyor, M., Wang, H. C., and Faust, R. (1992) *J. Macromol. Sci., Chem.* **A29**, 639.
3. Storey, R .F. and Choate, K. R., Jr. (1997) *Macromolecules* **30**, 4799.
4. Storey, R. F., Chisholm, B. J. and Brister, L. B. (1995) *Macromolecules* **28**, 4055.
5. Storey, R. F. and Shoemake, K. A. (1995) *ACS Div. Polym. Chem., Polym. Prepr.* **36**(2), 304.
6. Storey, R. F. and Donnalley, A. B. (1998) *ACS Div. Polym. Chem., Polym. Prepr.* **39**(1), 329.
7. Kaszas, G., Puskas, J. E, Chen, C. C., and Kennedy, J. P. (1998) *Polym. Bull.* **20**, 413.
8. Storey, R. F. and Lee, Y. (1992) *J.M.S.-Pure Appl. Chem.* **A29**, 1017.
9. Kaszas, G., Puskas, J., and Kennedy, J. P. (1988) *Makromol. Chem., Macromol. Symp.* **13/14**, 473.
10. Faust, R., Ivan, B., and Kennedy, J. P. (1991) *J. Macromol. Sci. Chem.* **A28**, 1.
11. Faust, R. and Kennedy, J. P. (1986) *Polym. Bull.* **15**, 317.

12. Puskas, J. E., Kaszas, G., Kennedy, J. P., Kelen, T., and Tüdös, F. (1982) *J. Macromol. Sci-Chem.* **A18**, 1229.
13. Fodor, Z., Bae, Y. C., and Faust, R. (1998) *Macromolecules* **31**, 4439.
14. Puskas, J. E., Kaszas, G., and Litt, M. *Macromolecules* **24**, 5278.
15. Kennedy, J. P. and Ivan, B. (1992) *Designed Polymers by Carbocationic Macromolecular Engineering: Theory and Practice*, Hanser, Munich.
16. Majoros, I., Nagy, A., and Kennedy, J. P. (1994) *Adv. Polym. Sci.* **112**, 1.
17. Fodor, Z., Bae, Y.C., and Faust, R. (1998) *Macromolecules* **31**, 5170.
18. Kaszas, G. and Puskas, J. E. (1994) *Polym. React. Eng.* **2**(3) 251.
19. Puskas, J. E. and Lanzendörfer, M. G. (1998) *Macromolecules* **31**, 8684.
20. Storey, R.F. and Donnalley, A.B. (1999) TiCl4 Reaction Order in Living Isobutylene Polymerization at Low [TiCl$_4$]:[Chain End] Ratios *Macromolecules* to appear.
21. Storey, R.F., Curry, C. L., Brister, L. B. (1998) *Macromolecules* **31**, 1058.
22. Held, D., Ivan, B., Muller, A. H. E., de Jong, F., and Graafland, T. (1996) *ACS Div. Polym. Chem., Polym. Prepr.* **37**(1), 333; (1997) *ACS Symp. Ser.* **665**, 63.
23. Bae, Y. C. and Faust, R. (1997) *Macromolecules* **30**, 7341.
24. Storey, R. F., Donnalley A. B., and Maggio, T. L. (1998) *Macromolecules* **31**, 1523
25. Matyjaszewski, K. (1993) *Macromolecules* **26**, 1787.
26. Puskas, J.E., Lanzendorfer, M., and Pattern, W. (1998) *Polym. Bull.* **40**, 55.
27. Weast R.C., Ed. (1984) *CRC Handbook of Chemistry and Physics, 65th Ed.*, CRC Press, Inc., Boca Raton, FL, p. E-50. Value for BCl$_3$ is that for closely related BBr$_3$.

STOPPED-FLOW TECHNIQUE AND CATIONIC POLYMERIZATION KINETICS

J.P. VAIRON, B. CHARLEUX, M. MOREAU
Laboratoire de Chimie Macromoléculaire, UMR 7610
Université Pierre et Marie Curie,
4, Place Jussieu, 75252 Paris cedex 05, France

1. Introduction

The major challenge concerning the cationic polymerization of alkenes still remains the identification of the propagating species and the determination of their concentration throughout the reaction. Very reactive and unstable carbocations have short lifetime (10 ms to a few minutes, depending on temperature) and exist at low concentrations ($< 10^{-4}$ M) in polymerization medium. A shophisticated technique is needed for styrenics with UV absorbing cations to directly follow their evolution with time, i.e. a device allowing a very short mixing time and a very fast and quantitative spectroscopic detection. The well-known stopped-flow method is the most suitable to observe in real time these short living species. Surprisingly, since the pioneering low temperature (-80 °C) work of Pepper [1-4], only two series of thorough studies from the Kyushu [5-10] and Kyoto [11-15] groups were reported using this technique, but owing to instrumentation they were limited to a very narrow range of temperatures (-1 to +30 °C). Furthermore the stability of carbocations depends on the experimental conditions, and experiments under vacuum in sealed vessels with highly purified reagents are required. The situation is particularly complex in rapid cationic polymerizations as the unstable ionic species can change in nature as well as in concentration during the short time of the reaction. Thus the validity of experiments strongly depends on the design of the instrument, the nature of materials in contact with reagents, the sources of leaks which can destroy active species, and on the available temperature range of the instrument.

We developed in the last ten years a one-block all quartz-glass stopped-flow device (S-F) working over a wide temperature range (i.e. −100 °C to r.t. or above), which can be filled up under vacuum with highly purified reagents and allows the sampling of a complete UV spectrum each 4 ms. We intend here to emphasize the outstanding capabilities of this technique through our recent studies devoted to rapid polymerization

177

J.E. Puskas et al. (eds.), Ionic Polymerizations and Related Processes, 177-204.
© *1999 Kluwer Academic Publishers. Printed in the Netherlands.*

kinetics: the reexamination of the "old" -but still unclear- cationic polymerization of styrene initiated in CH_2Cl_2 by triflic acid, the protonation of the trans-ethylenic dimer of styrene used as a model precursor of the styrene propagating species, and finally our last results about the ionization of cumyl chloride, an initiator for living cationic polymerization, in the presence of different Lewis acids.

2. Description and Operation of the Stopped-flow Instrument

Our custom-designed S-F apparatus [16] is of piston-driven and front-stopping type and a general diagram together with a simplified design of the sealed glass-quartz part is presented on Figure1. Valves of a classical S-F are replaced by PTFE taps (YOUNG Glassware Ltd.) to avoid air leaking in the instrument. The pistons of the syringes have each 2 high-friction PTFE rings which hold a static vacuum better than 5.10^{-4} torr. The quartz mixing chamber (4 tangential offset jets for each solution) is sealed to reference back-to-back cells (upstream) and to a 2 mm optical path observation cell (downstream) (Figure 2). and placed in a brass cooling box. Movement of the third syringe - and the flow - is stopped by a mobile pin equipped with a microswitch, which triggers the detection system. The deadtime, classically measured by using the reduction of 2,6-dichlorophenolindophenol by ascorbic acid, was found to be 4-6 ms, depending on the mode of pushing (manual, or pneumatic at 10 bars). Nevertheless, in this prototype, the mechanical microswitch is closed a few ms before stopping of the flow, and the first 10 ms recorded signals cannot be confidently used.

The continuous spectrum source is a 150 W Xenon arc, which may be used down to 270 nm. Optical fibers (Superguide G, Fiberguide Industries) bring the polychromatic beam to the cells and then to the monochromator (Chromex 250IS, holographic gratings). The light is dispersed, according to the wavelength, on a photodiode array (512 diodes), which is coupled with an optical multichannel analyzer (Princeton) controlled by a specific software (Winspec 1.6.1). All important parameters for the experiment like starting wavelength, spectrum width, time to record one spectrum, number of ignored spectra, delays, etc ..., can be defined. Spectra are obtained by substracting observation and reference records. Any UV-Visible wavelenght range can be observed. For instance, a spectrum from 250 to 550 nm with a resolution of 0.5 nm can be recorded in less than 5 ms and more than 1000 spectra can be stored during the course of the reaction. For all experiments monomers and solvents were purified on sodium mirrors and stored under vacuum as usual in our laboratory [17,18]. All reagents were distributed in graduated reservoirs equipped with high vacuum Teflon connections and taps (Young Glassware Ltd.). After adaptation of the reservoirs and suitable vessels for dilutions, the instrument was evacuated ($< 10^{-5}$ torr) and then filled up with the convenient solutions.

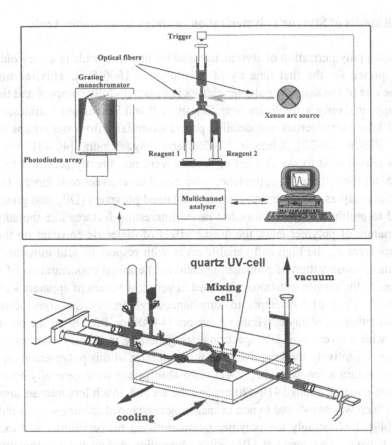

Figure 1. General diagram of the S-F instrument and details of the glass-quartz part.

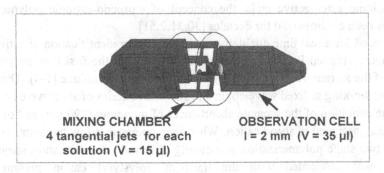

MIXING CHAMBER
4 tangential jets for each
solution (V = 15 µl)

OBSERVATION CELL
l = 2 mm (V = 35 µl)

Figure 2. Quartz mixing chamber and observation cell.

3. The Kinetics of Styrene Polymerization Initiated by Protonic Acids

The cationic polymerization of styrene initiated by protonic acids is a very old story as it was reported for the first time by M. Berthelot in 1866 [19]. This reaction which should be one of the simplest polymerizations is in fact awfully complex and the kinetic studies appeared only in the fifties from Szwarc [20] and Brown and Mathieson [21,22], followed later by numerous and detailed papers essentially from the groups of Pepper [23-28], Plesch [29-32], Chmelir [33-39] and Higashimura [40-43]. During this « blind » period -just to say that active species were not directly seen and followed in real time- all these thorough approaches were based on conventional kinetic techniques and careful analyses of conversion rates and obtained polymers (DP$_n$, end groups, etc..). They led to multiple interpretations but raised interesting features like the ethylenic or indanic nature of polymer ends, the drastic effect of dielectric constant on the overall rate or apparent k_p, the high and variable order with respect to acid initiator. As there was no direct observation of propagating cations, the initial concentration of acid was introduced in the kinetic equations which led to very low values of apparent k_p, typically 5-20 L.mol^{-1}.s^{-1} at r.t.. Attempts to simultaneously identify the styryl cation were performed either on phenylethylium precursors [44-48] or directly on the polymerizing medium with conventional UV spectrometers generally in several minutes time scale which was definitively too long owing to the high rates of this polymerization which is completed within a few seconds [32,47]. Two absorptions were generally observed, the main broad one in the range 410-420 nm and the second, much less intense, around 305-315 nm which was associated to non propagating rearranged species such as indanylium cations [49]. Interestingly the polymerization initiated by perchloric acid exhibited a peculiar behavior (no apparent UV-visible absorption during the reaction) which led Gandini and Plesch to propose a propagation mechanism involving monomer insertion into non ionic ester active ends, the concept of « pseudo-cationic polymerization » which has been controversial for decades [30,31,2,51].

Thus the need for a real time qualitative and quantitative identification of active species was obvious. This led Pepper and coworkers to study for the first time polymerization kinetics of the styrene-HClO$_4$ system by the stopped-flow technique [1-4]. They used an instrument working at fixed wavelength and successive series of shots were necessary to follow the evolution of the separate absorbances of monomer (290 nm) and of a peak at 340 nm they assigned to styryl cation. When they operated at low temperature (-60 to −80 °C) a two stage polymerization was clearly observed. The very short stage I (a few seconds) was associated with the transient polystyryl cation present at low concentrations. The corresponding apparent k_p, calculated by assuming similar extinction coefficient for styryl anion and cation $\varepsilon_{340} = 10^4$ L.mol^{-1}.cm^{-1} [50], were 2-5 10^3 L.mol^{-1}.s^{-1} in the -80 to −60 °C range, which are much more plausible values than those reported for the previous « blind » experiments. A salt effect led to an estimate of

$k_p^\pm$ and k_p^+, respectively 1-3 10^3 and 2-2.5 x 10^5 L.mol^{-1}.s^{-1} in same temperature range. The much slower stage II (minutes), dominant above –60 °C, was in the early analysis associated with pseudo-ionic propagation but it was recently reinterpreted in terms of ionic propagation, esters playing the role of ion-pairs reservoirs [51]. Later Kunitake and Takarabe [6-8] used the coupling of commercial rapid-scan S-F and rapid quenching devices (Union Giken - RA 1300 and RA 416) to simultaneously follow the evolution of the 340 nm styryl cation absorbance and the monomer conversion. They chose to study the styrene polymerization initiated by F_3CSO_3H (triflic acid), an expectedly much simpler system as the very unstable triflic esters have never been isolated [52]. They observed initiation first orders with respect to initiator and to monomer, and measured initiation, propagation and monomer transfer apparent rate constants with k_p^{app} in the range 5 to 50 10^4 L.mol^{-1}.s^{-1} between -1° and +30 °C. Again a salt effect allowed an estimate of $k_p^\pm$ and k_p^+, respectively 4-12 x 10^4 and 2.5-30 x 10^5 L.mol^{-1}.s^{-1} in the same temperature range, i.e. a ratio $k_p^+/k_p^\pm$ = 6-20 depending on temperature. Unfortunately the device was not adapted for lower temperature experiments and all results were obtained above –1 °C.

We decided to reexamine this triflic acid-initiated styrene polymerization in methylene dichloride solution. The goals were at the same time to test our sophisticated multi-scan instrument -high purity conditions, simultaneous recording of all absorbances, high number of spectra during the course of reaction, high resolution- with a confident reference system and to cover the missing temperature domain.

3.1. REEXAMINATION OF THE STYRENE POLYMERIZATION INITIATED BY TRIFLIC ACID [53,54]

The reaction was studied in methylene dichloride solution between –65 °C and room temperature within restricted concentration ranges $[M]_0$ = 1-40 x 10^{-3} mol.L^{-1} and $[I]_0$ = 0.5 to 3 x 10^{-3} mol.L^{-1}, due to optical density limitations. Monomer consumption (side peak at 292 nm, ε_{292} = 580 L.mol^{-1}.cm^{-1}) and evolution of cationic species (343 nm, ε assumed to be 10^4 L.mol.$^{-1}$cm^{-1}) were followed simultaneously.

3.1.1. *General Behavior and Initiation*

As reported by the previous authors, initiation is much faster and shorter than termination. Depending on temperature, a sharp increase of $[P_n^+]$ is observed at the onset of polymerization, followed by a stationary period (plateau or smooth maximum) and then by a long decay period (Figure 3). A very strong reverse dependence of $[P_n^+]_{max}$ with T is observed, and the maximum yield of carbocations with respect to acid remains low: it varies from 0.5% at r.t. to 35 % at –65 C for a system with $[M]_0/[I]_0$ = 20. Experiments at –65 °C with a large excess of TfOH (0.02<$[M]_0/[I]_0$<0.2) led also to incomplete yield of cations with respect to monomer (<30 %). In these latter conditions

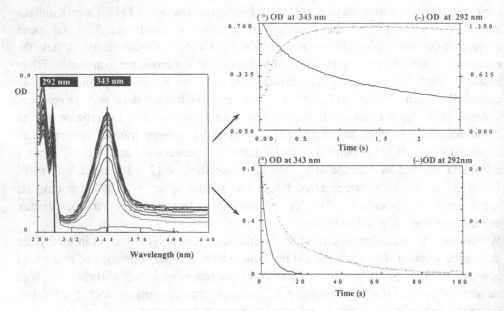

Figure 3. Simultaneous evolutions of monomer (292 nm) and cation (343 nm) absorptions
$[St]_o = 10.7$ mmol.L^{-1} ; $[TfOH]_o = 2.27$ mmol.L^{-1} ; T = -62 °C.

this can be explained by a propagation much more rapid than initiation but prevents any attempt of experimental verification of the ε of the styryl cation.

This could be explained by a propagation much more rapid than initiation but in fact such a behavior agrees much better with an equilibrium of protonation where the monomeric cation can either propagate or lose a β-proton and give back monomer and acid :

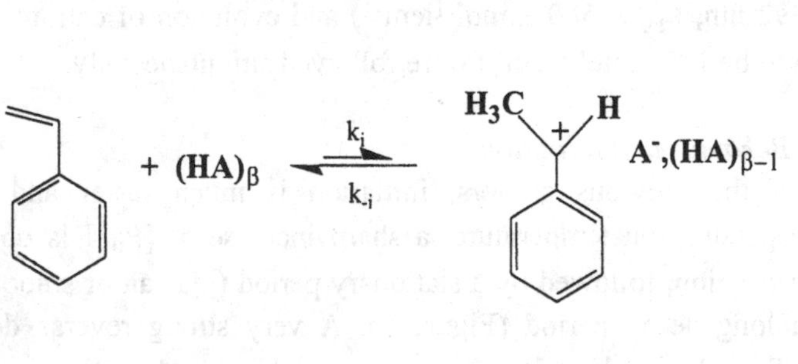

Scheme 1.

The drastic decrease of $[P_n^+]_{max}$ when increasing T could be explained by a higher activation energy for the deprotonation reaction. Measurements of initial rate of cation formation led to the expected external first order with respect to monomer for initiation (Figure 4). Situation is quite different for the external order with respect to acid as we observed very high values decreasing from 4.5 at -62 °C to 3.8 at -32 °C and 3.3 at -23 °C (Figure 5). It is thus difficult to give reliable k_i values as their units are changing with temperature. The most consistent explanation involves an equilibrium of aggregation of triflic acid in methylene dichloride which is shifted towards the larger aggregates when temperature is decreased. This also supposes that the higher aggregates are the most reactive. On the other hand, the maximum concentration of carbocations $[P_n^+]_{max}$

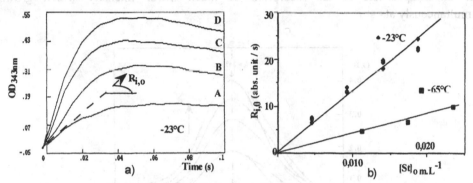

Figure 4. Initiation. First order with respect to monomer
a) : $[St]_o$=4.7 (A) ; 9.4 (B) ; 14.1 (C) ; 18.7 (D) mmol.L^{-1} ; $[TfOH]_o$=2.2 mmol.L^{-1}.

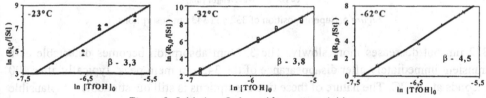

Figure 5. Initiation. Orders with respect to initiator
$[St]_o$= 10.7 mmol.L^{-1} ; $[TfOH]_o$= 0.77 to 3.37 mmol.L^{-1}.

observed at the stationary period appears proportional to $[TfOH]^2$ which agrees with an equilibrium of initiation involving dimeric acid. It might be considered that at the very beginning, when acid solution is suddenly diluted with the monomer one, initiation takes place essentially on large aggregates of TfOH, the higher the aggregate the higher the reactivity. Then the equilibrium of association is rapidly shifted towards the lower aggregates with a predominant initiation by average dimeric forms. These complex external orders with respect to acid, which are depending on temperature and concentration, may be related, even if the time scales of measurements are very different, with the results of Chmelir [34] who obtained an external order of 3 at − 15 °C for the same polymerization system and to those of Sigwalt [55,56] who showed the existence of 1:1 and 1:2 homoconjugates of TfO⁻ in CH_2Cl_2 solution. The only comparable stopped-flow measurements are those of Kunitake [5,7] which led to a first external order with respect to TfOH for initiation of the same polymerization. This obvious discrepancy might eventually be explained by the difference in purity conditions, as the commercial triflic acid directly used by this author could be hydrated. A second observation which should be pointed out relates to the exact nature of the 343 nm absorption. Even if this peak is undoubtedly associated with styryl cation and is rather sharp, it is clearly assymetric and an attempt of deconvolution (Figure 6) showed the superimposition of an absorption of much lower intensity which grows simultaneously at

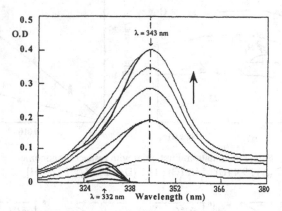

Figure 6. Superimposition of 332 and 343 nm absorptions.

332 nm but decreases more slowly. The 332 nm absorption becomes detectable as a transient immediately after disappearance of the 343 nm main peak, typically after 100 seconds at −60 °C. The nature of these two absorptions is still questionable . A plausible interpretation might assign the 332 nm peak to monomeric styryl cation wether the 343 nm peak could correspond to polymeric cation (D^+ to P_n^+) or even to solvated species. After a longer time, i.e. 25 mn, the spectrum of the reacting medium presents only the expected 311 and 417 absorptions resulting from indanylium cations generated by hydride abstraction from cyclized indanyl ends to triflic acid or to propagating cations (less probable owing to their concentration):

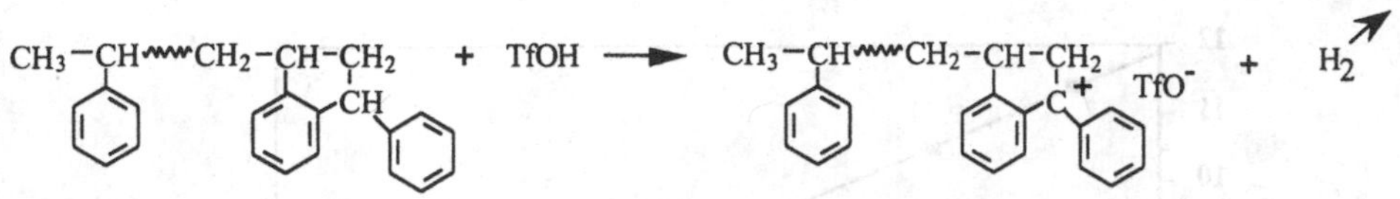

$$CH_3-\underset{\displaystyle \bigcirc}{\underset{|}{CH}}\text{\tiny wwww}CH_2-\underset{\displaystyle \bigcirc}{\underset{|}{CH}}-\underset{\underset{\displaystyle \bigcirc}{\underset{|}{CH}}}{CH_2} \;+\; TfOH \;\longrightarrow\; CH_3-\underset{\displaystyle \bigcirc}{\underset{|}{CH}}\text{\tiny wwww}CH_2-\underset{\displaystyle \bigcirc}{\underset{|}{CH}}-\underset{\overset{|}{\underset{\displaystyle \bigcirc}{\underset{|}{C^+}}}\; TfO^-}{CH_2} \;+\; H_2\nearrow$$

Scheme 2.

We recently observed that mixing 1-methyl-3-phenylindan with excess triflic acid in CH_2Cl_2 either at low or room temperature leads to exactly the same UV spectrum (311 and 417 nm abs.) as obtained with the styrene-TfOH polymerization system [57].

3.1.2. Propagation

Monomer consumption appears to be internally of second order with respect to monomer during the increase of carbocation concentration (which accounts for the first order dependence of initiation on monomer concentration), and to be first order during the period where the concentration of cations is close to stationary (0.5 to 4 sec, Figure 3). Values of the apparent k_p (ion pairs + free ions) at different temperatures (Table 1) are obtained from this stationary period and lead to an activation energy $E^\circ_{p,app.}$ ~39 kJ.mol^{-1} (Figure 7). Determination of k_p is limited to the –62 to –10 °C temperature range as, with the concentrations of monomer and acid we used, the concentration of carbocations is too low to be accurately measured above –10 °C. Such a determination was possible in the case of Kunitake who used a ten times higher $[M]_0$ with resulting larger $[P_n^+]$. Our values for -62 to –10 °C agree rather well with those of Kunitake and Pepper which are situated on both sides of our temperature range (Table 1).

TABLE 1. Comparison of our apparent k_p values with those previously reported [2,7,8].

T (°C)	k_p^{app} (l·mol^{-1}·s^{-1}) (this work)	k_p^{app} (l·mol^{-1}·s^{-1}) (Kunitake et al., 1978)	k_p^{app} (l·mol^{-1}·s^{-1}) (Pepper et al., 1974)
+ 30	-	$1.6 - 2.5 \cdot 10^5$	-
+ 20	-	$1.1 - 1.4 \cdot 10^5$	-
+ 10	-	$7.0 - 8.9 \cdot 10^4$	-
- 10	$9.0 \pm 1 \cdot 10^4$	$4.7 - 5.7 \cdot 10^4$	-
- 22	$3.7 \pm 1 \cdot 10^4$	-	-
- 32	$1.5 \pm 1 \cdot 10^4$	-	-
- 52	$2.9 \pm 1 \cdot 10^3$	-	-
- 62	$1.0 \pm 1 \cdot 10^3$	-	$4.5 \pm 0.5 \cdot 10^4$
- 80	-	-	$2.3 \pm 0.3 \cdot 10^4$

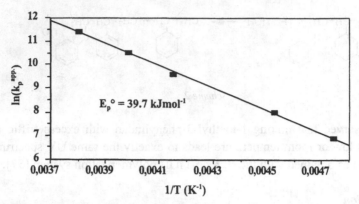

Figure 7. Activation energy for propagation (this work).

3.1.3. *Chain Transfer and Termination*

One of our most striking observations concerns termination. In most experiments carbocations are still remaining after polymerization, and decay of the 343 nm absorbance.

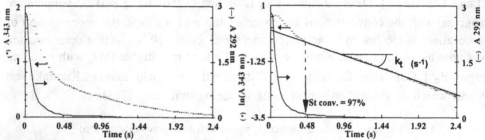

Figure 8. Termination . Simultaneous evolution of monomer and cations with first order second stage
$[St]_o = 21.9$ mmol.L^{-1}; $[TfOH]_o = 5.73$ mmol.L^{-1}; $T = -32$ °C .

exhibits systematically two periods, a rapid one which lasts as long as the monomer is present and a slower one which corresponds to termination of cations after complete monomer conversion and which is internally first order (Figure 8). It can reasonably be assumed that the profile of the 343 nm decay results from superimposition of this first order termination and of a second process related to monomer when it is still present.

TABLE 2. Influence of initial concentrations of monomer and initiator on k_t (spontan. term.).

T (°C)	$[St]_0$ (mmol·l⁻¹)	$[TfOH]_0$ (mmol·l⁻¹)	k_t (s⁻¹)
-	11.1	-	0.046 ± 0.006
-	11.1	-	0.045 ± 0.01
- 65	17.3	2.0	0.045 ± 0.015
-	23.4	-	0.05 ± 0.015
-	-	1.53	0.016 ± 0.002
- 62	10.7	2.27	0.037 ± 0.001
-	-	3.17	0.054 ± 0.003

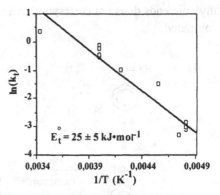

$E_t^\circ = 25 \pm 5 \text{ kJ·mol}^{-1}$

Figure 9. Activation energy for spontaneous termination.

The only secondary reaction which could be considered with polymer is a Friedel-Crafts alkylation either intermolecular or intra on the penultimate phenyl group. But it gives back the acid and obviously it is not a termination but a transfer to monomer and should not modify the cation concentration. A plausible explanation for the contribution of monomer to the decay of cations absorbance involves the equilibrium of initiation. In most cases this decay at 343 nm is observed for a period of polymerization with corresponds to 50-100 % of monomer conversion. Thus the rapidly decreasing monomer concentration shifts the protonation-deprotonation equilibrium to the left (Scheme 1), leading to lower $[P_n^+]$ values. This process is not a termination but the correlation of monomer concentration and disappearance of cations supports well the proposed equilibrium of initiation.

The true termination corresponding to the slowest stage is, as said above, a first order reaction and can reasonably correspond to the irreversible indanyl cyclization:

$$CH_3-CH\text{\scriptsize{wwww}}CH_2-CH-CH_2-CH^+ \ TfO^-,(TfOH)_x \longrightarrow CH_3-CH\text{\scriptsize{wwww}}CH_2-CH-CH_2$$

Scheme 3.

of the oligomers (DP about 20 at max. of SEC peak) which are obtained in our conditions of temperature and concentrations agrees with the superimposition of mono and polysubstituted phenyl rings vibrations. The collapse of counter anion could be considered but it was shown by Matyjaszewski [52] that this triflic ester is very unstable and, in our case, it will -if it exists- rapidly give back the cation. The formation of ethylenic ends does not correspond to a termination as the double bond can be reversibly protonated.

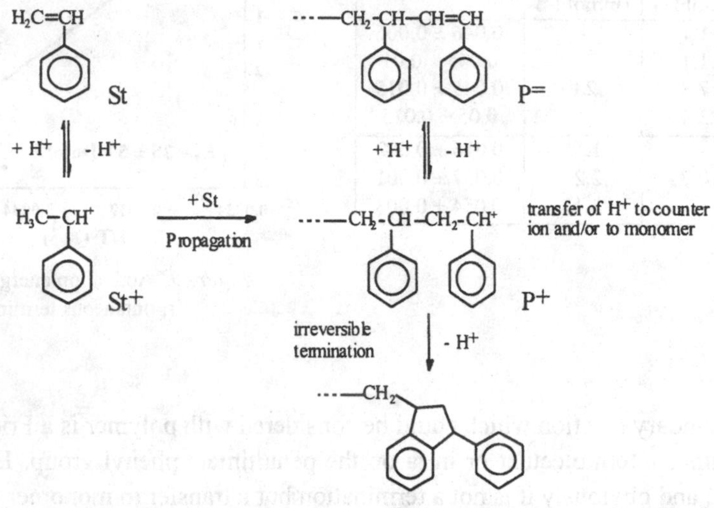

Scheme 4. Proposed mechanism for styrene polymerization initiated by triflic acid.

The internal Friedel-Crafts cyclization leading to indanyl end remains the most probable first order termination. It is irreversible and is the real driving force of the polymerization (Scheme 4). Alkylation of the chains of terminated polymer, the concentration of which is equal to $[M]_o$, could also be considered as it will lead to a pseudo-first order. This possibility, if it exists, is not the kinetically predominant termination process as the spontaneous termination rate constant k_t is independent of $[M]_o$ i.e. of polymer concentration (Table 3).

3.1.4. Conclusion

This detailed reexamination affords a new vision of the general mechanism of styrene polymerization initiated by triflic acid, particularly concerning the proposal of successive equilibria shifted by a unique driving force which is the irreversible indanic termination. The values of apparent k_p obtained between -60 and −10 °C agree well with the previous determinations of Pepper at lower temperatures and of Kunitake at higher temperatures. Nevertheless there is an obvious discrepancy with Kunitake's results concerning initiation kinetics and aggregation of triflic acid. Even if the values of k_p are coherent with those obtained by Pepper with ClO_4H as initiator during the first rapid stage of the polymerization, the overall mechanisms cannot be compared as it appears clearly that the termination in this latter case results from the reversible formation of the perchloric ester.

There are still persistent problems and questions :
- additional confirmation is needed for definitive assignment of the 343 nm absorption (mono or polystyryl cation?),
- the ε of the styryl cation remains to be determined,
- reversible protonation of polymer ethylenic chain ends remains to be proved,
- identification of secondary reactions linked to the evolution of UV-Vis absorption during the course of the reaction is still uncomplete,
- is aggregation of triflic acid the real explanation for observed high orders with respect to acid? The most aggregated, the most reactive?
- can polystyryl cations add to ethylenic chain ends?
- what is the real state of active ends? Salt effects performed by Pepper and Kunitake proved that propagation takes place simultaneously on ion pairs and free ions but the highest k_p' obtained at 30 °C are in the 10^6 L.mol^{-1}.s^{-1} range. Such value is close to the ones reported for γ-ray polymerizations of styrene [58,59] but is still far from the estimate of k_p obtained from recent model studies of addition of alkenes on cations indicating that monomer insertion should be close to diffusion controlled ($k_p = 6.\times10^8$ L.mol^{-1}.s^{-1} for isobutene) [60]. It is interesting to notice that the k_t we measured are dependent on $[TfOH]_o$ (Table 2) and this indicates a possible solvation of styryl cation by triflic acid. If this happens we should expect a slowing down of propagation with · respect to that with unsolvated ions.

Clearly, the exact behaviour of the unsaturated chain end is unknown, which led to consider the reaction with triflic acid of the styrene E-dimer as a model of this unsaturated end.

3.2. PROTONATION OF STYRENE E-ETHYLENIC DIMER. AN APPROACH TO THE STYRYL PROPAGATING CATION

The trans-1,3-diphenyl-1-butene, trans ethylenic dimer of styrene (D), has been chosen as a model for this approach because its reactions simulate the behaviour of the unsaturated polymer chain ends and provide information relevant to the real styrene polymerization system. Protonation of this molecule leads to the dimeric cation D$^+$ which can be considered as a better model for the polystyryl cation than 1-phenylethylium. Furthermore the homopolymerization of this sterically hindered dimer is limited to very short oligomers and this study should allow to answer, at least partly, to the questions we raised in the above section: Is the protonation reversible as for styrene monomer? Does it allow a confirmation of the assignment of the 343 nm absorption to polymeric PSt$^+$ alone or to both PSt$^+$ and monomeric St$^+$? If the constant of protonation equilibrium is high could we expect a more consistent estimate of the value of the ε_{343}?

This trans dimer, is obtained by cationic oligomerization of styrene under conditions where transfer is faster than propagation [64]. A few studies have been previously reported on its reaction with protonic and Lewis acids, involving spectroscopic analysis of the mixture [65] and products characterization [64,66-68]. Most of the analyzed species had indanic structure. It has been shown that not only cyclization of D into 1-methyl-3-phenylindan (I) was occurring but also formation of higher oligomers, the main fraction of which was composed of the indanic dimers of D (DI, Scheme 5).

Sawamoto and Higashimura [64] characterized these higher oligomers by ^{1}H NMR spectroscopy and size exclusion chromatography (SEC). They also found some styrene oligomers with an odd number of units (3 and 5) and concluded that, simultaneously to oligomerization, depolymerization of D was also occuring, leading to styryl cation and styrene monomer which could then be incorporated into oligomeric chains. Nevertheless none of these studies have been performed at T < 0 °C under the conditions for styrene cationic polymerization and a low temperature spectroscopic and kinetic analysis of the reactions was needed. We undertook a stopped-flow reexamination between –67 °C and room temperature.

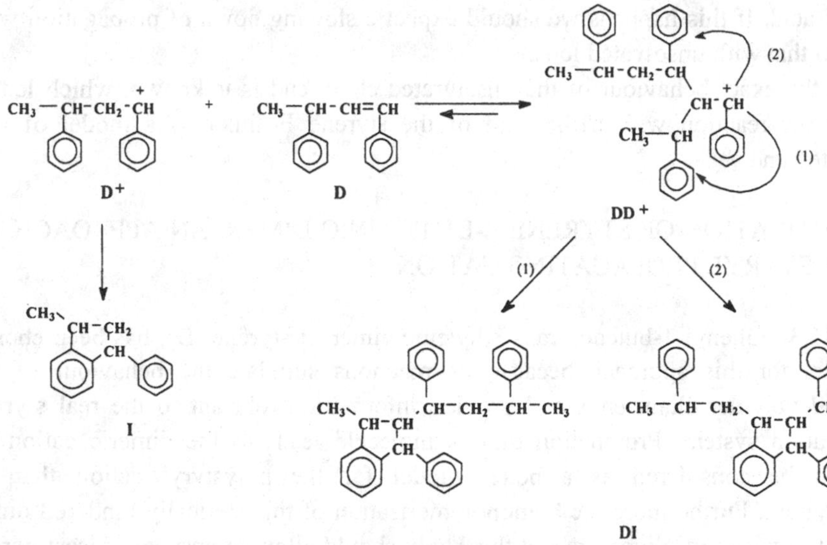

Scheme 5. Oligomerization of ethylenic dimer D of styrene.

3.2.1. *Low Temperature Range Stopped-flow Experiments (T < -30 °C)*

The reaction of dimer with triflic acid was studied in CH_2Cl_2 solution. Consumption of D was followed by the decrease of its 296 nm side-peak (ε_{296}=1160 L.mol^{-1}.cm^{-1}). It was always complete whatever the initial conditions. Between -67 and –30 °C a transient absorption at 340 nm appeared quickly, reached its maximum after about 1 second and then disappeared more slowly, within 50 s at lowest temperature (Figure 10). The intensity of this peak depended upon initial conditions: the lower the temperature and the higher the acid and dimer concentrations, the higher the OD at maximum. Time to reach the maximum (t_{max}) was fairly independent of reagents concentrations but very temperature dependent. The value of tmax decreased when temperature was increased: it was close to 1 s at – 67 °C and shorter than 0.1 s at –38 °C. The 340 nm absorption had a persistence time similar to that of D double bond and this peak has been assigned to the cation D$^+$.

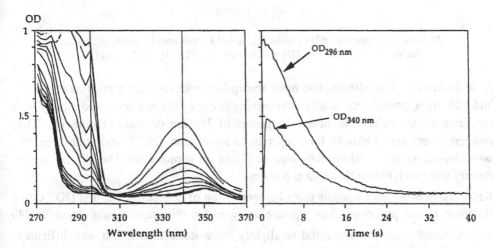

Figure 10. Protonation of D by triflic acid at –67 °C. Consumption of D (296 nm) and simultaneous evolution of the D$^+$ and D$_n^+$ cations (340 nm). [D]$_o$ = 6.7 mmol.L^{-1} ; [TfOH]$_o$ = 22.5 mmol.L^{-1}.

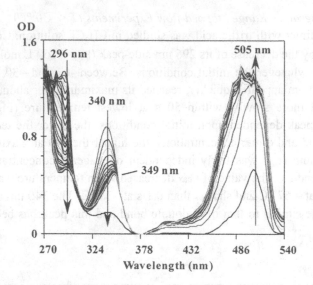

Figure 11. Protonation of D by triflic acid in CH_2Cl_2 solution. Evolution of spectra during 50 s. T = -64 °C ; $[D]_o = 22.9$ mmol.L^{-1}; $[TfOH]_o = 13.1$ mmol.L^{-1}.

In addition to peak at 340 nm, two other absorptions with respective maxima at 349 nm and 505 nm appeared immediatly after mixing the reagents and increased slowly with the same kinetic behaviour during the period of 340 nm decrease (Figure 11). They reached a very stable plateau for temperatures lower than –30 °C and their intensities were dependent on initial concentration of D and on temperature. The 505 nm optical density was much higher than that at 349 nm.

Knowing from product analysis that oligomerization of D occured, i.e. that DD^+ cation (Scheme 4) was produced, then a possibility was that this cation could absorb at 349 nm. Although such a cation could be slightly more stable than D^+, it was difficult to explain the large observed difference in stability (5 s vs. 20 mn at –67 °C),. Moreover there was no reasonable explanation for a strong absorption of a styryl type cation at 505 nm. Thus, it was more reasonable to consider that DD^+ and D^+ cations absorb at the same wavelength (340 nm), and have same stabilities. Since the two peaks at 349 nm and 505 nm have similar behaviour for all the temperatures studied, they could correspond to the same cationic species. This cation should have a strongly conjugated structure to agree with such high wavelength and high stability. Thus, it might correspond to an allylic cation we propose to be the 1,3-diphenyl-1-buten-3-ylium, D_i^+ (Scheme 6) produced by hydride transfer from D molecule to D^+ cation or even to triflic acid with H_2 formation.

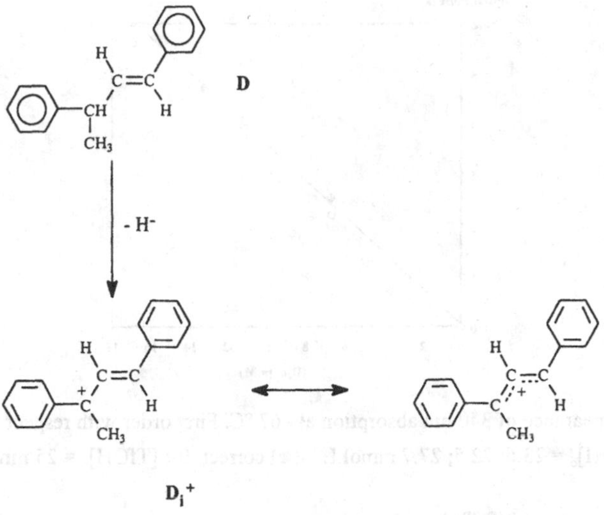

Scheme 6. Pathway for the formation of allylic D_i^+ cation.

Such a hydride transfer reaction would also occur from higher molecular weight unsaturated oligomers and also from the indanic oligomers that appear in the medium although these different products are in low concentration at the beginning of the reaction.

An allylic cation such as D_i^+ had been previously postulated by Gandini and Plesch [68] for styrene cationic polymerization initiated with perchloric acid at room temperature since they isolated some polyenes late after the end of the reaction. In their system, a very stable peak at 450 nm appeared within 1 hour after complete consumption of styrene and was assigned to corresponding allylic cations. In the light of our results it seems more reasonable to consider that this peak is related to final indanic species and that allylic cations are not stable at room temperature. Moreover, these authors observed a transitory peak at 510 nm which disappeared in less than 50 s but for which no assignement was proposed, and which could better correspond to the allylic species.

Concerning the quantitative and kinetic aspects of the protonation, we observed that initial rate of appearance of the 340 nm peak observed (Figures 10, 12, 13) was slower than for styrene and was expressed by

$$R_i = k_i . [D]_o . [TfOH]_o^{1.26}$$

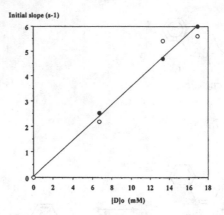

Figure 12. Appearance of 340 nm absorption at –67 °C. First order with respect to monomer (o): $[TfOH]_o$ = 23.6; 22.5; 27.7 mmol.L^{-1} (●) correct. for $[TfOH]_o$ = 25 mmol.L^{-1}.

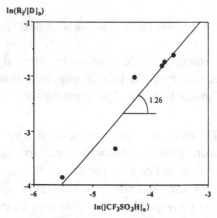

Figure 13. Appearance of 340 nm absorption at –67 °C. Order with respect to acid.

The complex order with respect to initiator indicates an initiation by monomeric and dimeric forms of acid in rapid equilibrium. The concentration of cations at t_{max} never exceeded 5% of $[D]_o$ at –64 °C, which here again prevented any attempt of ε_{340} determination.

At the maximum we found that

$$[D_n^+]_{tmax} \propto [D]_o \cdot [TfOH]_o^{1.38}$$

As at this point $R_i = R_t$ and, if disappearance of cations is unimolecular, we should found the same order with respect to acid as for initiation. The two experimental values are rather close which confirms that the disappearance is monoexponential. This disappearance is not a true termination but combination of the reversible proton transfer to counter-ion and of the irreversible indanic cyclization which is here again the driving force for dimer consumption and oligomerization.

3.2.2. *High Temperature Range Stopped-flow Experiments (T > -30 °C)*

When increasing T above –30 °C the assumed allylic D_i^+ cation becomes less stable. The linked 349 and 505 nm peaks progressively decreased, and two new absorptions appeared at 415 nm (Figure 14) and 316 nm. These two absorptions are similar to the ones observed when recording the UV spectrum of styrene cationic polymerization medium after a long evolution (minutes) at room temperature and, as proposed by Bertoli and Plesch [65], should correspond respectively to monoaryl (I_1^+) and diaryl (I_2^+) indanylium cations resulting from hydride shift from indanic end groups (here from indanic dimer I, **Scheme 5**). The same species might also be formed by protonation of the unsaturated indenic dimer resulting from cyclization of D_i^+ (**Scheme 7**).

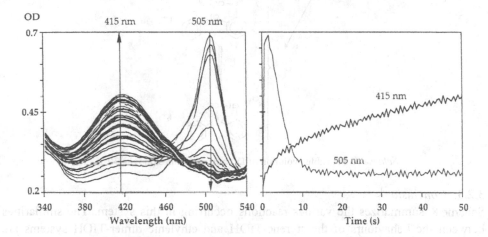

Figure 14. Decrease of 505 nm peak and appearance of 415 nm absorption at 0 °C.

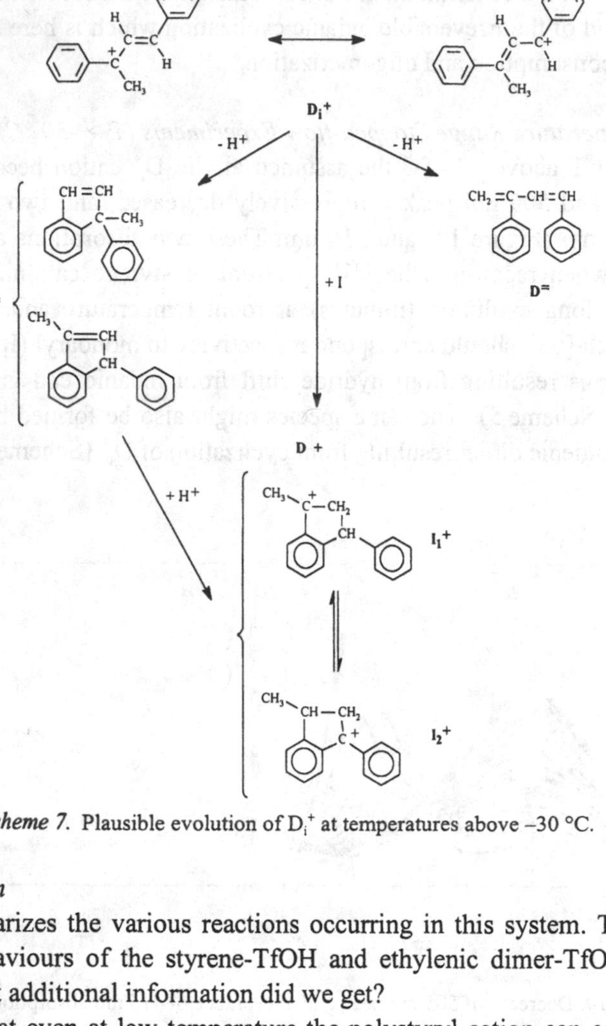

Scheme 7. Plausible evolution of D_i^+ at temperatures above $-30\,°C$.

3.2.3. *Conclusion*

Scheme 8 summarizes the various reactions occurring in this system. The similarities between the behaviours of the styrene-TfOH and ethylenic dimer-TfOH systems are obvious, but what additional information did we get?

- confirmation that even at low temperature the polystyryl cation can add to ethylenic chain ends.
- the 340 nm absorption can definitely be assigned to polymeric PSt$^+$ cation but we still do not have proof that the monomeric St$^+$ absorbs separately.
- the ε of the cation still could not be experimentally determined.
- the ethylenic chain end can be protonated but there is only strong evidence and not proof that this protonation is reversible.
- irreversible indanic cyclization appears again as the driving force of the reaction.
- the kinetic approach led to initiation by monomeric and dimeric forms of triflic acid in equilibrium, which is conflicting with the higher reactivity of large **aggregates we**

observed for styrene protonation. A tentative explanation might be that the hindered dimer molecule is accessible essentially by monomeric acid or smaller aggregates, which would not be the case for the styrene molecule.

- supplementary absorptions at 349 and 505 nm could be assigned to an allylic dimer cation stable at T lower than –30 °C.

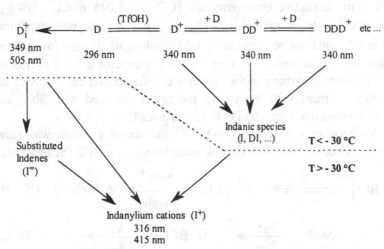

Scheme 8. Overall pathway for reaction of styrene ethylenic dimer with triflic acid.

4. Study of the Ionization of Cumyl Chloride by Some Lewis Acids [70]

We shall finish this presentation with our most recent approach of a system used as an initiator in controlled cationic polymerization (isobutylene, α-methylstyrene,..), the ionization of cumyl chloride by BCl_3. This preliminary study was performed in *polar solvent* (CH_2Cl_2) which is obviously different from conditions for « living » polymerization, and the objectives were to get an estimate of the rate of ionization and of the constant of equilibrium K_I :

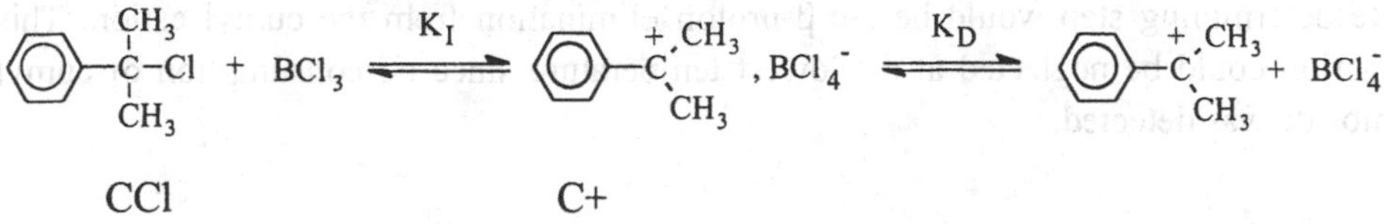

Scheme 9. Ionization / Dissociation of cumyl choride.

We first undertook a ^1H NMR study of the ionization of cumyl chloride (CCl) in CD_2Cl_2.

Chemical shifts have been previously reported for the cumyl cation (C^+, 2-phenyl-2-propylium) formed by the ionization of chloride in super acidic media at -30 °C [71-

73] and by the ionization of CCl by SbF_5 in CD_2Cl_2 at -70 °C [74]. In the latter case, it had been observed that cumyl chloride was almost quantitatively converted into the C^+ cation, the indanic dimer I of α-methylstyrene (1,1,3-trimethyl-3-phenylindan) and some oligomeric species were also identified as by-products.

In this work the ionization of CCl with a large excess of BCl_3 was first studied at -60 °C and -78 °C with respective concentrations: $[CCl]_o = 0.045$ mol.L^{-1}, $[BCl_3]_o = 0.53$ mol.L^{-1} and $[CCl]_o = 0.060$ mol.L^{-1}, $[BCl_3]_o = 2.3$ mol.L^{-1}. For both experiments the cumyl cation C^+ could not be detected by NMR indicating its very low concentration and therefore very weak ionization, even in the presence of a large excess of boron trichloride. At low temperatures in the presence of BCl_3, cumyl chloride was stable for hours and this contrasts with the strong ionization observed with SbF_5 and is in agreement with respective strength of both Lewis acids [75].

When the temperature was raised to $+13$ °C the cumyl chloride was slowly and quantitatively converted into I with simultaneous formation of HCl (**Scheme 10**).

$$CCl + BCl_3 \xrightleftharpoons{K} C^+ BCl_4^- \xrightarrow[\text{slow}]{k_{-H^+}} \alpha\text{-MeSt} + HCl, BCl_3$$

$$C^+ BCl_4^- + \alpha\text{-MeSt} \xrightarrow[\text{fast}]{k_{pl}} D^+ BCl_4^- \xrightarrow[\text{fast}]{k_c} I + HCl, BCl_3$$

Scheme 10. Evolution of cumyl chloride in presence of large excess of BCl_3 at $+13$ °C.

This reaction is slow at 13 °C as illustrated in Figure 15 where $\ln([CCl]_o/[CCl])$ is plotted versus time (80% conversion after 1 hour). Neither α-methylstyrene (α-MeSt) nor the monomeric C^+ and dimeric D^+ (4-methyl-2,4-diphenyl-2-pentylium) cations could be observed during the reaction although formation of dimers implies the existence of all of the three species. Formation of dimers can be explained by a slow deprotonation of the cumyl cation into α-methylstyrene followed by fast dimerization to form the D^+ cation which is rapidly converted into the indanic dimer I (see **Scheme 10**). Assuming that the equilibria of ionization and dissociation are rapidly established, the rate determining step would be the β-proton elimination from the cumyl cation. This reaction could be neglected at the lowest temperature since no consumption of cumyl chloride was detected.

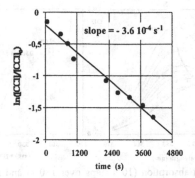

Figure 15. ^{1}H NMR determination of first order β-proton elimination rate constant k_{-H^+}.

$$[CCl] = [CCl]_0 \cdot e^{-2k_{H^+} \cdot t}; \, k_{-H^+} = 1.8 \times 10^{-4} \, s^{-1} \text{ at } +13 \, °C.$$

The stopped-flow study was performed with similar concentration range at –65 °C. Two transient absorptions at 348 and 468 nm appeared very rapidly and with same rate after the shot. These two correlated peaks reached their t_{max} within about 100-300 ms and then disappeared within 500 s (Figure 16) with a first order rate (k = 8.81 x 10^{-3} s^{-1} at –65 °C).

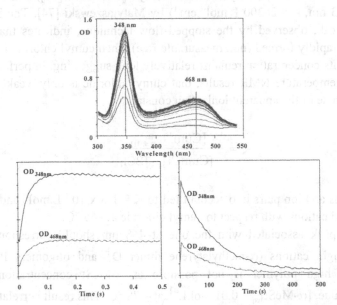

Figure 16. Ionization of cumyl chloride by BF$_3$ in CH$_2$Cl$_2$ at –65 °C. Evolution of 348 and 468 nm absorptions. [CCl]$_0$ = 0.025 mol.L-1 ; [BCl$_3$]$_0$ = 0.56 mol.L^{-1}.

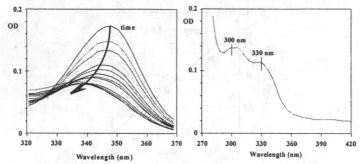

Figure 17. Decrease of 348 nm absorption (10 s steps over 130 s) and after 1000 s remaining stable cumyl cation peak at 330 nm. $[CCl]_o = 0.0113$ mol.L^{-1}; $[BCl_3]_o = 0.53$ mol.L^{-1}; T = -62 °C.

During the decrease of the 348 nm absorption the maximum was progressively shifted towards lower wavelengths, showing that this peak resulted from the superimposition of the absorption of an unstable species (348 nm) and of a stable species (330 nm) which remained after complete disappearance of the main one (Figure 17). The weaker peak at 330 nm was assigned to cumyl cation C^+ as close wavelengths have already been reported for its generation in super acid medium (326 nm, $\varepsilon = 11000$ L.mol^{-1}.cm^{-1}) by Olah [76], by laser flash photolysis (325 nm) by McClelland [77], and by ionization with SbF$_5$ (333 nm, $\varepsilon = 26300$ L.mol^{-1}.cm^{-1}) by Matyjaszewski [74]. The behaviour of the 330 nm peak, observed by the stopped-flow technique, indicates that the cumyl cation is very rapidly formed (non measurable rate) when cumyl chloride is mixed with BCl$_3$ and that its concentration remains relatively low, suggesting, in perfect agreement with the low temperature NMR results, that cumyl chloride is only weakly ionized by BCl$_3$. An estimate of the apparent ionization constant

$$K = \frac{[Cum^+]_{(ip + fi)}}{[CumCl]_o \cdot [BCl_3]_o}$$

taking free ions and ion pairs into account led to $K = 1\text{-}3 \times 10^{-3}$ L.mol^{-1} and to less than 0.1 % of cumyl cations with respect to cumyl chloride at –65 °C.

The 348 nm peak associated with the one at 468 nm should correspond to higher molecular weight cations (α-methylstyrene dimer D^+ and oligomers P$^+$), although formation of high polymer is not possible at α-MeSt concentration below its equilibrium value, $[\alpha\text{-MeSt}]_{eq} = 0.01$ mol.L^{-1} at -78 °C. This result correlates well with the first reported stopped-flow experiments where a peak at 350 nm was detected in the polymerization of α-methylstyrene initiated with perchloric acid in dichloromethane at –80 °C [1]. The evolution of the intensity of the 348 nm peak shows that the concentration of D^+ (and perhaps P$^+$) increases rapidly at the beginning of the reaction and then slowly decreases after reaching the maximum (Figure 16). This D^+ cation results from the reaction of C^+ with α-methylstyrene, and P^+ could be formed by the potential further oligomerization. According to the ^1H NMR results, the monomer cannot be formed in situ by β-proton elimination from cumyl cations since their amount is very low, especially at $-60\ °C$. Thus, D^+ is most probably formed by the reaction of C^+ with the small amount of monomer *initially* present in the medium (< 4 mol-% of $[CCl]_o$).

Owing to reversibility of the ionization reaction, the corresponding Cl-terminated dimer and oligomers of α-methylstyrene (respectively DCl and PCl) should also exist in the medium and be in dynamic equilibrium with the respective carbocations. The complete disappearance of these cations can only be explained by an irreversible reaction which is supposed to be the indanic cyclization since the indanic dimer I of α-methylstyrene (scheme 9) could be observed in the low temperature NMR spectra. After complete consumption of α-methylstyrene and disappearance of the oligomeric cations, the only remaining reaction is the equilibration between CCl / BCl$_3$ and C^+. The cumyl cation C^+ is then the only carbocationic species absorbing in the UV range. According to these experimental observations, a general scheme of the reactions occurring when mixing cumyl chloride and BCl$_3$ in CH$_2$Cl$_2$ at T < -60 °C is proposed (Scheme 11).

$$CCl + BCl_3 \xrightleftharpoons{K_i} C^+BCl_4^- \xrightleftharpoons{K_d} C^+ + BCl_4^-$$

$$C^+ (BCl_4^-) + \alpha\text{-MeSt} \xrightleftharpoons[?]{k_{pl}} D^+ (BCl_4^-) \xrightarrow{k_c} I + HCl + BCl_3$$

$$DCl + BCl_3 \xrightleftharpoons{K'_i} D^+ BCl_4^- \xrightleftharpoons{K'_d} D^+ + BCl_4^-$$

with :

K_i, K'_i : ionization equilibrium constants (for CCl and DCl respectively)

K_d, K'_d: dissociation equilibrium constants (for $C^+BCl_4^-$ and $D^+BCl_4^-$ respectively)

k_c : rate constant of irreversible termination of cations D^+

k_{pl} : rate constants of propagation for cumyl cation C^+

Scheme 11. Reactions of cumyl chloride with BCl$_3$ in CH$_2$Cl$_2$ at T < -60 °C.

Finally we very recently reexamined by the stopped-flow technique the behaviour of the system cumyl chloride - SbF$_5$ in CH$_2$Cl$_2$, for which a previous report indicated a strong ionization [74]. Experiments performed at $-62\ °C$ showed instantaneous (non measurable rate) and complete ionization (Figure 18).

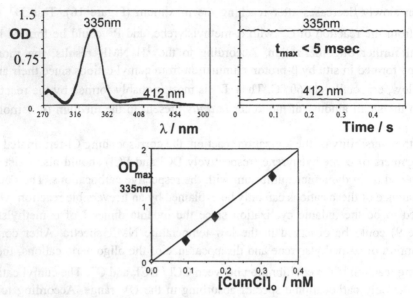

Figure 18. Complete ionization of cumyl chloride by SbF$_5$ in CH$_2$Cl$_2$.
[CCl]$_o$ = 0.34 mmol.L^{-1}; [SbF$_5$]$_o$ = 0.8 mmol.L^{-1}; T = -62°C.

This allowed an accurate determination of the extinction coefficient of the cumyl cation ε_{335} = 24300 L.mol^{-1}.cm^{-1}, confirming the value previously obtained by NMR and the much lower basicity of SbF$_5$ with respect to that of BCl$_3$. The value observed of ε for this α-methyl type styrenic cation appears about 2.5 higher than the value assumed for styryl cation (10^4 L.mol^{-1}.cm^{-1}) and obviously makes it questionable. Thus the concentrations of cations observed during the polymerizations of styrene are probably lower than calculated and the resulting kinetic constants underestimated.

5. General Conclusion

We do not intend to summarize here again the above results and simply hope that we succeeded in demonstrating the power of this stopped-flow technique.

6. References

1. De Sorgo, M., Pepper, D.C., and Szwarc, M. (1973) *J. Chem. Soc., Chem. Com.,* 419.
2. Pepper, D.C. (1974) *Makromol. Chem.* **175**, 1077.
3. Lorimer, J.P. and Pepper, D.C. (1976) *Proc. Roy. Soc. Lond.,* **A351**, 551.
4. Pepper, D.C. (1976) *J. Polym. Sci. Symp.,* **56**, 39.

5. Kunitake, T. and Takarabe, K. (1976) *J. Polym. Sci. Symp.*, **56**, 33.
6. Kunitake, T. and Takarabe, K. (1978) *Polymer J.*, **10**(1), 105.
7. Kunitake, T. and Takarabe, K. (1979) *Macromolecules*, **12**(6), 1061.
8. Kunitake, T. and Takarabe, K. (1979) *Macromolecules*, **12**(6), 1067.
9. Kunitake, T. and Takarabe, K. (1980) *Polymer J.*, **12**(4), 239.
10. Kunitake, T. and Takarabe, K. (1980) *Polymer J.*, **12**(4), 245.
11. Takarabe, K. and Kunitake, T. (1981) *Makromol. Chem.*, **182**, 1587.
12. Sawamoto, M. and Higashimura, T. (1978) *Macromolecules*, **11**(2), 328.
13. Sawamoto, M. and Higashimura, T. (1978) *Macromolecules*, **11**(3), 501.
14. Higashimura, T. and Sawamoto, M. (1978) *Polym. Bull.*, **1**, 11.
15. Sawamoto, M. and Higashimura, T. (1979) *Macromolecules*, **12**(4), 581.
16. Villesange, M., Rives, A., Bunel, C., Vairon, J.P., Froeyen, M., Van Beylen, M., and Persoons, A.
17. (1991) *Makromol. Chem;, Macromol. Symp.*, **47**, 271.
18. Sauvet, G., Vairon, J.P., and Sigwalt, P. (1970) *Bull. Soc. Chim. Fr.*, 4031.
19. Chéradame, H., Mazza, M., Hung, N.A., and Sigwalt, P. (1973) *Eur. Polym. J.*, **9**, 375.
20. Berthelot, M. (1866) *Bull. Soc. Chim. Fr.*, **6**, 294.
21. Throssell, J.J., Sood, S.P., Szwarc, M., and Stannett, V. (1956) *J. Am. Chem. Soc.*, **78**, 1122.
22. Brown, C.P. and Mathieson, A.R. (1957) *J. Chem. Soc.*, 3608.
23. Brown, C.P. and Mathieson, A.R. (1957)*Trans. Faraday Soc.*, **53**, 1033
24. Hayes, M.J. and Pepper, D.C. (1958) *Proc. Chem. Soc.*, 228.
25. Pepper, D.C. and Reilly, P.J. (1961) *Proc.Chem. Soc.*, 460.
26. Burton, R.E. and Pepper, D.C. (1961) *Proc.Roy. Soc. Lond.*, **A263**, 58.
27. Hayes, M.J. and Pepper, D.C. (1961) *Proc.Roy. Soc. Lond.*, **A263**, 63.
28. Pepper, D.C. and Reilly, P.J. (1961) *Proc.Chem. Soc.*, 200.
29. Pepper, D.C. and Reilly, P.J. (1966) *Proc.Roy. Soc. Lond.*, **A291**, 41.
30. Plesch, P.H. (1963) *The Chemistry of Cationic Polymerization*, Pergamon, Oxford.
31. Gandini, A. and Plesch, P.H. (1964) *Proc. Chem. Soc.*, 240.
32. Gandini, A. and Plesch, P.H. (1965) *J. Chem. Soc.*, 4826.
33. Gandini, A. and Plesch, P.H., (1965) *J. Chem. Soc.*, 4765.
34. Chmelir, M. (1973) *Int. Symp. Cat. Polym., Rouen (France)*, C7.
35. Chmelir, M. (1975) *Makromol. Chem.*, **176**, 2099.
36. Chmelir, M. (1976) *J. Polym. Sci. Symp.*, **56**, 311.
37. Chmelir, M., Cardona N., and Schultz G.V. (1977) *Makromol. Chem.*, 178, 169.
38. Cardona-Sütterlin, N. (1978) *Polym. Bull.*, **1**, 149.
39. Cardona-Sütterlin, N. (1978) *Polym. Bull.*, **1**, 308.
40. Cardona-Sütterlin, N. (1978) *Polym. Bull.*, **1**, 361.
41. Masuda, T., Sawamoto, M., and Higashimura, T. (1976) *Makromol. Chem.*, **177**, 2981.
42. Sawamoto, M., Masuda, T., and Higashimura, T. (1976) *Makromol. Chem.*, **177**, 2995.
43. Sawamoto, M., Masuda, T., and Higashimura, T. (1977) *Makromol. Chem.*, **178**, 389.
44. Sawamoto, M., Masuda, T., and Higashimura, T. (1977) *Makromol. Chem.*, **178**, 1497.
45. Grace, J.A. and Symons, M.C.R. (1959) *J. Chem. Soc.*, 958.
46. Inoue, T. and Mima, S. (1957) *Chem. High Polym. Japan*, **14**, 402.
47. Jordan, D.O. and Treloar, F.E. (1961) *J. Chem. Soc.*, 784.
48. Reilly, P.J. (1962) *Ph. D. Thesis*, Dublin.
49. Olah, G.A., Pittman, C.U., Jr., Waack, R., and Doran, M. (1966) *J. Am. Chem. Soc.*, **88**, 1488.
50. Bertoli, V. and Plesch, P.H. (1969) *Spectochimica Acta*, **25A**, 447.
51. Bhattacharyya, D.N., Lee, C.L., Smid, J., and Szwarc, M. (1965) *J. Phys. Chem.*, **69**, 612.
52. Pepper, D.C. (1995) *Macromol. Chem. Phys.*, **196**, 963.
53. Matyjaszewski, K. and Sigwalt, P. (1986) *Makromol. Chem.*, **187**, 2299.
54. Vairon, J.P., Rives, A., and Bunel, C. (1992) *Makromol. Chem., Macromol. Symp.*, **60**, 97.

55. Rives, A. (1994) *Thèse de Doctorat*, Université P. & M. Curie, PARIS.
56. Leborgne, A., Souverain, D., Sauvet, G., and Sigwalt, P. (1980) *Eur. Polym. J.*, **16**, 855.
57. Leborgne, A., Souverain, D., Sauvet, G., and Sigwalt, P. (1980) *Eur. Polym. J.*, **16**, 861..
58. Charleux, B., unpublished results.
59. Williams, F., Hayashi, K., Ueno, K., Hayashi, K., and Okamura, S. (1967) *Trans. Faraday Soc.*, **63**, 1501.
60. Hayashi, K., Hayashi, K., and Okamura, S. (1973) *Polymer J.*, **4**, 426.
61. Roth, M. and Mayr, H. (1996) *Macromolecules*, **29**, 6104.
62. 61 . Vairon, J.P., Charleux, B., and Rives, A. (1996) *Macromol. Symp.*, **107**, 307.
63. Charleux, B., Rives, A., Vairon, J.P., and Matyjaszewski, K. (1996) *Macromolecules*, **29**(18), 5777.
64. Charleux, B., Rives, A., Vairon, J.P., and Matyjaszewski, K. (1998) *Macromolecules*, **31**, 2403.
65. Sawamoto, M., Masuda, T., Nishii, H., and Higashimura, T. (1975) *J. Polym. Sci., Polym. Lett. Ed.*, **13**, 279.
66. Bertoli, V. and Plesch, P.H. (1968) *J.Chem. Soc. B*, 1500.
67. Sawamoto, M. and Higashimura, T. (1981) *Macromolecules*, **14**(3), 467.
68. Risi, J. and Gauvin, G. (1936) *Can. J. Res.*, **B14**, 255.
69. Barton, J.M. and Pepper, D.C. (1964) *J. Chem. Soc.*, 1573.
70. Gandini, A. and Plesch, P.H. (1968) *Eur. Polym. J.*, **4**, 55.
71. Russell, R., Moreau, M., Charleux, B., Vairon, J.P., and Matyjaszewski, K. (1998) *Macromolecules*,
72. **31**(12), 3775.
73. Olah, G.A. (1964) *J. Am. Chem. Soc.*, **86**, 932.
74. Farnum, D.G. (1964) *J. Am. Chem. Soc.*, **86**, 934.
75. Olah, G.A., Porter, R.D., and Kelly, D.P. (1971) *J. Am. Chem. Soc.*, **93**, 464.
76. Matyjaszewski, K. and Sigwalt, P. (1987) *Macromolecules*, **20**, 2679.
77. Satchell, D.P.N. and Satchell, R.S. (1969) *Chem. Rev.*, **69**, 251.
78. Olah, G.A., Pittman, C.U., Waack, R., and Doran, M. (1966) *J. Am. Chem. Soc.*, **88**, 1488.
79. McClelland, R.A., Chan, C., Cozens, F., Modro, A., and Steenken, S. (1991) *Angew. Chem., Int. Ed. Engl.*, **30**, 1337.

PHOTOINITIATED CATIONIC POLYMERIZATION

Y. YAGCI
Istanbul Technical University
Department of Chemistry
Maslak, Istanbul 80626, Turkey

Abstract. This paper describes the latest developments in the photoinitiating systems, based on onium salts, for cationic polymerization. The goal is, furthermore, to show differences as well as parallels between systems for different modes of photoinitiation. Besides their direct activation upon photolysis at appropriate wavelengths, these salts can undergo indirect decomposition to generate reactive cations. The indirect action can be based on electron transfer reactions between onium salt and photochemically generated electron donor radicals, photoexcited sensitizers and electron reach compounds in the photoexcited charge transfer complexes. Photoinitiation with allylic salts in conjunction with free radical sources is also discussed.

1. Introduction

Photoinitiated polymerization is an essential technique for many industrial applications such as uv curing of coatings, printing inks and photoresist technology [1]. In addition to widely used photoinitiated free radical polymerization, the corresponding cationic polymerization has also been of prime interest with regard to its inertness to oxygen and less hazardous properties of the monomers employed. Industrially important epoxides and alkyl vinyl ethers may be polymerized by photochemically generated Brønsted acids and carbocations as depicted below.

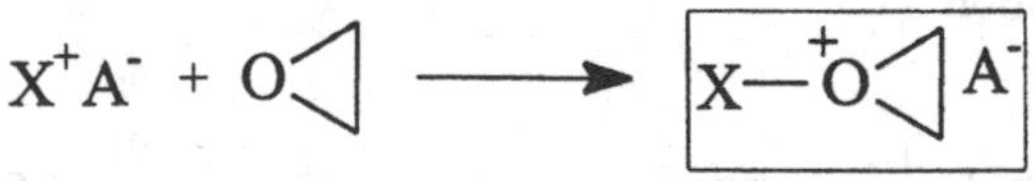

$$X^+A^- + O\triangleleft \longrightarrow \boxed{X-\overset{+}{O}\triangleleft\ A^-} \tag{1}$$

$$\boxed{X-\overset{+}{O}\triangleleft\ A^-} + NO\triangleleft \longrightarrow X\left[\!\!\left[O-CH_2-CH_2\right]\!\!\right]_N \overset{+}{O}\triangleleft\ A^- \tag{2}$$

J.E. Puskas et al. (eds.), Ionic Polymerizations and Related Processes, 205–217.

$$X^+A^- + CH_2=\underset{Y}{CH} \longrightarrow \boxed{X-CH_2-\underset{Y}{\overset{+}{C}H} \ A^-} \qquad (3)$$

$$\boxed{X-CH_2-\underset{Y}{\overset{+}{C}H} \ A^-} + n \ CH_2=\underset{Y}{CH} \longrightarrow X\left[CH_2-\underset{Y}{CH}\right]_n CH_2-\underset{Y}{\overset{+}{C}H} \ A^- \qquad (4)$$

This reaction scheme holds for all types of cationic photoinitiators. Notably, the counter anion has to be nonnucleophilic in order to prevent the termination of a growing chain. In general, for photoinitiated cationic polymerization with onium salts and metal salts, polymerization rate increases in the order of $BF_4^- \ll PF_6^- < AsF_6^- < SbF_6^-$.

What type of photoinitiators is applicable to generate cations depends on the chemical constitution of the system used. Regarding onium salts, which are the most prominent photoinitiators, direct and indirect acting systems can be applied [2]. In direct initiating systems, the energy absorbed by the onium salt and leads to its decomposition. In contrast to this, in indirect systems the energy is absorbed by an additional component. After absorbing energy, either the additives themselves or the species formed from the additives can react with the onium salt thus producing initiating cations. With changing the additives, one can often easily adjust to various wavelengths for specific applications.

The objective of this article is to present the main features of mechanisms involved for the major onium salt type photoinitiating systems namely, iodonium, sulphonium [2] and N-alkoxy pyridinium salts [3,4]. The aim is, furthermore, to show differences as well as parallels between these initiators for direct and indirect action. At the end of the paper, special emphasis will be given to the addition-fragmentation reactions [5] developped in the author's laboratory as a new photoinitiating system for cationic polymerization.

2. Direct Photoinitiation

All of the photoinitiators listed in Table 1 generate initiating species upon photolysis at appropriate wavelengths. In general, radical cations were found to be the primary species and play a key role for the initiation.

$$On^+ \xrightarrow{h\nu} X^{+\cdot} + product \qquad (5)$$

Two distinct mechanisms were proposed. Radical cations are highly reactive towards monomers and can directly initiate the polymerization.

$$X^{+\cdot} + M \longrightarrow \text{polymer} \tag{6}$$

The rate constants of the radical cations of the onium salts investigated are compiled in Table 2.

TABLE 1. Major onium salt photoinitiators and initiating species.

Onium Salt	Absorption maxima (nm)	Representative Compound	Initiating species
Iodonium salts	230-270	Ph_2I^+	$PhI^{+\cdot},\ H^+$
Sulphonium salts	230-300	Ph_3S^+	$Ph_2S^{+\cdot},\ H^+$
Alkoxy pyridinium salts	266-337		, H^+

TABLE 2. Bimolecular rate constants (l mol^{-1} s^{-1}) for the reaction of onium salt radical cations with monomers [3].

Monomer	PhI^+	Ph_2S^+		
Butyl vinylether	5.2 x 10^9	3.2 x 10^9	4.4 x 10^9	5.9 x 10^9
Cyclohexene oxide	7.2 x 10^6	1.3 x 10^5	6.7 x 10^6	6.8 x 105
Tetrahydrofuran	2.4 x 10^5	5.1 x 10^4	1.2 x 10^7	1.4 x 10^6

Alternatively, protons formed as a result of hydrogen abstraction of radical cations can add to monomers.

$$X^{+\cdot} + R\text{-}H \longrightarrow X\overset{+}{-}H + R\cdot \tag{7}$$

$$X\overset{+}{-}H \longrightarrow X + H^+ \tag{8}$$

$$H^+ + M \longrightarrow \text{polymer} \tag{9}$$

Although direct iniation requires less number of components i.e., onium salt is the only light absorber in the system and simple way in generating reactive cations, the spectral response of the simple onium salts are rarely acceptable for their practical application. Thus, it is the aim to find polymerization conditions at longer wavelengths in order to fulfill practical requirements. This may be achieved by chemical incorporation of additional chromophoric groups. But synthetic procedures often involved multistep routes and limited success was obtained. Therefore, indirect initiation turn out to be more technically useful pathway in starting the decomposition of onium salts.

3. Indirect Photoinitiation

Several systems were developed to extend the applicability of the onium salt photoinitiators towards longer wavelengths. In these cases additives are present which participate in the reaction sequences to yield reactive species capable of initiating the cationic polymerization. In the following sections, depending on the role played by the additives in the process, three modes of indirect initiation will be discussed.

3.1. FREE RADICAL PROMOTED CATIONIC POLYMERIZATION

Many photochemically formed radicals can be oxidized by onium salts according to the following reaction:

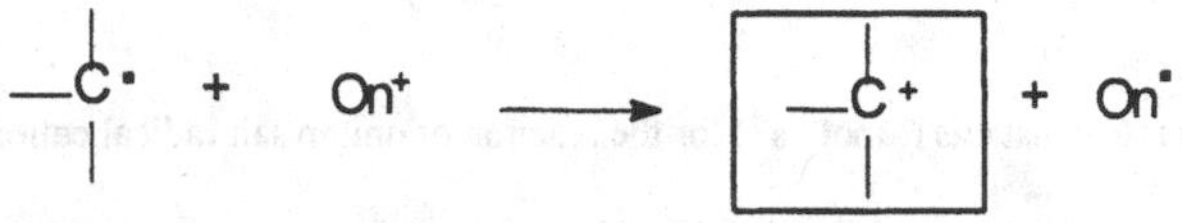

$$-\overset{|}{\underset{|}{C}}\cdot \; + \; On^+ \; \longrightarrow \; \boxed{-\overset{|}{\underset{|}{C}}{}^+} \; + \; On^\cdot \tag{10}$$

The cations thus generated are used as initiating species for cationic polymerization. This process is usually termed as the free radical primed cationic polymerization.

This so-called free radical promoted cationic polymerization is an elegant and fairly flexible type of externally stimulated cationic polymerizations [6,7]. Free radicals may be produced by various modes: photochemically, thermally or by irradiating the system with high energy rays. Suitable radical sources for all modes of stimulation are available. The photochemical generation of radicals can be applied even at low temperatures. In order to exploit the photon energy efficiently, a photolabile compound with an absorbency matching with the emission spectrum of the lamp has to be chosen. Usually, one works at a wavelength's region, where the onium salt itself is transparent. Tab. 3 shows photoinitiators generally used in promoted cationic polymerization and the way, electron donating radicals are generated.

TABLE 3. Initiating systems for free radical promoted cationic polymerization.

Free Radical Source	Onium Salt	Reference
Benzoin Derivatives	Iodonium /pyridinium salts	[8, 9]
Benzophenone	Iodonium /pyridinium salts	[8, 9]
Acylphosphine oxides	Iodonium /pyridinium salts	[10, 11]
Dye-Amine	Iodonium salt	[12]
Vinyl Halides	Sulphonium/pyridinium salts	[13, 14]
o-Phtaldehyde	Pyridinium salts	[15]
Azo compounds	Iodonium /pyridinium salts	[8, 9]
Polysilanes	Iodonium /pyridinium salts	[16, 17]

Being photolized with fairly high quantum yields (0.41 for benzoin [18]), benzoin derivatives are so far the most effective photoinitiators. The photolysis of benzoin salts results in the generation of strong electron donor radicals. The influence of factors, such as light intensity, onium salt concentration or the type of radical source onto the rate of radical induced cationic polymerization with benzoin derivative/onium salt systems has been carefully investigated [19]. In addition to direct generation of electron donating radicals, non-nucleophilic radicals, like $PhCO^{\cdot}$, $(R_1R_2)PO^{\cdot}$ and $Ph^{\cdot}$ formed upon photolysis of the photolabile compound may react with monomer molecules producing electron donating radicals, as described on the example of $PhCO^{\cdot}$ in reactions (11) and (12).

$$\text{(11)}$$

$$\text{(12)}$$

Thus formed, these radicals can easily be oxidized by onium salts yielding initiating species. The efficiency of onium salts as oxidizing agents is related to their electron affinity. The higher the oxidation power of the onium salt, the higher (more positive) is the reduction potential $E_{red}^{1/2}$ (On^{+}).

The efficiency of onium salts in this mode of polymerization rises in the order of trialkyl sulphonium sats ($E_{red}^{1/2}=-1.2V$) < alkoxy pyridinium salts ($E_{red}^{1/2}=-0.7V$) < diaryliodonium salts ($E_{red}^{1/2}=-0.2V$) < aryldiazonium salts ($E_{red}^{1/2}=0.3V$) [20].

Aryldiazonium salts are most suitable for the oxidation of radicals. However, their practical application is hampered by the lack of thermal stability. Diphenyliodonium salts have also a relatively high reduction potential. Being very suitable for the oxidation of free radicals, these salts have been most frequently used for the oxidation of photogenerated free radicals. On the other hand, triphenylsulphonium salts have only limited potential for radical induced cationic polymerizations due to their low reduction potential. However, some highly nucleophilic radicals could be oxidized with sulphonium salts.

Provided the oxidation and reduction potentials of the free radical and the onium ion, respectively, are known, it can be estimated on the bases of the Rehm-Weller equation (32) whether a radical can be oxidized by a given onium salt or not.

$$\Delta G = F\,[E_{ox}^{1/2}\,(R^{\cdot}) - E_{red}^{1/2}\,(On^{+})] \tag{13}$$
F: Faraday constant

However, the calculation of ΔG is usually not feasible since the exact oxidation potentials $E_{ox}^{1/2}\,(R^{\cdot})$ of most radicals involved in radical promoted polymerizations are unknown.

3.2. PHOTOINITIATION BY ELECTRON TRANSFER BETWEEN PHOTOEXCITED MOLECULE AND ONIUM SALT

Many aromatic hydrocarbons are able to sensitize the decomposition of onium salts via electron transfer in an excited complex referred to as exciplex [21, 22]. For this type of cationic initiation, the following general scheme holds:

$$S \xrightarrow{h\nu} S^{*} \xrightarrow{On^{+}X^{-}} \left[S^{*}\cdots On^{+}X^{-}\right] \longrightarrow \boxed{S^{+\bullet}\,X^{-}} + \overset{\bullet}{On} \tag{14}$$

$$S^{+\bullet}\,X^{-} + R{-}H \longrightarrow HS^{+}X^{-} + \overset{\bullet}{} \tag{15}$$

$$HS^{+}X^{-} \longrightarrow \boxed{H^{+}X^{-}} + \tag{16}$$

The excitation of the sensitizer is followed by the formation of a complex between excited sensitizer molecules and ground state onium salt. In this complex, one electron is transferred from the sensitizer to the onium salt giving rise to the generation of sensitizer radical cations. These can by themselves initiate the polymerization of appropriate polymers or, alternatively, interact with hydrogen containing constituents of the polymerization mixture (solvent, monomer) resulting in the release of Brønsted acid. In the case of alkoxy pyridinium salts, an additional mechanism has to be taken into account. Alkoxy radicals, which are generated by the decomposition of alkoxy pyridinium salts, react with sensitizer radical cations yielding initiating sulfur centered cations [23].

$$S^{+ \bullet} \; X^- \; + \; \overset{\bullet}{O}\diagdown\diagup \longrightarrow \boxed{S^+ {\diagdown}O{\diagup}\diagdown} + X^- \qquad (17)$$

The electron transfer (right part in (14)) is energetically allowed, if ΔG calculated by (18) (extended Rehm-Weller equation) is negative.

$$\Delta G = F\,[E_{ox}^{1/2}\,(S) - E_{red}^{1/2}\,(On^+)] - E\,(S^*) \qquad (18)$$
F: Faraday constant
E (S*): Excitation energy of the sensitizer (singlet or triplet)

Since the oxidation potentials of sensitizers, $E_{ox}^{1/2}$ (S), are easy to determine (in contrast to that of radicals), the calculation of ΔG can indeed be applied in order to predict whether or not an oxidation would take place as presented for the example of pyridinium salt in Table 4.

TABLE 4. Parameters related to photoinduced electron transfer reactions of photoexcited sensitizers and pyridinium salt.

Sensitizer	E(S*) (kJmol.L^{-1})	E^{ox}1/2 (V)	ΔG (kJmol^{-1})	λ_{max} , S$^{+\cdot}$ (nm)	Photosensitizatio
Benzophenone	290	2.7	+39.8	-	No
Acetphenone	308	2.9	+41.2	-	No
Thioxanthone	277	1.7	-44.2	430[a]	Yes
Anthracene	319	1.1	-144.4	715	Yes
Perylene	277	0.9	-121.8	535	Yes
Phenothiazine	239	0.6	-112.9	514	Yes

[a] Observed with iodonium salts [22]

However, not all sensitizers are suitable in conjunction with onium salts. According to (18), the requirements are low oxidation potentials, $E_{ox}^{1/2}$(S), and relativley high excitation energies, E (S*), of the sensitizer. Besides that, only onium salts with high (low negative) reduction potentials $E_{red}^{1/2}$ (On$^+$) , such as diphenyliodonium or alkoxy pyridinium salts are easily reduced by the sensitizer.

3.3. PHOTOINITIATION BY CHARGE TRANSFER COMPLEXES

Pyridinium salts are capable of forming ground state CT complexes with electron-rich donors such as methyl- and methoxy-substituted benzene [24]. Notably, these complexes absorb at relatively high wavelengths, where the components are virtually transparent. For example, the complex formed between N-ethoxy-4-cyano pyridinium hexafluorophosphate and 1,2,4-trimethoxybenzene possesses an absorption maximum at 420 nm. The absorption maxima of the two constituents are 270 nm and 265 nm for the pyridinium salt and trimethoxybenzene, respectively. It was found that the CT complexes formed between pyridinium salts and methyl- and methoxy-substituted benzene act as photoinitiators for the cationic polymerization of cyclohexene oxide and 4-vinyl cyclohexene oxide [24]. The following mechanism for the initiation of the cationic polymerization has been suggested:

(19)

Since the proton scavenger 2,6-di-tert-butylpyridine did not noticeably influence the polymerization, the initiation by Brønsted acid that could be formed after an interaction with hydrogen containing components can be excluded. Notably, the CT complexes described above are applicable for the photoinitiation of epoxide monomers but not for the photoinitiation of vinyl ethers and N-vinyl carbazol. The latter monomers are already polymerized in a dark reaction upon addition of these complexes.

It should be noted that iodonium and sulphonium salts either do not form charge transfer complexes with electron donating compounds or the absorption band of the complexes overlap with that of the components.

4. Photoinitiation by Addition-Fragmentation Reactions

The use of addition fragmentation reactions for photoinduced cationic polymerization has been subject of recent investigations in our laboratory [25-29]. Being not based on easily oxidizable radicals, addition fragmentation reactions are indeed a very versatile method to adjust the spectral response of the polymerization mixture with the aid of free radical photoinitiators. The allylic salts that have so far been applied for addition fragmentation type initiations are compiled in Table 5.

The advantage of allylic salts that can undergo addition fragmentation reactions derives from the fact that virtually all sorts of thermal and light-sensitive radical initiators may be utilized for cationic polymerization, which enables an adoption to most initiation conditions. In contrast to radical promoted cationic polymerization based on the oxidation of radicals, one is not limited to oxidizable radicals.

TABLE 5. The allylic salts used in addition fragmentation type initiations.

4.1. ALLYLIC SULPHONIUM SALTS

The allyl sulphonium salt, ETM, was shown to efficiently initiate the cationic polymerization of various monomers in the presence of suitable free radical sources [25]. As can be seen from Table 6, CHO was polymerized quite effectively with benzoin and 2,4,6 trimethyl benzoyl diphenylphosphonyloxide (TMDPO) as free radical sources. Since the sulphonium salt does not absorb the light at $\lambda > 300$ nm, all irradiations were performed at $\lambda_{inc} > 340$ nm. In the absence of free radical sources the photolysis of the sulphonium salt solution in methylene chloride failed to produce any precipitated polymer.

The mechanism involves the production of radicals from the radical initiator and the subsequent addition of these radicals to the allylic double bond. The energy rich intermediates thus produced undergo fragmentation. The resulting sulphinium radical cations may directly initiate the cationic polymerization. Protons generated by hydrogen abstraction may also participate in the initiation process. The overall process is shown in the reaction (20). Support for the initiation process was also obtained by the detection of the unsaturated compound originating from the addition fragmentation steps.

In addition to CHO, (BVE), N-vinyl carbazole (NVC) and 4-vinyl cyclohexene dioxide (4-VCHD) were also examined (Table 7). These monomers also polymerized readily in solutions containing suphonium salt and TMDPO. In the case of 4-VCHD, which possesses two epoxide groups, an insoluble polymer was readily formed indicating the possibility of practical applications of this type of initiation for industrially important UV curing of epoxy coatings.

(20)

TABLE 6. Photoinitiated polymerization of CHO using free radical photoinitiators (PIs) in the presence of the allylic salt (ETM) in methylene chloride[a].

PI	λ_{inc} (nm)	Conversion (%)	M_n[b]	M_w / M_n
TMDPO	380	27.4	9 099	2.08
Benzoin	340	71.1	15 023	1.99

[a]Polymerization time = 30 min., [CHO] = 6.46 mol.L^{-1}, [PI] = 5 x 10^{-3} mol.L^{-1}, [ETM] = 5 x 10^{-3} mol.L^{-1}
[b]Determined by GPC according to polystyrene standards

TABLE 7. Photoinitiated polymerizations of various monomers using TMDPO in the presence of allylic salt (ETM) in methylene chloride.

Monomer	Time (min.)	[M] (mol.L^{-1})	[ETM$^+$]x10^3 (mol.L^{-1})	[TMDPO]x10^3 (mol.L^{-1})	Conversion (%)	M_n[a] g mol^{-1}	M_w / M_n
CHO	30	6.46	5	5	27.4	9 099	2.08
CHO	30	6.46	5	-	0	--	--
BVE	30	2.2	5	5	74.02	13 964	1.79
BVE	30	2.2	5	-	52.4	19 111	2.46
NVC	5	1	5	5	49.07	45 803	2.53
4-VCHD	20	5.2	5	5	100	--	--

[a]Determined by GPC according to polystyrene standards

Besides free radical photoinitiators, certain photosensitizers namely, benzophenenone, anthracene, thioxanthone, perylene or phenothiazine can activate the allylic sulphonium salt (Table 8). As in the case of conventional onium salts, the sensitization action is based on electron transfer from the excited photosensitizer to the allylic salt, resulting in the formation of the radical cation of the sensitizer (reaction 14).

TABLE 8. Sensitized photopolymerization of CHO (8.24 mol.L^{-1}) in methylene chloride containing the allylic salt (ETM) (7 x 10-3 mol.L^{-1}) at λ = 380 nm.

Sensitizer	[Sensitizer] 10^{-3} mol.L^{-1}	Irradiation time (min)	Conversion (%)	M_w[a] g.mol^{-1}
Benzophenone	5	30	94	6 900
Thioxanthone	5	30	4.1	11 500
Anthracene	5	45	6.8	8 900
Perylene	5	30	11	6 800
Phenothiazine	5	30	375	6 589

[b]Determined by GPC according to polystyrene standards

4.2. ALLYLOXY PYRIDINIUM SALTS

Allyloxy-pyridinium salts with various substitiuents at the allylic moiety are shown to be very efficient coinitiators in radical promoted cationic polymerization. Depending upon the radical initiator chosen cationic polymerizations may be initiated by heat or light according to the mechanism depicted below:

$$\text{Radical Source} \xrightarrow{\Delta \text{ or } h\nu} \dot{R} \quad (21)$$

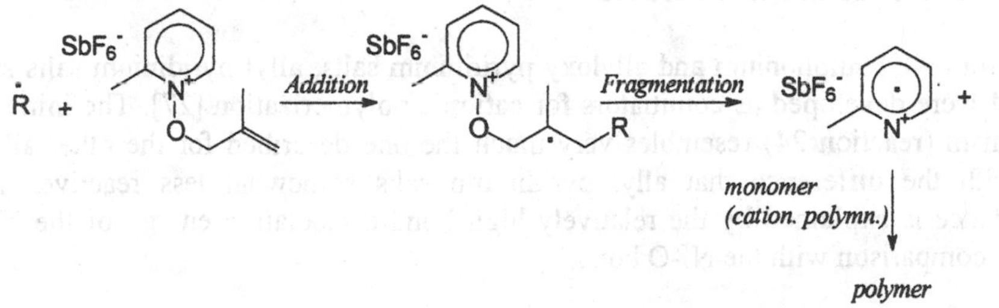

Benzoin has been chosen as photochemical initiator since it absorbs light in the near UV region, thus, tuning the initiating system to wavelength applicaple for practical purposes. The high efficiency in this case is again understood as a proof of the efficiency of addition fragmentation reaction (Figure 1) . Interestingly the polymerization rate for the initiation with EAP is higher than for all other allyloxy pyridinium salts.

The mechanism involving the oxidation of hydroxy benzyl radicals stemming from benzoin by EAP according to the reactions [22-23] probably accounts for the faster initiation in the case of EAP.

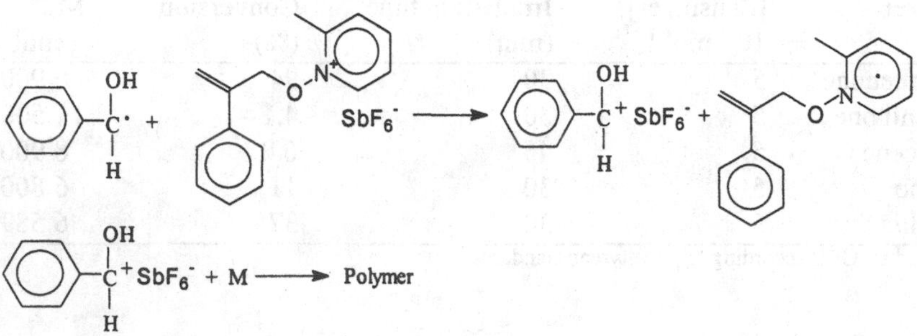

$$(22)$$

$$(23)$$

Figure 1. Photopolymerization of CHO with Benzoin. $\lambda = 367$ nm, 20 °C, O.D.= 0.1, [Pyridinium Salt] = 5 x 10^{-3} mol. L^{-1}, [Benzoin] = 2.4 x 10^{-2} mol. L^{-1}

It is clear that the mechanism of the initiation by allyloxy pyridinium salts follows either just addition fragmentation polymerization or involves both addition fragmentation and oxidation of free radicals, depending on the free radical formed and the pyridinium salt. As far as addition fragmentation is concerned, there is no influence of the substituent giving rise to the conclusion that fragmentation rather than addition is that rate determined.

4.3. ALLYLIC PYRIDINIUM SALTS

Apart from allyl sulphonium and allyloxy pyridinium salts, allyl pyridinium salts such as EPM were developed as coinitiators for cationic polymerization [27]. The initiation mechanism (reaction 24) resembles very much the one described for the other allylic salts with the difference that allyl pyridinium salts somewhat less reactive. This circumtance is explained by the relatively high bond dissociation energy of the N^+-C bond in comparison with the N^+-O bond.

$$(24)$$

4. References

1. Fouassier, J.-P. (1995) *Photoinitiation, Photopolymerization and Photocuring*, Hanser Verl., München
2. Crivello, J.V. (1991) *Chemistry & Technology of UV & EB Formulations for Coatings, Inks & Paints,* (Dietliker, K. Ed.) SITA Technology Ltd., London
3. Yagci, Y. and Schnabel, W. (1994) *Macromol. Symp.*, 85, 115
4. Yagci, Y. and Endo, T. (1997) *Adv. Polym. Sci.*, 127, 59
5. Yagci, Y. and Reetz, I. (1998) *Macromol. Symp.*, in press
6. Yagci, Y. (1989) *J Rad. Curing*, 16, 9
7. Yagci, Y. and Schnabel, W. (1992) *Makromol. Chem., Macromol. Symp.*, 60, 133
8. Yagci, Y. and Ledwith, A. (1988) *J. Polym. Sci., Polym.Chem.Ed.*, 26, 1911
9. Böttcher, A., Hasebe, K., Hizal, G., Yagci, Y., Stellber, P., Schnabel, W., (1991)*Polymer*, 32, 2289
10. Yagci, Y. and Schnabel, W. (1987) *Makromol. Chem., Rapid Commun.*, 8, 209
11. Yagci, Y., Borbely, J., and Schnabel, W. (1989) *Eur. Polym. J.*, 25, 129
12. Bi, Y. and Neckers, D.C. (1994) *Macromolecules*, 27, 3683
13. Johnen, N., Koboyashi, S., Yagci, Y., and Schnabel, W. (1993)*Polym. Bull.*, 30, 279
14. Okan, A., Serhatli, I.E., und Yagci, Y. (1996) *Polym. Bull.*, 37, 723
15. Yagci, Y. and Denizligil, S. (1995) *J. Polym. Sci., Polym. Chem.*, 33, 1461
16. Yagci, Y., Kminek, I., and Schnabel, W. (1992) *Eur. Polym. J.*, 28, 387
17. Yagci, Y., Kminek, L., and W. Schnabel, *Polymer*, 34, 426 (1993)
18. Baumann, H. and Timpe, H.J. (1984) *Z. Chem.*, 24,
19. Timpe, H.J. and Rajendran, A.G. (1991) *Eur. Polym. J.*, 27, 77
20. Yagci, Y. and Schnabel, W. (1992) *Makromol. Chem., Macromol. Symp.*, 60, 133
21. Yagci, Y., Lukac, I., and Schnabel, W. (1993)*Polymer*, 34, 1130
22. Manivannan, G. and Fouassier, J.P. (1991*) J.Polym.Sci., Polym.Chem.Ed.*, 29, 1113
23. Dossow, D., Zhu, Q.Q., Hizal, G., Yagci, Y., and Schnabel W. (1996) *Polymer*, 37, 2821
24. Hizal, G., Yagci, Y., and Schnabel, W. (1996) *Polymer*, 37, 2821
25. Denizligil, S., Yagci, Y., and Mc Ardle, C. *Polymer*, (1995)36, 3093
26. Denizligil, S., Resul, R., Yagci, Y., McArdle, C., and Foussier, J. P. (1996) *Macromol. Chem. & Phys.*, 197, 1233
27. Yagci, Y. and Önen, A. (1996) *J..Polym.Sci., Polym.Chem.Ed.*, 34, 3621
28. Reetz, I., Bacak, V., and Yagci, Y. (1997) *Macromol.Chem. & Phys.*, 198, 19
29. Reetz, I., Bacak, V., and Yagci, Y. (1997) *Polym. Int.*, 43, 27

A FORGOTTEN CLASS OF HIGH T_G THERMOPLASTIC MATERIALS: ANIONIC COPOLYMERS OF STYRENE AND 1,1-DIPHENYLETHYLENE

Synthesis, toughening and block copolymerization with butadiene

K. KNOLL, M. SCHNEIDER AND S. OEPEN

Polymers Laboratory, ZKT - B 1
BASF Aktiengesellschaft, D-67056 Ludwigshafen, Germany

Abstract: Anionic copolymerization of 1,1-diphenylethylene (DPE) and styrene yields random copolymers containing a molar excess of styrene units or alternating polymers. Depending on the DPE content their glass transition temperature can be as high as 180°C. The toughness of the brittle copolymers can be enhanced with grafted rubber particles. Block copolymerization with minor amounts of butadiene yields tough and transparent materials. Thermoplastic elastomers with an enhanced softening point having the block sequence S/DPE-b-Bu-b-S/DPE and S/DPE-b-EB-b-S/DPE are presented.

In order to upgrade polystyrene towards an engineering plastic material like ABS the long-term service temperature needs to be raised above 100 °C. Consequently a stiffening of the polymer chain becomes necessary which can be accomplished by copolymerization of styrene with a variety of different monomers like α-methylstyrene. However, in many cases the influence of the co-monomer content on the polymer glass temperature is not very pronounced. Using our anionic polymerization technology know how, we succeeded to develop a new styrenic copolymer class based on styrene and 1,1-diphenylethylene (DPE) which not only exhibits a higher glass temperature but also an overall advanced performance compared to GPPS without sacrificing the typical polystyrene property profile.

Although these copolymers have, in principle, been known for quite a long time[1-3], they have never been commercialized, probably due to the high monomer price of DPE. In order to obtain a cost-effective monomer supply, we developed a process for the zeolite catalyzed alkylation of benzene with styrene on a pilot plant scale yielding the precursor 1,1-diphenylethane. A process to dehydrogenate 1,1-diphenylethane to DPE in the gas phase has also been worked out.

The copolymerization kinetics have been studied in detail. Since DPE cannot homopolymerize, maximally alternating copolymers are accessible. The addition of DPE to poly(styryllithium) in cyclohexane is generally preferred compared to an addition of styrene monomer. Consequently, an azeotropic temperature exists for each copolymer composition (r_s changes from 0.47/50°C to

J.E. Puskas et al. (eds.), Ionic Polymerizations and Related Processes, 219–221.
© 1999 *Kluwer Academic Publishers. Printed in the Netherlands.*

0.72/70°C; the polymerization rate decreases with DPE concentration) and copolymers with excellent chemical homogeneity along the polymer chain can be synthesized.

The incorporation of additional phenyl rings into the polystyrene backbone results in a stiffened polymer chain. Therefore, depending on the content of 1,1-diphenylethylene the glass temperature of the copolymer can be adjusted between 104 and 180°C, corresponding to polystyrene homopolymer and a strictly alternating copolymer containing 63 wt.% DPE. The glass transition temperature of poly(S/DPE) was found to increase linearly by 1.26°C per weight% of DPE.

S/DPE copolymers with a molecular weight and MWD similar to general-purpose polystyrene (GPPS) were synthesized in a combination of a CSTR with a tube reactor. Compared to GPPS, the heat distortion temperature of this material, as well as the modulus, hardness, thermal stability of melt under air, UV stability, and resistance toward various polar and nonpolar solvents and media is improved with rising DPE content. S/DPE copolymers are about as brittle as GPPS; their toughness depends mainly on the molecular weight. S/DPE copolymers containing up to 15% DPE are miscible with GPPS. Consequently, block copolymers S/DPE-b-S and S/DPE-b-S-b-S/DPE having more than 15% DPE in the S/DPE block are phase-separated. This can be shown by electron microscopy of RuO_4 stained samples. Two glass transitions for the PS phase and Poly(S/DPE) phase are observed.

A major drawback of S/DPE copolymers is their brittleness. As in GPPS, deformation is governed by crazing processes. We have elaborated several concepts for impact modification. One way is to blend random S/DPE polymer with butadiene-rich high-impact polystyrene and a small amount of S/DPE-b-S diblock copolymer as compatibilizer. A more straightforward approach for impact-modified S/DPE with a still higher softening point was found by *in situ* generation of a Bu-b-S/DPE block rubber followed by polymerization of the S/DPE matrix.

Transparent, impact-resistant resins consisting of a S/DPE hard phase have been made by repeated initiator and monomer addition and final coupling. In the first step a long S/DPE block is formed, followed by further addition of butyllithium, DPE and styrene, yielding short S/DPE blocks. The molar ratio of short to long chains is significantly larger than 1. Finally a randomizer and then butadiene are added, resulting in a butadiene block with incorporated residual DPE from the foregoing steps. This mixture, consisting of short, butadiene-rich and long, S/DPE rich diblocks, is coupled with an oligofunctional coupling agent giving on average an unsymmetical star polymer with about 4 arms.

Thermoplastic elastomers (TPE) S/DPE-b-Bu-b-S/DPE having a block ratio of 16/68/16 are advantageously synthesized by coupling of the living diblock copolymers S/DPE-b-Bu-Li. This route is particularly useful for DPE-rich S/DPE blocks, which can be polymerized initially under high-concentration conditions. DPE may even serve as a solvent in the first reaction phase. Styrene is slowly added together with cyclohexane to ensure that virtually all DPE has been consumed at the end of the polymerization. Butadiene blocks with about 40% 1,2-vinyl microstructure are best suited for hydrogenation due to the low crystallinity of the resulting ethene/butene block having a T_g around -60°C. In this case 0.25 vol% THF with respect to cyclohexane was added after the S/DPE polymerization had been completed. In comparison with SBS[4] or SEBS[5] the TPEs produced by this route have improved mechanical properties. The softening point of the TPEs increases with the T_g of the S/DPE block.

SEBS-type TPEs are usually not used as neat materials but often blended with thermoplastic resins to improve toughness. For rubber applications, e.g. two component injection molding, the are compounded with aliphatic oil and small amounts of polypropylene in order to reduce hardness and cold flow, and to optimize processability and last but not least economics. The properties of the compounds and blends usually improve with the molecular weight of the TPE component. Therefore high-molecular-weight TPEs having the structure S/DPE-b-EB-b-S/DPE and block lengths 28,800/122,400/28,800[6] have been synthesized and compounded together with aliphatic oil and polypropylene in a ratio of 100/100/34 to give materials with a continuous PP phase. The resulting elastomers are soft and highly flexible and exhibit elevated softening points compared to SEBS compounds.

References

1. Yuki, H., Hotta, J., Okamoto, Y. and Murahashi, S. (1967), *Bull. Chem. Soc. Jap.*, **40**, 2659

2. Trepka, W. J. (1970), *Polymer Letters*, **8**, 499

3. Asahi, *JP Patent Appln. Disclosure No.* 79613/91, *Appln. No.* 214104/89

4. Kraton D 1102 of Shell

5. Kraton G 1652 of Shell

6. Block structure comparable to Kraton G 1651 of Shell

CONTROL OF ACTIVE CENTERS REACTIVITY IN THE HIGH TEMPERATURE BULK ANIONIC POLYMERIZATION OF HYDROCARBON MONOMERS

A. DEFFIEUX, P. DESBOIS, M. FONTANILLE,
*Laboratoire de Chimie des Polymères Organiques, ENSCPB-CNRS,
UMR 5629, Université Bordeaux 1, BP 108, 33402 Talence-cedex, France*
V. WARZELHAN, S. LÄTSCH, C. SCHADE
BASF AG, Polymer Laboratory, D-67056 Ludwigshafen, Germany

Abstract. The control of the reactivity in the anionic polymerization of styrene at elevated temperature (100 °C) and in the bulk has been achieved by the use of dialkylmagnesium additives to polystyryllithium. The rate of polymerization was found to decrease with increasing dialkylmagnesium/ polystyryllithium ratio ® in the range 0 to 20. In addition, chain initiation by the magnesium derivatives allows keeping low metal/chain ratios (<2). The results are interpreted by the formation of "ate" complexes in dynamic equilibrium.

1. Introduction

A study devoted to the control of the reactivity in the anionic polymerization of hydrocarbon monomers, typically styrene, at elevated temperature and in concentrated monomer conditions, approaching the bulk, has been undertaken.

The living anionic polymerization technique presents several important advantages in comparison to styrene thermal polymerization presently used in industrial processes. Contrarily to the thermal free radical-prepared ones, polystyrenes obtained by a living type anionic mechanism are expected to be free of residual monomer, thermal dimers and trimers. Moreover they should exhibit a higher thermal stability in particular during the processing step since defects along the chains such as head-to-head monomer placement are absent. It would become also possible to benefit of the living polymerization characteristics to elaborate polymers and copolymers with sophisticated chain structures such as telechelics, branched polymers, block copolymers, etc... .

To make the anionic process economically competitive, it is however necessary to avoid polymerization in solution and therefore found conditions which allow high temperature bulk anionic polymerization.

In non polar solvents, which exhibit physical properties close to the bulk hydrocarbon monomers, alkyllithiums are the most commonly used initiators. They yield

223

J.E. Puskas et al. (eds.), Ionic Polymerizations and Related Processes, 223–237.

polystyryllithium (PSLi) species in the form of aggregates in equilibrium with a low fraction of monomeric ion-pairs [1-2] (Scheme 1).

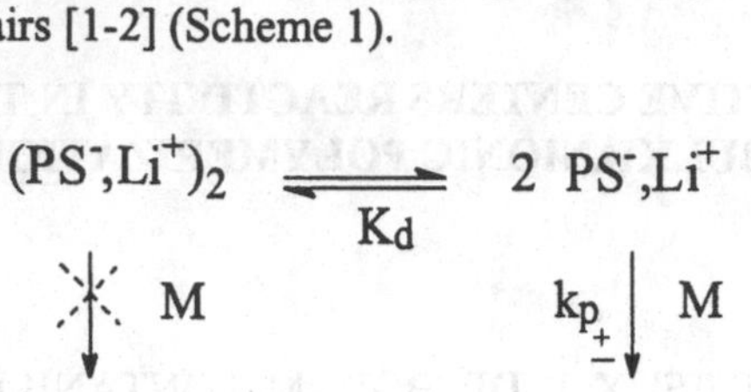

$$(PS^-,Li^+)_2 \underset{K_d}{\rightleftharpoons} 2\ PS^-,Li^+$$

$$\cancel{\ \ }\ M \qquad\qquad k_{p_+}\Bigg\downarrow\ M$$

Scheme 1.

It is assumed that propagation proceeds mainly via monomeric PSLi with a negligible contribution of PSLi dimers, although, finally, each initial lithium species yield the formation of a polystyrene chain through rapid and reversible exchange reactions.

The absence of highly reactive free ions as well as the very small instantaneous fraction of active PSLi explains that, in diluted hydrocarbon solutions, styrene polymerization remains controlled up to relatively high temperature, about 50 to 80°C.

In this temperature range, the bulk styrene polymerization results already in a run-away reaction. The increase in the polymerization rate resulting from the elevated monomer concentration is rapidly worsened by problems of heat removal associated to high viscosity; very high temperatures (> 250°C) can be reached in a few seconds, yielding a deactivation of the carbanionic polymer ends [3-4].

To preserve the livingness of the styrene polymerization, the overall reactivity of the propagating species should be, at first, drastically reduced. The control of the kinetics should permit the control of the polymerization temperature and hence preserve the stability of the propagating species. A decrease of the overall rate of polymerization can be obtained, in principle, by two complementary approaches; reduction of the intrinsic reactivity of propagating active centers, and/or lowering of the fraction of active species by transformation of monomeric polystyryl ends into "dormant" ones.

Several routes, reported in the abundant literature devoted to the styrene and diene anionic polymerizations, can be considered to reach this goal.

The substitution of alkali metal counterions by alkaline-earth ones [5-10] leads to a significant decrease of the styrene and dienes polymerization rates in THF, in relation with the much lower ionic dissociation constants of alkaline-earth systems. In low polar media, however, the polymerization rate with alkaline-earth systems is of the same order of magnitude or even higher than with the lithium one [10], since alkaline-earth species are almost completely in the form of active monomeric ion-pairs. Therefore these systems are not suitable for a control of the reactivity in low polarity media.

The effect of a broad series of additives on the polystyryllithium reactivity has been investigated in detail. Complexing agents, such as ethers [11] and pluri-amines [12-13], significantly reduce the intrinsic reactivity of polystyryl lithium ion-pairs, see Table 1. However, since these additives also lower the portion of inactive aggregated species, the reduction of the overall polymerization rate is only observed at low active centers concentration for which most of the polystyryllithium ion-pairs are monomeric.

TABLE 1. Effect of solvent and/or additive on the absolute rate constant on PSLi contact ion-pairs, kp_+.$(l.mol^{-1}.s^{-1})$ [11-13].

dioxane	benzene			cyclohexane		
pure	pure	2 diox/Li[a)	2 THF/Li[a)	PMDT[b)	TMEDA [b)	HMTT [b)
0.94	15.5	0.4	0.9	0.08	0.15	2.3

a) molar ratio; b) molar ratio additive/Li $\geq$ 1
PMDT: pentamethyldiethylenetriamine; TMEDA: tetramethylethylenediamine; HMTT:
hexamethyltriethylenetetramine

Alkalimetal alkoxides, M_tOR (Li, Na, K) [14-17], alkylmetals ($ZnEt_2$ [18-19], $AlEt_3$ [18,20], MgR_2 [21-22],....) have been used as modifiers for anionic polymerization systems. In particular, the influence of dialkylmagnesium on the styrene and butadiene polymerizations initiated by alkyllithium and alkylsodium (0 < Mg/Li (or Na) < 1) has been investigated by Hsieh and Wang [21] and Liu et al.[22]. In cyclohexane at 50°, n,s-dibutylmagnesium, [n,s-DBuMg], was found to retard the polymerization by a factor 2-3 at the highest Mg/Li ratio examined (r = 0.8). Although n,s-DBuMg is not an active initiator alone, when it is complexed with the propagating polymer-lithium molecules, styrene insertion takes place into a fraction of the Mg-butyl bonds yielding the formation of new polymer chains.

We report in this paper, an investigation on the influence of n,s-DBuMg additive on the styrene anionic polymerization initiated by alkyllithium in hydrocarbon media at temperatures ranging from 50 to 100 °C. The influence of n,s-DBuMg on the kinetics of styrene anionic polymerization, on the polystyrene molar masses and on the UV-visible spectra of polystyryl species is first examined in cyclohexane, at 50 °C, over a broad domain of Mg/Li ratios ranging from 0 to 20. The mechanism of polymerization, the nature and structure of the mixed species formed, as well as their behavior during the high temperature styrene anionic polymerization will then be examined.

2. Experimental

2.1. MATERIALS

Sec-Butyllithium (1.3M in cyclohexane from Aldrich), n,s-dibutylmagnesium ether-free (1.0M in heptane from Aldrich) were used as received. Cyclohexane (RPE. 99.5% from Aldrich) was degassed over freshly crushed CaH_2, stored over polystyryllithium and distilled before use. Styrene (RPE. 99% from Aldrich) was degassed over freshly crushed CaH_2 and distilled over dibutylmagnesium.

2.1.1. UV Visible Spectroscopy

The absorption spectra of the polystyryllithium/n,s-dibutylmagnesium solutions were recorded on a UV-Vis spectrometer Varian-Cary 3E, using a quartz cell (0.01 cm path-length) attached to the glass reactor used, ε_{PSLi} = 13000 mol.dm^3.cm^{-1} at 326 nm, $\varepsilon_{styrene}$ = 450 mol.dm^3.cm^{-1} at 290 nm.

2.2. POLYMERIZATION

They were carried out under dry nitrogen in cyclohexane, at 50°C, in glass flasks equipped with quartz cell and fitted with PTFE stopcocks. Addition of five equivalents of styrene in cyclohexane to sec-BuLi under nitrogen leads to the formation of polystyryllithium seeds ($\overline{DP}_n$ = 5, [PSLi]= 3-7x10^{-3} M). A known amount of n,s-dibutylmagnesium was added to the polystyryllithium seeds solution to obtain the correct ratio Mg/Li (Mg/Li = 0 to 20). After addition of styrene, consumption of the monomer during the polymerization was followed by UV-visible spectrometry.

Investigation of the polystyryl ends stability was performed on living polystyrene prepared in a glass reactor fitted with a UV-Visible cell and thermostated. Measurements were carried out by UV-visible spectroscopy by following the decrease with time of their concentration at the given temperature.

2.2.1. Polymer Characterization

The average molar masses and polydispersity of the resultant polymers were measured by size exclusion chromatography. Measurements were performed using a JASCO HPLC-pump type 880-PU, a Varian apparatus equipped with refractive index/UV detection and 3 TSK Gel columns (HXL 2000, 3000 and 4000) calibrated with polystyrene standards.

The ^{1}H NMR polymer end groups analysis was performed on a Bruker AC 250 using d-chloroform as solvent.

The MALDI-TOF measurements were performed on a BIFLEX III (Bruker-Franzen AnalytiK GmBH) in reflection mode using a dithranol matrix in THF and Silver trifluoroacetate.

3. Results and Discussion

The kinetics of styrene polymerization in presence of PSLi/n,s-DBuMg systems were investigated first in cyclohexane, at 50°C, in a series of experiments performed at approximately constant initial concentrations in monomer and in polystyryllithium. Styrene polymerization does not take place in the presence of n,s-DBuMg alone.

As it may be seen in Figure 1, the increase of the n,s-Bu$_2$Mg proportion with respect to PSLi, from 0 to 20, is accompanied by a strong decrease of the overall reactivity of the system. Nevertheless, polymerization remains quantitative and the semi-logarithmic plots of the styrene concentration versus time give straight lines up to high styrene conversions. Although this is not shown in Figure 1, because of the very low polymerization rate, this linear relationship is verified even for the high Mg/Li ratios, supporting for the whole domain examined the absence of significant chain termination.

The corresponding values of ln(Rp/M) versus increasing r are plotted in Figure 2. As already stressed, a strong and continuous decrease of the styrene polymerization rate is observed in the whole range of r (factor 10^3 and 6.10^3 for r = 4 and 20 respectively).

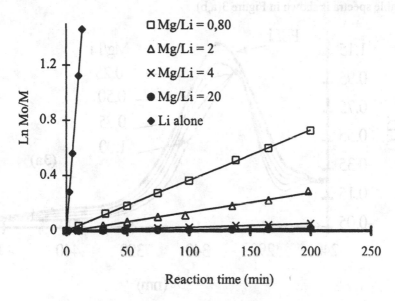

Figure 1. Influence of the ratio *n,s*-DBuMg /PSLi on the kinetics of styrene polymerization; [PSLi] = 3-8x10^{-3}M; [styrene] = 0.3-0.5M; cyclohexane, 50°C.

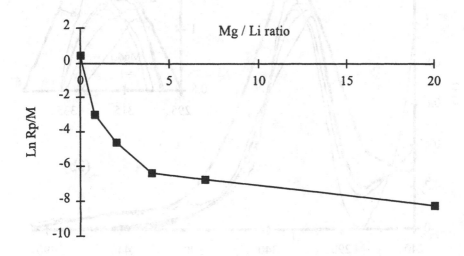

Figure 2. Influence of the ratio *n,s*-DBuMg /PSLi on the styrene propagation rate; [PSLi] = 3-8x10^{-3}M; [styrene] = 0.3-0.5M; cyclohexane, 50°C.

228

This strong effect on kinetics can be related to the formation of mixed «ate» complexes PSLi/*n,s*-DBuMg of various stoichiometry, as shown above by UV-visible spectroscopy. The influence of an increasing proportion of *n,s*-DBuMg on the PSLi UV-visible spectra is shown in Figure 3(a,b).

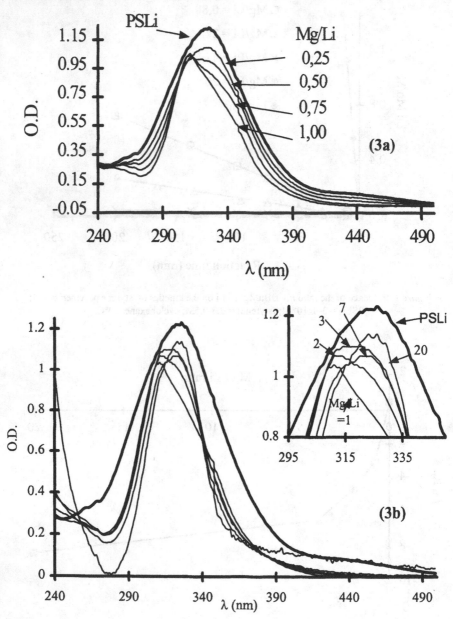

Figure 3. a,b Influence of increasing amounts of *n,s*-DBuMg on the UV-visible absorption spectrum of PSLi; a) Mg/Li = 0-1; b) Mg/Li = 1-20; cyclohexane, 20°C.

Addition of n,s-DBuMg on PSLi in cyclohexane leads almost instantaneously to a decrease of the main PSLi absorption band (λ_{max} = 326 nm). This initial signal decreases at the expense of a new peak located at 310 nm and almost completely vanishes for r equal to 1-1.2. Another peak centered approximately at 325 nm, clearly appears for r > 1. On further addition of n,s-DBuMg, up to r = 20, the 325 nm peak continues to increase to the detriment of the peak located at 310 nm.

These numerous spectral changes suggest that several types of PSLi/n,s-DBuMg «ate» complexes (likely more than two) are formed. The first one (absorption maximum at 310 nm), might correspond to a PSLi:n,s-DBuMg complex with a 1:1 stoichiometry. For higher r values, the spectral evolution suggests the transformation of the 1:1 complex into new ones of different stoichiometry and higher proportion in n,s-DBuMg. A possible complexation pathway is illustrated in Scheme 2.

$$PSLi \ + \ MgBu_2 \ \rightleftharpoons \ PSLi : MgBu_2$$

$(\lambda_{max} = 326 \ nm)$ $(\lambda_{max} = 310 \ nm)$

$$PSLi : MgBu_2 + (n-1) \ MgBu_2 \ \rightleftharpoons \ (PSLi) : (MgBu_2)n$$

$(\lambda_{max} - 310 \ nm)$ $(\lambda_{max} = 325 \ nm)$

Scheme 2.

For high r values, it is likely that the « ate» complexes are in equilibrium with magnesium aggregates free of lithium.

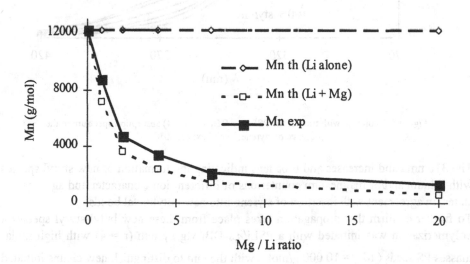

Figure 4. Influence of the ratio n,s-DBuMg /PSLi on the polystyrene experimental molar masses; polymerization in cyclohexane, at 50°C.

For the different r ratios investigated, the $\overline{M}_n$ of polystyrene increases linearly with conversion and distribution remains narrow. However, the experimental $\overline{M}_n$ do not match the theoretical values calculated assuming initiation only by lithium species. $\overline{M}_n$ continuously decreases with increasing r indicating beyond initiation by PSLi seeds a second mechanism of chain formation The variation of the experimental polystyrene molar masses with an increasing portion of n,s-DBuMg in the initiating system is presented in Figure 4.

The contribution of n,s-DBuMg to the formation of new polystyrene chains (active or dormant) is further supported by the signal increase of polystyryl chromophores, observed by UV-visible spectroscopy, upon addition of styrene on the PSLi/n,s-DBuMg system, Figure 5 (r = 4).

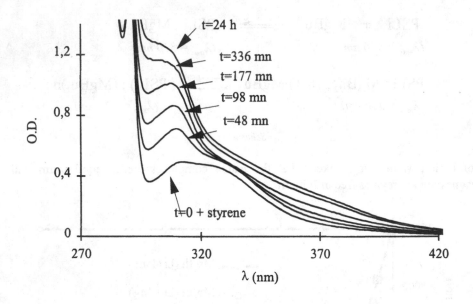

Figure 5. Evolution with time of the PSLi/n,s-DBuMg (r = 4) absorption spectrum in the presence of styrene; cyclohexane; 50°C.

The 310 nm band increases and broadens indicating the formation of new styryl species with lithium and/or magnesium counterions of different ionic character and aggregation states, in agreement with insertion of styrene into new butylmetal bonds.

To further confirm that propagation takes place from these new butyl-styryl species a polymerization was initiated with a PSLi/n,s-DBuMg system (r = 4) with high molar masses PS seeds ($\overline{M}_n$ = 10 000 g/mole) with the aim to distinguish new chains initiated from butyl groups of the magnesium compound. As it may be seen in Figure 6, a bimodal distribution is observed in agreement with the simultaneous growth of PS chains from PS seeds ($\overline{M}_n$ = 13 100 g/mole) and from butyl groups ($\overline{M}_n$ = 3 400 g/mole), each one exhibiting a narrow molar mass distribution.

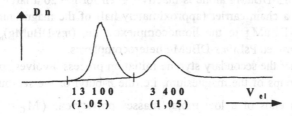

Figure 6. Bimodal GPC trace of polystyrene obtained in the presence of the PSLi ($\overline{M}_n$ = 10 000 g/mole) /*n,s*-DBuMg system (r = 4); cyclohexane; 50°C.

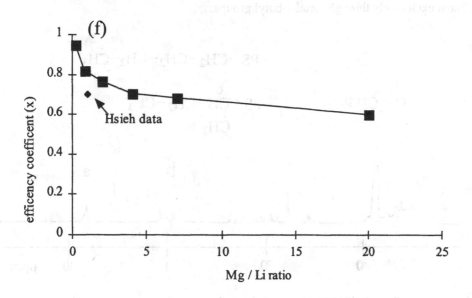

Figure 7. Influence of the Mg/Li ratio on the efficiency factor of *n,s*-DBuMg toward styrene initiation; styrene polymerization in cyclohexane at 50°C.

In experiments initiated with low molar masses polystyrene seeds ($\overline{M}_n$ = 500-1000) or s-butyllithium and *n,s*-DBuMg polystyrene with monomodal narrow distribution are obtained. The experimental polystyrene molar masses are in agreement with the formation of one PS chain by lithium derivative and 0.8 to 0.6 (f) by magnesium species, i.e.

$$\overline{M}_n \ (GPC) = [S]/ \ ([RLi]+ f[n,s\text{-}DBuMg]).$$

The efficiency (f) with respect to magnesium slightly decreases for increasing values of r, see Figure 7.

However, although n,s-DBuMg alone is inactive, even for r = 20 a large proportion of n,s-DBuMg acts as a chain carrier (approximately half of the magnesium compound) suggesting that n,s-DBuMg in the homocomplexes, i.e, $(n,s$-DBuMg)$_n$ also rapidly exchanges with the mixed PsLi/n,s-DBuMg heterocomplexes.

To determine whether the secondary styrene initiation process involves specifically one of the two butyl groups of the magnesium, i.e. the n-butyl or the sec-butyl, or is non selective, the chain ends of a low molar masses polystyrene ($\overline{M}_n$ = 1500 g/mole) obtained from a PsLi/n,s-DBuMg system (r = 10) were analyzed by [1]^{13}C NMR (Figure 8). The observed signals at δ = 11,5 ppm (a), 19 ppm (b) and 32 ppm (c) can be assigned respectively to the two -CH$_3$ and the =CH- of s-butyl ends. The absence of any detectable resonance at 14 ppm (a'), expected for the methyl of primary butyl groups, indicates that styrene insertion into PSLi/n,s-DBuMg is highly selective and proceeds almost exclusively through metal-s-butyl groups.

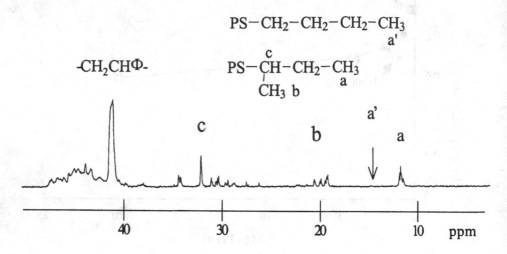

Figure 8. ^{13}C NMR spectrum of a low molar mass polystyrene prepared with PSLi/n,s-DBuMg system (r = 10).

The strong reactivity difference between secondary and primary alkyl groups in mixed complexes towards styrene insertion is further demonstrated by the MALDI-TOF analysis of polystyrenes prepared with the n-hexylLi/n,s-DBuMg initiating system (r = 1). The polystyrene spectrum is shown in Figure 9. Most of the chains (>90%) contains a butyl endgroup, likely the s-butyl one, whereas the presence of hexyl groups of n-hexyllithium is less than 10%. This shows that the nature of the alkyl group is predominant over its initial location on the lithium or the magnesium. The alkyl structure determines the possibility of first styrene insertion and the consecutive formation of new PS chains. These results also stress the role of ligand exchanges between the two metals.

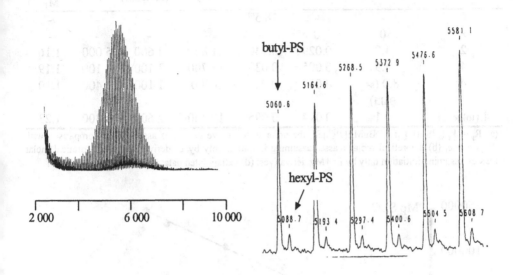

Figure 9. MALDI-TOF spectrum of a polystyrene prepared with n-hexylLi/n,s-DBM(1:1).

The influence of higher temperatures on the kinetics of styrene polymerization and on the polymer characteristics have been investigated.

At 100 °C, in cyclohexane n,s-DBuMg alone is still inactive and no significant traces of polystyrene is detected after several days. As shown from kinetic data collected in Table 2 the reactivity is still drastically lowered by increasing the n,s-DBuMg /Li ratio. The k_p value for the RLi system, given as reference, is extrapolated from data obtained at low temperature using the Arrhenius equation (E_a = 50KJ/mol; A = 1.83x10^8). For a ratio Mg/Li equal to 4, the polymerization reactivity is decreased by a factor 600 compared to the PSLi system. The polystyrene molar masses are also in agreement with an initiation involving both the lithium derivative and n,s-DBuMg. At 100°C the efficiency (f) of the magnesium derivative (r = 4) get close to one whereas the polystyrene molar masses (M_p, SEC peak molecular weight) increases almost linearly with conversion. However, a broadening of MWD is noticed with samples of increasing molar masses (from 1.2 to 1.9) suggesting that the living character of the process is not completely preserved. This could be attributed to a slow contribution of primary alkyl groups of DBuMg during the polymerization.

It is worthy to note that the substitution of cyclohexane by toluene yield an increase of k_p of a factor 3 but do not substantially modify the kinetic profile of the polymerization.

TABLE 2. Anionic styrene polymerization in the presence of PsLi/n,s-DBuMg in cyclohexane at 100°C.

n,s-DBuMg / Li	[PSLi] 10^3.M	Rp/[M] mn$^{-[1]}$	kp$_{app}$$^{a)}$ $M^{0.5}$mn$^{-[1]}$	$\overline{M}_n$ th$^{b)}$ (Li)	$\overline{M}_n$ th$^{c)}$ (Li+DBM)	$\overline{M}_n$ GPC	$\dfrac{\overline{M}_W}{\overline{M}_n}$
0	-	-	18.5$^{d)}$	-	-	-	-
∞	0	0	-	-	-	-	-
2	4.8	0.029	0.240	11 800	4 600	5 000	1.16
4	4.9	0.005	0.033	11 700	3 100	3 100	1.19
4	8.0 (s-BuLi)	0.004	0.022	5 200	1 100	2 400	1.10
4 (toluene)	4.1	0.014	0.095	10 400	2 600	5 300	1.23

(a) $R_p = k_p.([PSLi]+[\,n,s\text{-DBuMg}])^{[1]/2}.[S]$; the order ½ for active centers is assumed for comparison with PSLi alone; (b) theoretical molar masses assuming initiation only by Li derivatives; (c) theoretical molar masses assuming initiation only by Li +Mg derivatives; (d) extrapolated data; $[n,s\text{-DBuMg}] = 8 \times 10^{-3}$M

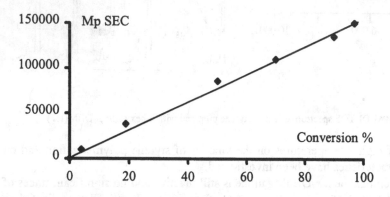

Figure 10. Evolution of SEC peak molar masses with conversion in the styrene polymerization initiated with PSLi/ n,s-DBuMg (r = 5), at 100°C in cyclohexane. [Styrene]$_0$= 3,6 mol.L^{-1}.

Similar study in decaline at 150 °C shows the limit of stability of these mixed systems. In these conditions a rapid evolution (within 30mn) of the active polymer solution from yellow to red and dark is observed indicating that isomerization of styryl ends is taking place. Polymerizations still go to completion but yield colored polymers with broad distribution (>2).

The stabilizing effect of n,s-DBuMg on polystyryllithium species was investigated by UV-visible spectroscopy.

The mechanism of living polystyryl species degradation at high temperature has been investigated by several authors [3,4,24,25]. Two distinct reactions take place as illustrated in Scheme 3: a spontaneous deactivation of the chain terminus (a), yielding metal hydride and an unsaturated polystyrene chain, and a bimolecular process (b) involving the previously formed unsaturated chain end and a new polystyryl chain yielding an isomerized polystyryl terminus. This latter reaction can be followed by monitoring the increase of the absorption band of the new carbanionic species formed which absorbs at high wavelength (λ_{max} = 440nm). Indeed in the case of the

DBuMg/PSLi systems the first deactivation path is extremely slow whereas the second reaction is not observed even at high degradation rate (50%) and therefore can be neglected.

(a)

$RCH_2CH\Phi CH_2 \bar{C} H\Phi, Met^+ \rightarrow RCH_2CH\Phi CH=CH\Phi + MetH$

$\lambda max = 326$ nm

(b) Met^+

$RCH_2CH\Phi CH=CH\Phi + RCH_2 \bar{C} H\Phi Met^+ \rightarrow RCH_2 \bar{C} \Phi CH=CH\Phi + RCH_2CH_2\Phi$

$\lambda max = 440$ nm (Li^+)

Scheme 3.

TABLE 3. Influence of n,s-DBuMg on the stability of polystyryl species at 100 °C.

nature of end groups [a]	[PSLi] 10^3 M	$t_{1/2}$ [b] (H)	R_t [c] min$^{-[1]}$	R_p/M [d] min$^{-[1]}$	k_p / k_t
PSLi (50°C)	6.2	166	$1.2.10^{-5}$	1.56	135 000
PSLi	6.2	3	$6.4.10^{-4}$	18.6	29 000
DBuMg/PSLi (r = 2)	7.2	102	1.9×10^{-5}	0.241	13 000
DBuMg/PSLi (r = 4)	6.2	290	6.6×10^{-6}	0.033	5 000

(a) PSLi, $\overline{DP}n$ = 50-100
(b) half life time of polystyryl species
(c) R_t : assumed deactivation rate $R_t = -d[PS^-]/dt = k_t.[PS^-]^\alpha$
(d) R_p : assumed propagation rate $R_p = -d[PS^-]/dt = k_p.[M][PS^-]^\alpha$

The rate of deactivation of polystyryl ends in cyclohexane in the case of lithium and RLi/n,s-DBuMg systems was determined by monitoring the intensity decrease of the absorption band of the polystyryl end groups versus time. Only a very limited decrease of the UV signal is observed over a period of several days. Data for different r values and various temperatures are collected in Table 3.

The corresponding propagation rate is also given for comparison. As it may be seen a drastic reduction of R_t is observed both at 50°C and 100°C. The rate of deactivation is also drastically lowered by increasing the ratio n,s-DBuMg/Li at a given temperature. If we assume that the propagation and the spontaneous termination reactions involve the same active polystyryl species and that the contribution of « dormant » ones may be neglected in the two elementary proceesses, a kp/kt value can be calculated. Despite the important stabilization of polystyryl species observed in presence of n,s-DBuMg, the more drastic effect of the magnesium derivative on the propagation rate finally results in an important decrease of kp/kt with increasing r. This is in apparent contradiction with the better control of the high temperature styrene polymerization in the presence of n,s-DBuMg/PSLi than with PSLi systems alone. This could indicate that this is the capacity to regulate the inside polymerization temperature, by lowering the overall reactivity of the system in presence of DBuMg, which is the crucial parameter in these controlled n,s-DBuMg/Ps Li anionic polymerization reactions.

In conclusion, n,s-DBuMg used as an additive in the styrene anionic polymerization initiated by lithium derivatives, allows a drastic reduction of the reactivity of propagating active species in hydrocarbon media. The rate decrease depends upon the initial ratio n,s-DBuMg/lithium species. The living character of the polymerization, especially the control of molar masses is preserved over the entire range Mg/Li studied up to elevated temperatures. The kinetics of polymerization and polymer molar masses obtained are consistent with the formation of « ate » complexes PSLi/n,s-DBuMg of various stoichiometry in which rapid and reversible ligand exchanges between lithium and magnesium species take place.

Polymerization may proceed either directly via the mixed Li/Mg aggregates which may exhibit, according to their structure, different intrinsic reactivities towards styrene or, via a rapid pre-equilibrium between lithium derivatives and n,s-DBuMg yielding a drastic reduction of the concentration of non aggregated, reactive polystyryllithium ion-pairs. (Scheme 4 for illustration).

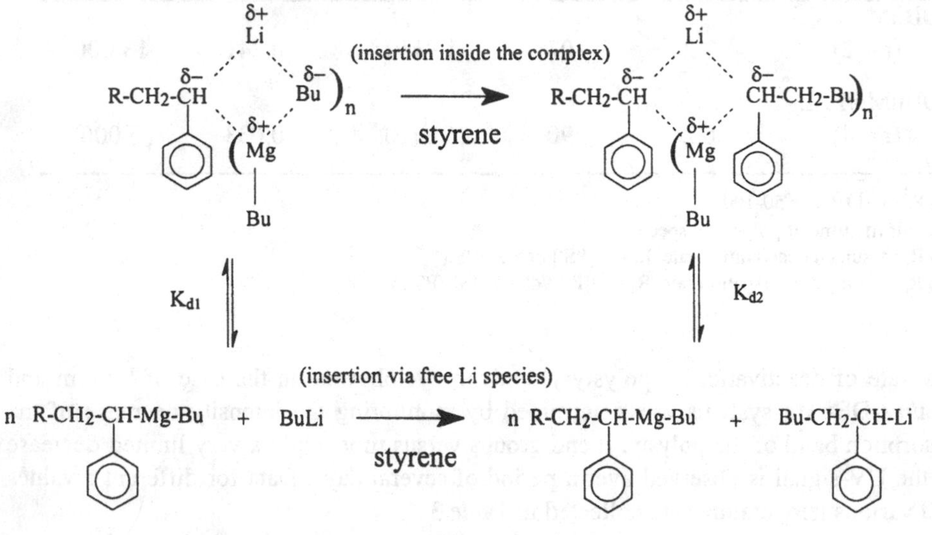

Scheme 4.

The elementary mechanisms involved in dialkylmagnesium/poly(styryl)lithium retarded anionic polymerization of hydrocarbon monomers as well as the possibility to use these initiating systems at higher polymerization temperatures and more concentrated monomer solutions are under investigation.

Acknowledgements. We would like to thank W. Schrepp (BASF) for running the MALDI-TOF experiments.

4. References

1. Szwarc, M. and Van Beylen, M. (1993) 'Ionic Polymerization and Living Polymers', Chapman and Hall, New York, London, ; Hsieh, H.L. and Quirk, R.P. (1996), 'Anionic Polymerization principles and practical applications', M. Dekker, Inc, New York
2. Bywater, S. (1994), *Prog. Polym. Sci.* **19**, 287
3. Kern, W.J., Anderson, J.N., Adams, H.E., Bouton, T.C., and Bethea, T.W. (1972) *J. Appl. Polym. Sc.* **16**, 3123 ; Glasse, H.D. (1983) *Prog. Polym. Sci.* **9**, 133
4. Michaeli, W., Hocker, H., Berghaus, U., and Frings W. (1993), *J. Appl. Polym. Sc.* **48**, 871
5. Mathis, C. and François, B.J. (1978), *Polym. Sci. Chem. Ed.* **16**, 1297.
6. Van Beylen, M., Jappens-Loosen, P., and Huyskens P. (1993) *Makromol. Chem.* **194**, 2949
7. De Groof, B., Van Beylen, M., and Szwarc, M. (1977) *Macromolecules* **10**, 598
8. De Groof, B., Van Beylen, M., and Szwarc, M. (1975), *Macromolecules* **8**, 396
9. De Smedt, C. and Van Beylen, M. (1989) *Makromol. Chem.* **190**, 653
10. Arest-Yakubovich, A.A. (1994) *Chem. Reviews* **19** (4), 1
11. Bywater, S., (1968) *J. Pol. Sci.* **A1**(6), 3407
12. Fontanille, M. and Hélary, G. (1977) *Eur. Pol. J.* **14**, 345 ; Fontanille, M., Adès, D., Léonard, J., and Thomas, M. (1983) *Eur. Pol. J.* **19**, 305
13. Fontanille, M., Szwarc, M., and Hélary, G. (1988) *Macromolecules* **21**, 1532.
14. Roovers, J.E. and Bywater, S. (1966) *Trans. Faraday. Soc* **62**, 1876.
15. Cazzaniga, L. and Cohen, R.E. (1989) *Macromolecules* **22**, 4125
16. Ndebeka, G., Caubère, P., Raynal, S., and Lécolier, S. (1981) *Polymer* **22**, 347
17. Ndebeka, G., Caubère, P., Raynal, S., and Lécolier, S. (1981) *Polymer* **22**, 356
18. Welch, F. J. (1960) *J. Am. Chem. Soc.* **82**, 6000
19. Hsieh, H. L.(1976) *J. Pol. Sci.* **A1**(4), 379
20. Arest-Yakubovich A.A. (1994) *Macromol. Symp.* **85**, 279
21. Hsieh, H.L. and Wang, I.W. (1986) *Macromolecules* **19**, 299
22. Liu, M., Kamiensky, C., Morton, M., and Fetters, L. J. (1986) *J. Macromol. Sci.* **A23**, 1387
23. Kaminski, C. and Elroy B. J. (1986) *Organomet. Synth.* **3**, 395
24. Spach, G. J. (1962) *Chem. Soc.*, 351
25. Schué, F., Nicol, P., and Azuai, R. (1993) *Makromol. Chem., Macromol. Symp.* **67**, 213

ARBORESCENT POLYMERS: HIGHLY BRANCHED HOMO- AND COPOLYMERS WITH UNUSUAL PROPERTIES

M. GAUTHIER

Institute for Polymer Research, Department of Chemistry, University of Waterloo, Waterloo, ON, N2L 3G1, Canada

Abstract. High molecular weight dendritic polymers were synthesized using anionic polymerization and grafting. Aborescent polystyrenes with branching functionalities varying from 5E4-5E8, weight average molecular weights of 14-22 000 and low polydispersity indices (PDI less than 1.2) were obtained. Unusual physical properties, consistent with a rigid sphere morphology, resulted from the very compact structure of these materials. The grafting technique was extended to the preparation of amphiphilic copolymers incorporating an aborescent polystyrene core with end linked poly(ethylene oxide) segments by a *grafting from* scheme. Copolymers with a high polyisoprene content were also obtained by a *grafting onto* scheme.

1. Introduction

Dendritic polymers are a class of macromolecules with a highly branched structure. Two main types of dendritic molecules can be identified, based on their structure and the synthetic methods used to obtain them [1]. *Dendrimers* are generally prepared from protection-condensation-deprotection reaction sequences of AB_n monomers, starting from a tri- or tetrafunctional core molecule. The strictly controlled branching process used to prepare these compounds leads to a well-defined structure and a very narrow molecular weight distribution (MWD). Since small molecule building blocks are used in the reactions, the rate of increase in molecular weight is generally low, and many reaction cycles (generations) are necessary to attain relatively high molecular weights. Steric effects, significant for higher generation dendrimers, can lead to structural defects affecting the topology of the molecules. *Hyperbranched* polymers are dendritic molecules obtained from controlled, single-pot condensation reactions of AB_n monomers. This approach enables the easy synthesis of high molecular weight randomly branched polymers. Control over the branching process is much more limited under these conditions, however, and the branching densities attained are typically much lower than for dendrimers. Because these syntheses rely on random condensation reactions, the MWD of the products generally corresponds to the Flory most probable distribution, with a polydispersity index (PDI = M_w/M_n) of *ca.* 2.

A different approach, based on successive grafting reactions of *polymeric* building blocks, was introduced to solve some of the problems encountered in dendrimer and

239

J.E. Puskas et al. (eds.), Ionic Polymerizations and Related Processes, 239–257.

hyperbranched polymer syntheses [2]. The generic scheme used in the preparation of *arborescent polymers* is shown in Figure 1. The so-called 'graft-on-graft' method relies on anionic polymerization and grafting to prepare well-defined, high molecular weight dendritic polymers in a few steps. A linear core polymer with a narrow MWD is randomly functionalized with grafting sites, and coupled with "living" macroanions. The comb (generation G0) polymer obtained in high yield (> 95%) is purified by fractionation, functionalized and reacted further with macroanions to produce subsequent generations G1, G2, *etc.*

A few requirements need to be satisfied to obtain well-defined graft polymers by this approach. The monomer must yield macroanions with good "living" character, and ideally with a narrow MWD. The side chains grafted in each generation must further allow the introduction of chemical functionalities suitable for anionic grafting. The functionalization reaction used to introduce the grafting sites is required to be free of side reactions leading to intermolecular cross-linking. Finally, the yield of the grafting reaction should ideally be high, to simplify sample purification by fractionation if the removal of non-grafted material from the product is required.

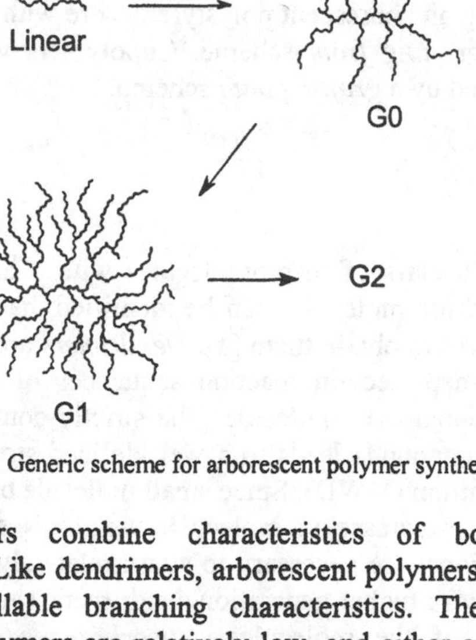

Figure 1. Generic scheme for arborescent polymer synthesis.

Arborescent polymers combine characteristics of both dendrimers and hyperbranched polymers. Like dendrimers, arborescent polymers have a narrow MWD and well-defined, controllable branching characteristics. The branching *densities* attained in arborescent polymers are relatively low, and either comparable to or lower than in the hyperbranched systems. The utilization of polymeric chains rather than small molecules as building blocks has major advantages. The increase in molecular weight and branching functionality per generation is typically 10- to 15-fold for arborescent polymers, as compared to 2- or 3-fold for dendrimers. An additional advantage of the 'graft-on-graft' method is that the branching density and the side chain molecular weight can be varied independently for each generation, providing further control over the structure of the molecules. For the simplest case of an arborescent polymer synthesis using branches with an identical molecular weight M_b for each generation and a constant grafting site density, the molecular weight of a generation G polymer is given by the equation

$$M = M_b + M_b n + M_b n^2 + \ldots = \sum_{x=0}^{G+1} M_b n^x \qquad (1)$$

where n represents the average number of grafting sites introduced per side chain in the functionalization reaction. An exponential increase in molecular weight (and branching functionality) is thus expected for successive generations, if all grafting sites are consumed.

2. Arborescent Polystyrenes

2.1. SYNTHESIS

Styrene is one of the monomers satisfying the requirements stated above regarding good "living" characteristics and ease of functionalization. Furthermore, polystyrene molecules with different structures (star-branched, comb-branched, etc.) have been extensively investigated, enabling direct comparison with the physical properties of arborescent polystyrenes. The grafting reaction used for the synthesis of arborescent polystyrenes is depicted in Scheme 1.

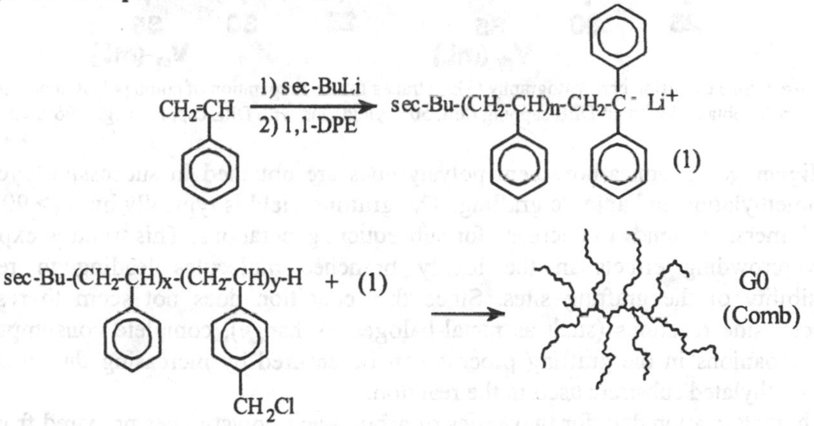

Scheme 1. Arborescent polystyrene synthesis

The synthesis of comb-branched polymers by grafting polystyryllithium onto a partially chloromethylated polystyrene backbone was reported in the 1960's [3]. One of the main problems encountered in these reactions was their low grafting yield (typically ca. 50%). The low grafting efficiency was linked to the occurrence of a metal-halogen exchange reaction competing with the coupling process. It was recognized in the work on arborescent polystyrene synthesis [2] that this problem could be virtually eliminated, and the coupling efficiency increased to 96% by "capping" the polystyryllithium chains with a single 1,1-diphenylethylene (DPE) unit prior to the grafting reaction (Figure 2). Another advantage of the DPE addition is that it facilitates monitoring the stoichiometry of the grafting reaction: The dark red macroanion solution can be simply titrated with the chloromethylated polymer solution until the color fades. Non-grafted material is easily removed from the raw product by precipitation fractionation from toluene-methanol mixtures.

An additional problem in the graft polymer synthesis is that chloromethylation with chloromethyl methyl ether and a Lewis acid is prone to cross-linking reactions by methylene bridge formation [4]. This leads to MWD broadening in the functionalization reaction, and consequently also for the grafting product. *Intermolecular* cross-linking reactions can be eliminated in the arborescent polymer synthesis by chloromethylation under dilute solution conditions, with the selection of an appropriate solvent (CCl$_4$), a relatively mild Lewis acid (either SnCl$_4$ or an AlCl$_3$/1-nitropropane complex) and employing a large (10-fold) excess of chloromethyl methyl ether. While the occurrence of *intramolecular* cross-linking still cannot be excluded under these conditions, its effect on the MWD of the polymer is negligible.

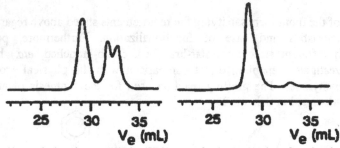

Figure 2. Size exclusion chromatography (SEC) traces for the preparation of comb polystyrenes: graft polymers obtained without DPE capping (left, 50% yield) and with DPE capping (right, 96% yield)

Higher generation arborescent polystyrenes are obtained in successive cycles of chloromethylation and anionic grafting. The grafting yield is typically high (> 90%) for G1 polymers, but tends to decrease for subsequent generations. This trend is explained by overcrowding effects in the highly branched molecules leading to reduced accessibility of the grafting sites. Since this condition does not seem to result in enhanced side reactions (such as metal-halogen exchange), complete consumption of the macroanions in the grafting process can be ensured by increasing the amount of chloromethylated substrate used in the reaction.

Characterization data for two series of arborescent polystyrenes prepared from side chains with a weight-average molecular weight of either $M_w^{br} \approx 5\,000$ (S05 series) or 30 000 (S30 series) [5] are summarized in Table 1. The absolute weight-average molecular weight of the graft polymers (M_w^{AGP}) was determined from light scattering measurements. The branching functionality was calculated from the equation

$$f_w = \frac{M_w^{AGP}(G) - M_w^{AGP}(G-1)}{M_w^{br}} \tag{2}$$

where $M_w^{AGP}(G)$, $M_w^{AGP}(G-1)$ and M_w^{br} refer to the weight-average molecular weight of polymers of generation G, of the preceding generation and of the side chains, respectively. The molecular weight and branching functionality of the graft polymers increase in an approximately exponential fashion as predicted by Equation (1), at least for the lower generations. For example, in the S05 series polymers M_w^{AGP} increases roughly 10-fold for each generation up to G2. In contrast, grafting onto a G3 substrate to synthesize the G4 polymer only leads to a 2.2-fold increase in molecular weight, because of the overcrowding effects discussed earlier. The range of molecular weights and branching functionalities attainable by the graft-on-graft technique is very broad, with M_w^{AGP} ranging from 6.7×10^4 to 5×10^8 and f_w from 14 to over 20 000. The apparent molecular weight, $(M_w)_{app}^{AGP}$, and PDI, $(M_w/M_n)_{app}^{AGP}$, of all the branched polymers that could be characterized by SEC analysis (using a linear polystyrene standards calibration curve) are also included in Table 1. Comparison of the apparent and absolute molecular weight values shows that SEC analysis strongly underestimates the molecular weight of arborescent polymers. This is expected, because of the very compact structure of these molecules. The apparent PDI of the graft polymers remains low over successive generations, and is below 1.2 in most cases.

TABLE 1. Characterization results for two series of arborescent polystyrenes with different side chain molecular weights. The indices br and AGP refer to the side chains and the graft polymers, respectively.

G	$M_w^{br}/10^3$	$(M_w/M_n)^{br}$	S05 series M_w^{AGP}	$(M_w)_{app}^{AGP}$	$(M_w/M_n)_{app}^{AGP}$	f_w
0	4.3	1.03	6.7×10^4	4.0×10^4	1.07	14
1	4.6	1.03	8.7×10^5	1.3×10^5	1.07	170
2	4.2	1.04	1.3×10^7	3.0×10^5	1.20	2 900
3	4.4	1.05	9×10^7	4.5×10^5	1.15	17 500
4	4.9	1.08	2×10^8	—	—	22 000

G	$M_w^{br}/10^4$	$(M_w/M_n)^{br}$	S30 series M_w^{AGP}	$(M_w)_{app}^{AGP}$	$(M_w/M_n)_{app}^{AGP}$	f_w
0	2.8	1.15	5.1×10^5	2.1×10^5	1.12	18
1	2.7	1.09	9.0×10^6	5.9×10^5	1.22	310
2	2.7	1.09	1×10^8	—	—	3 400
3	2.8	1.09	5×10^8	—	—	14 300

2.2. PHYSICAL PROPERTY INVESTIGATIONS

Among all arborescent polymers, arborescent polystyrenes have been investigated most thoroughly, partly because they were synthesized first. Furthermore the synthesis and characterization of polystyrenes with various architectures have been reported in the literature, making the comparison of structural effects easier. The physical properties of arborescent polymers have been investigated in solution and in the pure state by different techniques. The main results of these investigations are summarized, and a morphological model for arborescent polymers based on these findings is presented.

2.2.1. Solution Properties

Intrinsic Viscosity. The viscosity of arborescent polystyrenes in solution was examined by capillary viscometry [5]. Toluene (at 25°C) and cyclohexane (at 34.5°C) were selected for the measurements as typical good and poor polystyrene solvents, respectively. The intrinsic viscosity [η] was determined for two series of arborescent polymers of successive generations incorporating side chains with either $M_w \approx 5\,000$ (S05 series) or 30 000 (S30 series). The results obtained are summarized in Figure 3. For comparison, the variation in [η] with M_w for linear polystyrene samples in both solvents is also depicted.

To a first approximation, all arborescent polymer curves are relatively flat, *i.e.*, [η] remains almost independent of M_w over 4-5 orders of magnitude. Remarkably, the intrinsic viscosity of upper generation polymers is comparable to or even slightly lower than that of the linear core polymers. The intrinsic viscosity of a dispersion of rigid spheres is described by the Einstein equation

$$[\eta] = \lim_{c \to 0} \frac{\eta_{sp}}{c} = \frac{5}{2} N_A \frac{V_H}{M} = \frac{10\pi}{3} N_A \frac{R_H^3}{M} \tag{3}$$

where N_A, V_H, R_H, and M are Avogadro's number, the hydrodynamic volume, the hydrodynamic radius, and the mass of the particles, respectively. According to Equation (3) the intrinsic viscosity of rigid spheres should remain independent of their size as long as the ratio M/V_H, the so-called *hydrodynamic density*, remains constant. Based on Figure 3, it can be concluded that arborescent polymer molecules behave approximately like rigid spheres in dilute solution.

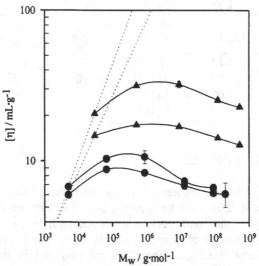

Figure 3. Intrinsic viscosity of arborescent polystyrenes of successive generations. The curves, from top to bottom, are for S30-series polymers in toluene and in cyclohexane, and for S05-series polymers in toluene and in cyclohexane.

The hydrodynamic radii calculated for the arborescent polystyrene molecules using Equation (3) are represented in Figure 4. Expansion of the molecules with $M_w \approx 30\ 000$ side chains from a poor solvent (cyclohexane) to a good solvent (toluene) is clearly observable. The dimensions of molecules with shorter side chains ($M_w \approx 5\ 000$), in contrast, remain essentially unchanged in both solvents. This shows that it is possible to control the properties of arborescent polymers in solution by varying the structure of the molecules.

Light Scattering Investigations. The variation in certain parameters such as the second virial coefficient (A_2), the radius of gyration (R_g), and the translational diffusion coefficient (D_z) with molecular weight typically follows a scaling relation of the type

$$Y = kM_w^b \tag{4}$$

where b is a structure-sensitive scaling factor. The investigation of scaling relations is thus useful in obtaining information on the structure of polymer molecules. This approach was used by comparing arborescent polystyrenes incorporating side chains with either $M_w \approx 5\ 000$, 10 000, or 20 000 [6]. Each scaling exponent was obtained from the slope of the corresponding log-log plot of Y *vs.* M_w. The scaling behavior of the molecules was found to be independent of the dimensions of the side chains for all three series of polymers investigated. The scaling factors obtained for the graft polymers are compared in Table 2 with the values expected for flexible (coiled) molecules and for rigid spheres. The scaling behavior of arborescent polymers in dilute solution is consistent with that of molecules behaving like rigid spheres, with limited interpenetration ability.

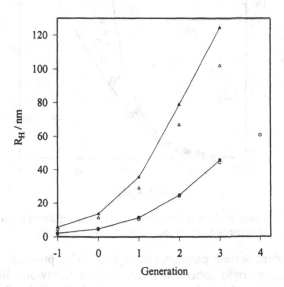

Figure 4. Hydrodynamic radii of arborescent polymer molecules in solution.
The order of the curves is the same as in Figure 3.

TABLE 2. Structure-dependent scaling factors for the second virial coefficient, the radius of gyration and the translational diffusion coefficient.

Parameter	Coils	Rigid spheres	Graft polymers
D_z	-0.6	-0.33	-0.34
R_g	0.6	0.33	0.2
A_2	-0.2	-1	-0.9

The viscometry and light scattering experiments in dilute solution, while hinting at rigid sphere-like behavior, are inappropriate to examine the effect of variables such as the branching functionality on the properties of arborescent polymers. The influence of these variables should become much clearer when the molecules are forced to interact with each other, *i.e.*, in the semidilute concentration range [6]. Measurements of the repulsive intermolecular forces in solution, expressed as the *osmotic modulus*, have been suggested to compare the rigidity of molecules in solution. Experimentally, the osmotic modulus is determined as the ratio of the weight-average molecular weight at infinite dilution (M_w) to the apparent molecular weight at a finite concentration (M_{app}). The osmotic modulus values determined for successive generations of arborescent polymers with $M_w \approx 5\,000$ side chains are given in Figure 5 as a function of a

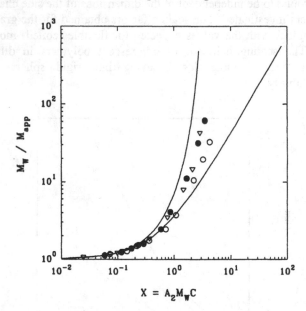

Figure 5. The osmotic modulus to characterize the structural rigidity of arborescent. polystyrenes of generations G0 (o), G1 (•), and G2 (∇).

concentration-dependent scaling parameter $X = A_2 M_w c$. Also provided on the plot are curves corresponding to rigid sphere and flexible coil behaviors. Repulsive forces increase with concentration for each sample, as expected. The data collected for the G0 polymer is closer to what is expected for the random coil case than the rigid sphere case. The results for the G1 and G2 polymers, while closer to the rigid sphere case, are very similar. An increase in branching functionality (from G0 to G1) leads to a decreased interpenetrability, but even higher generation molecules apparently retain the ability to interpenetrate each other to some extent.

Fluorescence Quenching. In an effort to investigate the morphology of arborescent polystyrenes, the molecules were labeled with low concentrations of pyrene for use in fluorescence quenching experiments [7]. Nitrobenzene, a small molecule quencher, and nitrated polystyrene, a polymeric quencher, served to probe the accessibility of the pyrene labels. Nitrobenzene quenching was shown to follow Stern-Volmer kinetics given by the equation

$$\frac{I_o}{I} = 1 + k_q \tau_o [Q] \tag{5}$$

where I_o and I represent the fluorescence emission intensities without and with quencher, respectively, k_q is the quenching constant, τ_o is the lifetime of pyrene in the absence of quencher, and [Q] is the quencher concentration. The quenching constants obtained in these experiments (Table 3) tend to decrease for higher generation polymers, reflecting an increase in average segmental density with branching functionality.

TABLE 3. Quenching constants (k_q) for nitrobenzene quenching, and fraction of accessible material (f_a) for nitrated polystyrene quenching of pyrene-labeled arborescent polystyrenes.

Sample	$k_q / 10^{11}$ L·mol^{-1}·s^{-1}	f_a
Linear	6.7 ± 0.6	0.73
G1	1.9 ± 0.2	0.73
G2	2.9 ± 0.3	0.57
G3	1.3 ± 0.1	0.46

Similar quenching experiments using nitrated polystyrene as a large molecule quencher yielded Stern-Volmer plots with downward curvature. This effect is typically attributed to the inaccessibility of a fraction of the chromophores in the quenching process. A fractional quenching model [8] was shown to be more appropriate to describe the fluorescence quenching of pyrene-labeled arborescent polystyrenes by nitrated polystyrene. The fraction f_a of accessible chromophores determined using the fractional quenching model decreases for higher generation polymers (Table 3). According to the results from the nitrated polystyrene quenching experiments, only a fraction of the chromophores attached to arborescent polymer molecules are easily accessible to large quenchers, while the balance are less or not accessible.

2.2.2. Solid State and Melt Properties

Atomic Force Microscopy. The film formation behavior of arborescent polystyrene molecules was investigated by atomic force microscopy, after spreading the molecules as monolayers on a mica substrate [9]. Only granular films are obtained for a G1 polymer sample with short ($M_w \approx 5\,000$) side chains, but the analogous G3 polymer is characterized by long-range lattice-like ordering of the molecules on the surface (Figure

6). This difference in behavior is easily explained, since molecules with a higher branching functionality are expected to have a higher segmental density and a lower interpenetrability.

The thickness of the monolayer is invariably lower than the diameter of the molecules, due to adsorption forces at the mica surface. Flattening is least pronounced for molecules of the higher generations with short side chains. For example, a G3 sample with 13 400 side chains each with $M_w \approx 30\ 000$ (S30-3) yields a film with a thickness of 80 nm and an average molecular diameter of 170 nm. For a G3 sample (S05-3) with short side chains ($M_w \approx 5\ 000$) and $f_w = 5\ 650$, in comparison, a film thickness of 35 nm and a molecular diameter of 50 nm are observed. Annealing of the films at 150°C also reveals very different behaviors depending on structure: The dimensions of the molecules in sample S30-3 are unaffected by annealing, but those in sample S05-3 relax to a perfectly spherical geometry.

Figure 6. Atomic force microscopy pictures of monomolecular films of generation G1 (left) and G3 (right) arborescent polystyrenes. The width of each picture is 500 nm

Melt Rheology. The zero-shear viscosity (η_o) of arborescent polymers in the molten state was examined as a function of branching functionality and side chain molecular weight [10]. Branched polymers of generations G0 and G1 incorporating side chains with M_w ranging from 5 000 to 20 000 have η_o values *ca.* 20 to over 1 000 times lower than linear polystyrene samples of comparable molecular weight (Figure 7). This behavior is consistent with a very compact structure and a limited entangling ability in the melt. The lack of entanglement formation is also evidenced by the first power scaling of η_o with M_w. The second-generation (G2) samples, while still characterized by a relatively low viscosity, display enhanced η_o values relative to the G0 and G1 polymers. This effect is most likely linked to the high segmental density of the molecules rather than to the entanglement of individual side chains, which have a M_w significantly below the critical molecular weight for entanglement formation ($M_c \approx 30\ 000$ for linear polystyrene). Longer terminal relaxation times are also observed for molecules with increasing side chain molecular weight and branching functionality.

2.3. MORPHOLOGICAL MODEL

The physical property investigations of arborescent polymers in solution, in the solid state and in the molten state consistently emphasized the rigid sphere-like behavior of these materials. However, even for upper generation polymers a limited ability for interpenetration is apparently maintained. Successive grafting reactions should lead to a dense core structure characterized by a high segmental density. The chains grafted in the last reaction, on the other hand, are unhindered and expected to form a diffuse, more accessible outer shell on the molecules. A model comprising a rigid core surrounded by a soft shell (Figure 8) therefore seems most appropriate to represent the morphology of arborescent polymers. Based on this model, the amount of accessible shell material is expected to decrease relative to the core material as the branching functionality (generation number) of the polymer increases, in agreement with all experimental results. This happens because when new polymer layers are added in successive grafting reactions, the volume fraction of the molecule occupied by the last layer progressively decreases, as illustrated on Figure 8.

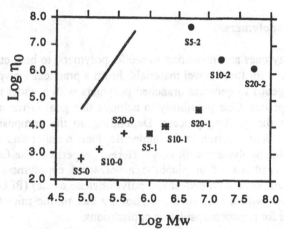

Figure 7. Zero-shear melt viscosity (η_o) of arborescent polystyrenes at 170°C. The labels (Sx-y) specify the side chain molecular weight (x) and generation number (y)

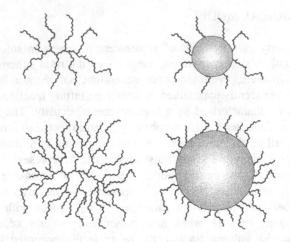

Figure 8. Morphological model for arborescent polymers of generations G0 (top) and G1 (bottom).

3. Amphiphilic Copolymers

Arborescent polystyrenes are interesting as model polymers, to help establish structure-property correlations for these novel materials. From a practical viewpoint, however, it would be advantageous to generate branched polymers with a wider range of physical and chemical properties. One possibility to achieve this goal is *via* inclusion of other monomer units in the grafting process. Depending on the composition of the new polymer segments, their location in the molecule, their number and size, it should be possible to generate copolymers with very different properties. The first example to be discussed is the synthesis of amphiphilic arborescent copolymers incorporating a branched polystyrene core with end-linked poly(ethylene oxide) (PEO) segments. This results in molecules with a core-shell morphology that mimic micellar behavior, and may thus be useful for microencapsulation applications.

3.1. SYNTHESIS

The synthetic technique used in the preparation of amphiphilic arborescent copolymers is an extension of the basic method developed for arborescent polystyrenes. The synthesis of a copolymer incorporating a G1 polystyrene core with end-linked PEO segments, based on a *grafting from* method, is described in Scheme 2. The same approach can be used for the preparation of copolymers based on higher generation

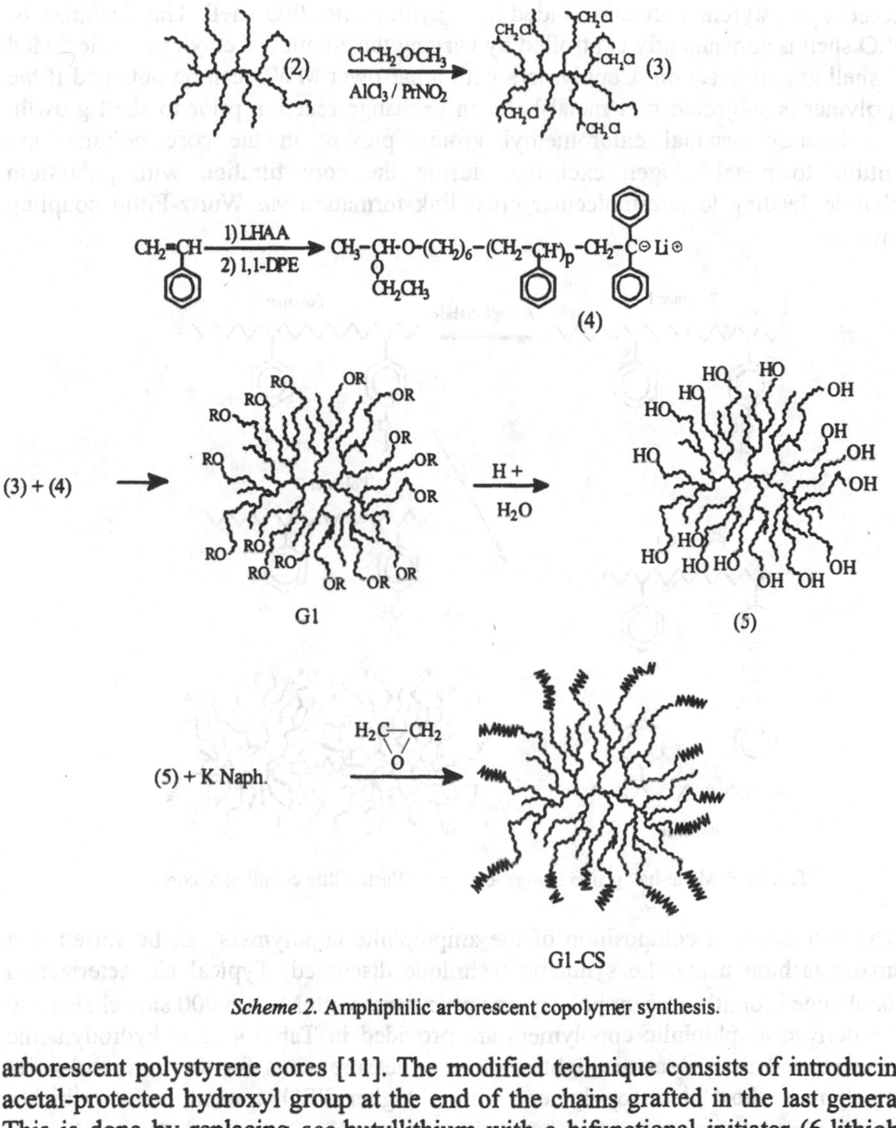

Scheme 2. Amphiphilic arborescent copolymer synthesis.

arborescent polystyrene cores [11]. The modified technique consists of introducing an acetal-protected hydroxyl group at the end of the chains grafted in the last generation. This is done by replacing *sec*-butyllithium with a bifunctional initiator (6-lithiohexyl acetaldehyde acetal) in the preparation of the side chains. The acetal groups are easily cleaved under mildly acidic conditions to yield hydroxyl chain ends. Hydrolysis is most conveniently achieved by adding a small amount of a mineral acid (HCl) during the fractionation process. Non-grafted material removal and deprotection of the hydroxyl functionalities are thus achieved simultaneously.

The introduction of the PEO segments involves titration of the hydroxyl chain ends with a strong base (potassium naphthalide), and addition of purified ethylene oxide monomer. The branched polymer serves as a polyfunctional initiator, and the hydrophilic PEO segments are obtained in a chain extension reaction of the arborescent polymer chains. This gives rise to a core-shell morphology with a hydrophobic arborescent polystyrene core surrounded by a hydrophilic PEO shell. The thickness of the PEO shell is conveniently controlled by varying the amount of ethylene oxide added in the shell growth reaction. Copolymers with a narrower MWD can be obtained if the core polymer is subjected to a metal-halogen exchange reaction prior to shell growth. This is because residual chloromethyl groups present in the core polymer are susceptible to metal-halogen exchange during the core titration with potassium naphthalide, leading to intermolecular cross-link formation *via* Wurtz-Fittig coupling (Scheme 3).

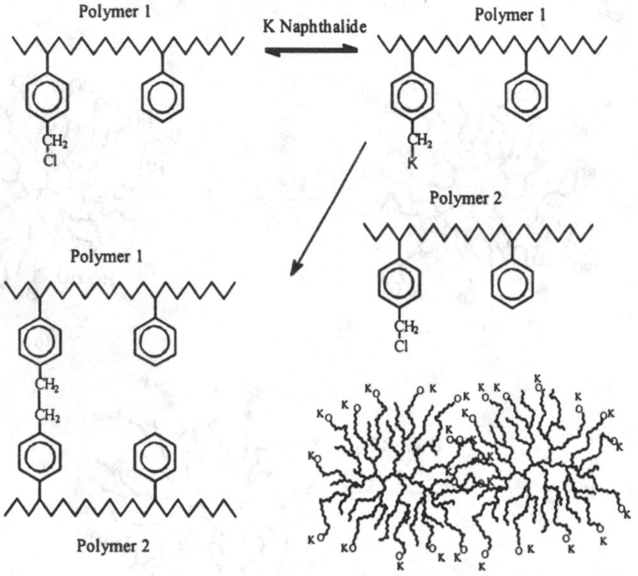

Scheme 3. Metal-halogen exchange leading to Wurtz-Fittig coupling of cores.

The structure and composition of the amphiphilic copolymers can be varied in a systematic fashion using the synthetic technique discussed. Typical characterization results obtained for arborescent polystyrene precursors with $M_w \approx 5\,000$ side chains and for the derived amphiphilic copolymers are provided in Table 4. The hydrodynamic radii determined from dynamic light scattering measurements for the core and core-shell polymers show that copolymers with a higher PEO content have a thicker hydrodynamic shell layer [$\delta = R_h(\text{core-shell}) - R_h(\text{core})$], as expected. Comparison of the hydrodynamic shell thickness with the PEO chain dimensions calculated for fully extended and coiled conformations (D_h^{PEO}) clearly demonstrates that the PEO segments at the surface of the molecules essentially adopt a coiled conformation.

TABLE 4. Dimensions of amphiphilic core-shell copolymers from dynamic light scattering measurements.

Sample	R_h/nm	δ/nm	PEO chain M_w	Extended PEO chain length/nm	D_h PEO chain/nm
G1 core	12.9	—	—	—	—
G1, 19% PEO	16.8	3.9	3.7×10^3	30	2.1
G1, 66% PEO	23.3	10.4	3.1×10^4	250	6.1
G4 core	63.7	—	—	—	—
G4, 36% PEO	66.0	2.3	4.8×10^3	39	2.4

3.2. PHYSICAL PROPERTIES

Evidence for a core-shell morphology can be found in the solubility behavior of the amphiphilic molecules. Dilute solutions (1-3%) of the copolymers in THF, a relatively good solvent for the polystyrene and PEO components, yields either clear (G1) or opalescent (G4) solutions when added dropwise to water or methanol. This is only possible if the molecules have a core-shell morphology, enabling shielding of the polystyrene core from the solvent by the PEO segments [11].

The properties of the copolymers in solution correspond to those of classical micelles. For example, a copolymer solution added to a dilute pyrene solution in water/methanol leads to a gradual increase in excimer formation, demonstrating the ability of the copolymers to extract and concentrate non-polar species in their core.

The amphiphilic character of arborescent copolymers was further investigated in a series of measurements using the Langmuir-Blodgett force balance technique [12]. Chloroform solutions of copolymers with different structures were spread on the water surface. After evaporation of the solvent, compression isotherms were recorded, and samples of films obtained at different surface pressures were transferred to silicon substrates for investigation by atomic force microscopy.

Compression isotherms for arborescent polymers based on the same polystyrene core (G1, $M_w \approx 30\,000$ side chains) with different PEO contents (or chain lengths) can display very different characteristics (Figure 9). Samples with short PEO segments yield a single transition from a "gas-like" to a "solid-like" state. For copolymers with longer PEO segments, in contrast, a "liquid-like" phase appears in the isotherm.

The origin of the new transition observed for molecules with longer PEO segments can be explained by comparing AFM pictures obtained for films of both copolymers (Figure 10). The copolymer with short PEO segments forms large island-like aggregates with a broad size distribution, characteristic of a dewetting process (Figure 10, left). At higher surface pressures, the aggregates are simply forced in closer proximity to each other, eventually giving rise to a close-packed "solid-like" state. Molecules with a higher PEO content have either a low aggregation level or are non-aggregated at low surface pressures (Figure 10, center). At higher surface pressures, association takes place to yield ribbon-like aggregates (Figure 10, right). The width of the ribbons is comparable to the diameter of the individual molecules, showing that aggregation involves strictly end-to-end association. Most interestingly, the pressure-induced phase transition is completely reversible: The monolayer, when allowed to expand back to its

254

pre-compression state, yields a film with an aggregation level essentially identical with the sample obtained prior to compression. Based on these results, the new transition in the compression isotherm can be unambiguously attributed to the reversible formation of ribbon-like aggregates for polymers with sufficiently long PEO side chains. It is obviously possible to control the association behavior of amphiphilic arborescent copolymers by varying the structure of the molecules.

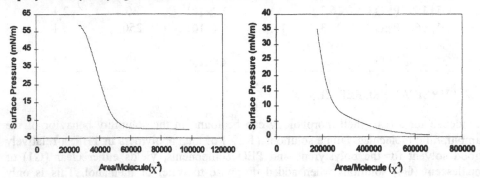

Figure 9. Compression isotherms for styrene-ethylene oxide amphiphilic arborescent copolymers based on a G1 30K core with 6.5% w/w PEO content (left), 31% w/w PEO content (right).

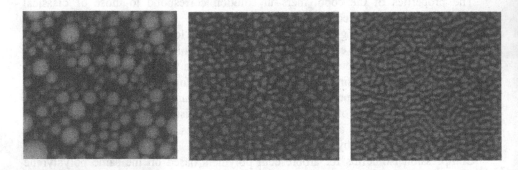

Figure 10. AFM pictures for amphiphilic copolymer monolayers removed for a 6.5% w/w PEO content sample at $\pi = 0.8$ mN/m (left), and for a 31% w/w PEO content sample at $\pi = 1$ mN/m (center), $\pi = 8.8$ mN/m (right). The width of each picture is 2 μm.

4. Isoprene Copolymers

The amphiphilic styrene-ethylene oxide copolymers previously discussed are synthesized using a *grafting from* scheme. An alternate approach used for the synthesis of arborescent copolymers containing polyisoprene segments is based on a *grafting onto* method. It allows the preparation of branched copolymers containing elastomeric chains with only a minor change in the synthetic technique used for arborescent polystyrenes [13]. The introduction of polyisoprene segments into arborescent polymers should yield materials with very interesting elastomeric properties, in analogy to star-branched polymers. Furthermore the residual double bonds of the polyisoprene chains are easily functionalized, allowing the preparation of reactive polymers useful as polyfunctional cross-linkers, elastomeric network components, *etc.*

4.1. SYNTHESIS

The method used for the synthesis of an isoprene copolymer by grafting polyisoprenyllithium onto a randomly chloromethylated linear polystyrene substrate is depicted in Scheme 4. This approach was employed to graft onto chloromethylated linear, G0, and G1 branched polystyrenes. As with the arborescent polystyrene synthesis, capping of the polyisoprenyl anions with DPE was found beneficial in avoiding metal-halogen exchange reactions and increasing the grafting yield.

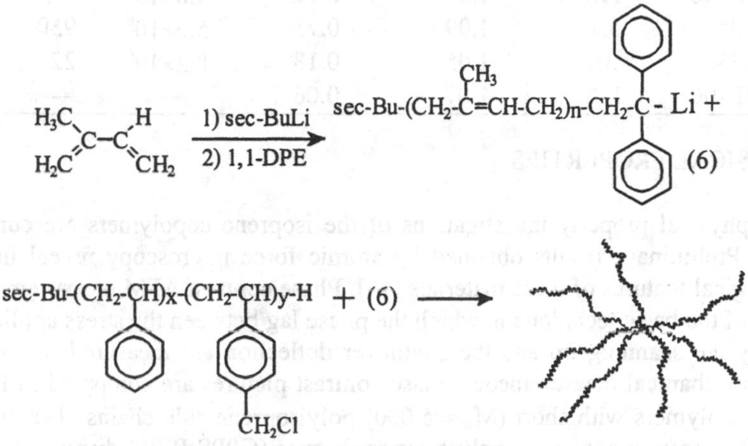

Scheme 4. Arborescent isoprene copolymer synthesis.

Characterization results for the polystyrene grafting substrates and for the copolymers obtained are provided in Tables 5 and 6, respectively. The substrates are chloromethylated linear or arborescent polystyrenes with a narrow MWD, and side chains with $M_w \approx 5\,000$ containing about 10 grafting sites each.

The grafting efficiency decreases for higher generation grafting substrates, as well as for higher molecular weight side chains (Table 6). Polyisoprene side chains with molecular weights M_w^{br} ranging from $ca.$ $5\text{-}100\times10^3$, when grafted onto chloromethylated polystyrene substrates, yield copolymers with a high content of elastomeric material: Even for $M_w^{br} \approx 5\,000$ side chains, the polyisoprene content of the copolymers is at least 86 mol%. For longer side chains ($M_w^{br} \approx 30\text{-}100\times10^3$), the polyisoprene content varies from 93 to > 99 mol%. Depending on the polymerization solvent used (cyclohexane or THF) the microstructure of the polyisoprene side chains can also be varied, providing further control over the properties of the materials.

TABLE 5. Characteristics of arborescent polystyrene grafting substrates

Polymer	M_w^{br}	$(M_w/M_n)^{br}$	M_w	f_w
Linear	—	—	4.4×10^3	—
G0PS	4 430	1.04	5.0×10^4	10
G1PS	5 750	1.03	8.0×10^5	130

TABLE 6. Characterization data for the isoprene arborescent copolymers synthesized

Sample	$M_w^{br}/10^3$	$(M_w/M_n)^{br}$	Grafting efficiency	M_w^{AGP}	f_w	PIP /mol%
PS-PIP5	4.5	1.07	0.80	5.0×10^4	11	93
PS-PIP30	35	1.03	0.66	4.2×10^5	12	> 99
PS-PIP100	118	1.08	0.43	—	—	> 99
G0PS-PIP5	4.7	1.06	0.65	5.8×10^5	112	89
G0PS-PIP30	32	1.06	0.63	2.8×10^6	85	96
G0PS-PIP100	112	1.03	0.35	4.0×10^6	35	> 99
G1PS-PIP5	5.1	1.09	0.71	5.6×10^6	950	86
G1PS-PIP30	35	1.05	0.18	8.2×10^6	225	96
G1PS-PIP100	118	1.07	0.06	—	—	> 99

4.2. PHYSICAL PROPERTIES

Detailed physical property investigations of the isoprene copolymers are currently in progress. Preliminary results obtained by atomic force microscopy reveal interesting morphological features of these materials [14]. Phase contrast AFM measurements are a variation of the basic technique in which the phase lag between the stress applied on the sample by the scanning tip and the cantilever deflection are measured, in analogy to dynamic mechanical measurements. Phase contrast pictures are compared in Figure 11 for two copolymers with short ($M_w \approx 5\ 000$) polyisoprene side chains. The copolymer obtained by grafting onto a G0 polystyrene substrate (G0PS-PIP5) displays little phase contrast between the glassy polystyrene "core" and the rubbery polyisoprene "shell" portions of the molecule. The copolymer derived from a G1 core (G1PS-PIP5), however, has more clearly defined morphological features, and a more pronounced phase contrast (*ca.* 10-12 degrees) between the core and shell.

5. Future Developments

The results obtained so far highlight some of the very unusual properties of arborescent polymers that make them interesting for a range of specialty applications. To realize the full potential of arborescent polymers it is necessary to further explore these materials in terms of new synthetic methods as well as more detailed physical characterization.

The inclusion of new monomers in the grafting reaction, and the chemical modification of the arborescent polymers already available will result in a broader range of physical and chemical characteristics than for the basic systems. Both *grafting onto* and *grafting from* schemes are potentially applicable to a wide range of anionically polymerizable monomers, depending on the reactivity of the monomers and of the macroanions generated. Chemical modification of polystyrene and polyisoprene is easily achieved with different reagents. The polarity, reactivity and physical properties of arborescent polystyrenes and of the isoprene copolymers can thus be varied.

The techniques used so far have relied solely on chloromethyl groups as grafting

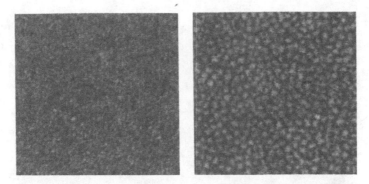

Figure 11. Phase contrast AFM pictures obtained for samples G0PS-PIP5 (left) and
G1PS-PIP5 (right). The width of each picture is 500 nm.

sites. While this approach works very well, it is not viable from a practical viewpoint because it uses chloromethyl methyl ether, a known carcinogen. It is important to develop new grafting methods for the preparation of arborescent polymers that do not rely on toxic reagents. These new approaches are currently being developed in our laboratory.

Because polymer research is mainly applications-driven, it is very important to understand the effect of the high branching functionalities encountered in arborescent polymers on the physical properties of these materials. A number of physical characterization techniques have already been applied to the investigation of arborescent polymers. It will be important to expand the range of techniques used in these investigations, to establish reliable structure-property correlations. For the same reason, it is important that as many techniques as possible be applied to the investigation of all the materials generated.

6. References

1. Newkome, G.R., Moorefield, C.N., and Vögtle, F. (1996) *Dendritic Molecules, Concepts, Syntheses, Perspectives*, VCH, New York.
2. Gauthier, M. and Möller, M. (1991) *Macromolecules* **24**, 4548-4553.
3. Altares, T., Jr., Wyman, D.P., Allen, V.R., and Meyersen, K.J. (1965) *J. Polym. Sci.: Part A*, **3**, 4131-4151.
4. Green, B. and Garson, L.R. (1969) *J. Chem. Soc (C)*, 401-405.
5. Gauthier, M., Li, W., and Tichagwa, L. (1997) *Polymer*, **38**, 6363-6370.
6. Gauthier, M., Möller, M., and Burchard, W. (1994) *Macromol. Symp.*, **77**, 43-49.
7. Frank, R.S., Merkle, G., and Gauthier, M. (1997) *Macromolecules*, **30**, 5397-5402.
8. Lakowicz, J.R. (1983) *Principles of Fluorescence Spectroscopy*, Plenum, New York, p. 279.
9. Sheiko, S.S., Gauthier, M., and Möller, M. (1997) *Macromolecules*, **30**, 2343-2349.
10. Hempenius, M.A., Zoetelief, W.F., Gauthier, M., and Möller, M. (1998) *Macromolecules*, **31**, 2299-2304.
11. Gauthier, M., Tichagwa, L., Downey, J.S., and Gao, S. (1996) *Macromolecules*, **29**, 519-527.
12. Cao, L. (1997) *Amphiphilic Arborescent Core-Shell Polymers – Synthesis and Interfacial Properties*, M.Sc. Thesis, University of Waterloo, Waterloo.
13. Kee, R.A. and Gauthier, M. (1997) *Polym. Mater. Sci. Eng.*, **77**, 176-177.
14. Sheiko, S.S., Gauthier, M., and Möller, M. (1998) To be submitted to *Macromolecules*.

SIMILARITIES AND DISCREPANCIES BETWEEN CONTROLLED CATIONIC AND RADICAL POLYMERIZATIONS

K. MATYJASZEWSKI
Department of Chemistry, Carnegie Mellon University,
4400 Fifth Avenue, Pittsburgh, PA 15213 USA

Abstract. Comparison of conventional and controlled/"living" carbocationic and radical systems indicate many differences but also some similarities. Conventional cationic and radical polymerizations are typically carried out under dramatically different conditions, use very different monomers, solvents and reaction conditions. However, recently developed controlled/"living" carbocationic and radical polymerizations employ the same mechanistic principle based on dynamic equilibration between dormant covalent species and growing species. In contrast to anionic polymerization of styrene and dienes, which provides polymers with predetermined molecular weights, low polydispersities and controlled end-functionalities, conventional carbocationic and radical polymerizations have been very difficult to control at a similar level. Thus, in carbocationic and radical polymerizations, the concentration of growing species (carbocations or free radicals) must be very low, otherwise cationic polymerization would proceed too rapidly and/or radicals would recombine too quickly. This is accomplished by using slow initiation. In both cationic and radical polymerization the control of molecular weights, polydispersities and functionalities is satisfactory only for relatively short chains. To prepare these polymers while maintaining a low concentration of active species, it is necessary to establish a dynamic equilibrium between growing species and dormant species. Some approaches toward the controlled/"living" carbocationic and radical polymerizations are based on the concept of reversible activation of alkyl halides as dormant species. They are similar in some aspects but also different since they include heterolytic vs. homolytic cleavage, different concentrations of counteranions and counterradicals and different rules for the stabilization of growing species.

1. Introduction

Cationic and radical polymerizations are typically carried out under dramatically different conditions and use very different monomers, solvents and reaction conditions. Thus, there are many differences between these two systems. However, recently developed controlled/"living" carbocationic and radical polymerizations employ the same mechanistic principle based on either spontaneous, catalyzed or degenerative equilibration between dormant covalent species and active species. In contrast to anionic

259

J.E. Puskas et al. (eds.), Ionic Polymerizations and Related Processes, 259–268.
© 1999 *Kluwer Academic Publishers. Printed in the Netherlands.*

polymerization of styrene and dienes, which provides polymers with predetermined molecular weights, low polydispersities and controlled end-functionalities, conventional carbocationic and radical polymerizations have been very difficult to control at the similar level. Several reasons, related to the very fundamental nature of carbocations and radicals, are responsible for such a behavior but probably the most important one has been the lack of efficient initiation system, which would yield quantitatively growing chains. In addition, carbocations react very rapidly with alkenes and readily participate in transfer reactions by loss of β-protons, especially in the presence of basic impurities [1]. On the other hand, radicals recombine and disproportionate with rates close to the diffusion controlled limit [2]. Thus, in typical carbocationic and radical polymerizations, the concentration of growing species (carbocations or radicals) must be very low ($<10^{-6}$ M), otherwise cationic polymerization would proceed too rapidly and/or radicals would recombine too quickly. However, if the total concentration of chains would be so low, the molecular weight of resulting polymers would be very high ($M_n >10^8$) and most likely limited by transfer. In both cationic and radical polymerizations the control of molecular weights, polydispersities and functionalities is satisfactory only for relatively short chains ($M_n <10^5$). To prepare these polymers while maintaining a low concentration of active species, it has been necessary to establish a dynamic equilibrium between growing species and dormant species. Some approaches toward the controlled/"living" carbocationic and radical polymerizations based on the concept of reversible activation of dormant species are surprising similar [1,3].

2. Conventional Carbocationic Polymerization [1]

The active center in the carbocationic polymerization is the electron deficient sp^2 hybridized carbenium ion. In order to achieve sufficient stability, it should have either resonance or inductively stabilizing substituents such as α-aryl, alkyl or alkoxy groups. Therefore, a typical range of monomers for carbocationic polymerization includes styrene, isobutene and vinyl ethers. The latter two monomers can be efficiently polymerized only by cationic pathway, although they have been successfully copolymerized by radical or coordination mechanisms. Electrophilic addition of carbocations to alkenes is very fast ($k_p>10^{5\pm1}$ mol^{-1}·L·s^{-1}) and competes with the elimination processes occurring by the loss of p-protons. Since, the latter transfer reaction has higher activation energy than propagation; it is necessary to use very low temperatures (T<-80 °C) to achieve high molecular weights by cationic polymerization. Moreover, since elimination occurs by the nucleophilic (basic) attack on the partially charged β-protons, it is necessary to eliminate all basic components from the reaction mixture. Therefore, many cationic polymerizations require very high purity of reagents and nearly complete water removal.

Reactivity of monomers vary over a range of a few orders of magnitude and scale reciprocally to carbocations and are very strongly affected by substituents. Thus, it is difficult to prepare random/statistical copolymers and therefore homopolymerization of vinyl ether can be carried out in solution of styrene or isobutene without their incorporation to the polymer chains.

Typical initiators for conventional carbocationic polymerization include protonic acids, which are often generated in situ from various Lewis acids and residual moisture. Since protonation of alkenes is not very fast, initiation is slow and molecular weights are typically limited by transfer reactions. Transfer is the most important chain breaking reaction and it can occur to monomer, to counterion or to any other basic component of the reaction mixture. Because activation energy for transfer is higher than that for propagation, carbocationic polymerizations are carried out at low temperatures. On the other hand, many cationic reactions are nearly terminationless, at least at lower temperatures and without moisture.

Carbocations are accompanied by counterions (charge must be balanced) and they can form ion pairs or stay as free ions. It seems that the reactivities of free and paired carbocations are similar which may have origin in sp^2 hybridization and relatively large ionic radiuses. It is not clear what is a typical proportion of free ions and ion pairs in polymerization, since reactions are run at very low concentrations (favoring free ions) but also in non-polar media (favoring ion pairs).

3. Conventional Radical Polymerization [2]

The active center in the radical polymerization is the sp^2 hybridized free radical. It has been assumed for a long time that radicals are relatively unselective which is manifested in a facile copolymerization. However, recently a concept of nucleophilic and electrophilic radicals has been widely accepted which is illustrated by a tendency of radicals to form alternating rather than blocky structures. Substituents with multiple bonds such as vinyl, phenyl, carbonyl or cyano stabilize radicals and also increase monomers reactivity. Therefore, dienes, styrenes, (meth)acrylates and acrylonitrile are readily radically polymerized and copolymerized.

Rate constants of radical polymerizations are a few orders of magnitude smaller than that of carbocationic process. Therefore, much higher temperatures are necessary to achieve a sufficient polymerization rate. Thus, at 60 °C $k_p \approx 10^{2\pm1}$ mol$^{-1}\cdot$L$\cdot$s^{-1} which is at least 1000 times less than for the carbocations. Relatively slow polymerization enables reactions in bulk, either in homogeneous or heterogeneous systems (e.g. suspension or emulsion).

Transfer reactions are relatively insignificant (unless a special transfer agent is added) and it is possible to achieve high molecular weights at temperatures exceeding 100 °C. Transfer requires groups or atoms which are easily homolytically cleavable. This would include multiple halogenated compounds (CBr_4, CCl_4) or weakly bound hydrogen atoms, for example on tertiary carbons, benzylic, allylic, on carbon atom next to carbonyl and cyano group as well as bound to some heteroatoms such as S-H, P-H, Sn-H, etc.

The main chain breaking reaction in the radical systems is termination, not transfer. Termination between two growing radicals is very fast, nearly diffusion controlled, and can not be avoided. However, since termination is second order but propagation is first order in respect to growing radicals, the ratio of rates of propagation to termination increases with reduction of the concentration of growing radicals. Therefore, to achieve a high molecular weight polymer, it is necessary to work at very low radical concentrations ($<10^{-7}$ mol/L). Fortunately, radicals are insensitive to water and the only purification required is deoxygenation. In conventional radical systems, the extremely low concentration of growing free radicals is very simply approached by using slow initiation. The steady state concentration of radicals is achieved by balancing rates of initiation and termination. Thus, at higher initiator concentration, faster rates but lower molecular weight polymers are formed. Typical initiators for conventional radical polymerization include compounds which cleave homolytically in the presence of light or higher temperatures (diazo or peroxy species) but can also include redox chemistry.

4. Comparison of Conventional Carbocationic and Radical Systems

Thus, at the first glance, carbocationic and radical polymerizations are very different. They include different monomers, initiators, solvents, and temperatures. Although both active centers are sp^2 hybridized, there are several types of carbocationic species depending on the degree of ionization (free ions, contact and separated ion pairs), whereas only free radicals have been so far observed in radical systems. Rate constants of propagation are much higher for cationic systems and require very low carbocations concentrations to achieve a reasonably low polymerization rate. Significant charge on β-protons in carbocations resulting from hyperconjugation leads to a significant elimination (transfer), unless at very low temperatures. On the other hand, in radical systems, a small proportion of transfer is overbalanced by the very fast termination ($k_t > 10^7$ mol$^{-1}\cdot$L$\cdot$s^{-1}). Here again, although for the different reason, a low concentration of growing species is required. Finally, initiators for both processes are very different, acids in cationic polymerization and thermally dissociating species in radical systems but both initiate slowly. Thus, among all discrepancies between cationic and radical systems, there are two similarities: slow initiation and low concentration of growing species. Additionally, both reactions often exhibit first order kinetics in respect to monomer and provide polymers with degrees of polymerization very different from the ratio of concentrations of reacted monomer to the introduced initiator (DP $\neq \Delta[M]/[I]_0$) and with high polydispersities, $M_w/M_n >> 1.5$.

5. Controlled/"Living" Polymerization [4-7]

The classic living anionic polymerization provides polymers with degrees of polymerization predetermined by the ratio of concentrations of reacted monomer to the introduced initiator (DP $= \Delta[M]/[I]_0$) and with low polydispersities, $M_w/M_n << 1.5$ [8]. This is in a big contrast to conventional cationic and radical systems. The control of molecular weights and polydispersities in living polymerization is achieved by quantitative initiation. It just happens that commercially available alkyl lithium compounds are fast initiators for styrene and dienes. Because they give many side reactions for (meth)acrylates, it was much more difficult to develop proper conditions and initiation systems for this monomers. In addition, because anionic polymerization is much slower than the cationic polymerization and transfer reactions are much less significant, it was easy to prepare polymers with the convenient range of molecular weights and even chain extend them to make many useful block copolymers.

Thus, the main challenge for the radical and cationic systems was to develop initiating systems, which would be quantitative. The range of molecular weights affected by transfer is approximately $M_n \approx 100,000$ for these systems. This value corresponds for bulk or 10% solution polymerization to the concentration of growing chains $[P] \approx 10^{-3}$ M and is significantly larger than concentration of free radicals or carbocations in conventional systems, $[P^*] \approx 10^{-7}$ M.

The solution to this problem was the development of reversibly deactivating systems. Total concentration of growing chains could still be high, $[IP] > 10^{-3}$ M, but only a small fraction of them should be active at the same time, $[P] \approx 10^{-7}$ M. This means that the equilibrium between growing radicals/carbocations and some kind of dormant species should be established. If exchange between this species is fast, then all chains appear to grow at the same rate. This approach allows also quantitative initiation with species, which may closely resemble the dormant chains.

6. Controlled Carbocationic Systems [1,9]

Several approaches have been successful for controlling carbocationic polymerization. They include use of alkyl halides in combination with relatively weak Lewis acids to assure dynamic equilibration with growing carbocations, use of nucleophiles that scavenge growing carbocations and use of salts that convert rapidly free ions to the corresponding ion pairs. The dormant species dominate in all of these systems as shown in Scheme 1.

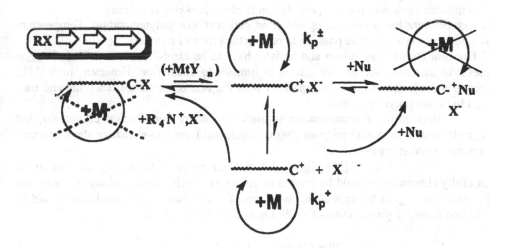

Scheme 1. Approaches towards controlled carbocationic polymerization.

Dormant species preserve their functionalities after quenching the polymerization. For example, when alkyl halide/Lewis acids are used as initiators in polymerization of styrene, vinyl ethers and isobutene, polymers with terminal halogen atoms and with high degrees of functionality have been isolated. This is very important for the subsequent transformation process.

Controlled carbocationic polymerizations are multicomponent systems and it is useful to understand a role of the major components and compare them subsequently with the controlled radical polymerization [10].

1. Alkyl halides, RX, should ionize at least as efficiently as the corresponding dormant species, ...-C-X. Thus, it is useful to use groups R, which stabilize carbocations such as those with alkoxy substituents or tertiary rather than secondary halides. In addition, a leaving group, X in the initiator should be at least as nucleofugic as that in dormant species. Since the leaving groups at the chain end and those attached to Lewis acid exchange, it is useful to use alkoxy derivatives, to accelerate the initiation process.

2. Lewis acid, MtYn should ionize both the RX and ...-C-X species reversibly and dynamically. Neither too strong nor too weak Lewis acids are efficient. Some additives may adjust the strength of the Lewis acid.

3. The role of the nucleophiles, Nu, is not yet fully understood. Nucleophiles may act in different ways. They can scavenge carbocations by reversible formation of onium ions but can also interact with and reduce strength of Lewis acids. In both ways they accelerate exchange reactions by reducing lifetime of carbocations. Some nucleophiles, such as hindered pyridines, may also trap protons.

4. Salts act also in several different ways. They suppress dissociation of carbocationic ion pairs by common ion effect. Sometimes they also increase degree of ionization by ionic strength effect and by exchanging leaving groups / counterions by the special salt effect. This way they may also modify structure of Lewis acids.

5. Solvents play an important role as in any ionic reactions. They affect dissociation constant of ion pairs and they may accelerate deactivation of carbocations. They may also influence ratios of rates of propagation to chain breaking reactions.

6. Temperature has a very strong effect on carbocationic polymerization. Temperature should be always as low as possible to reduce the effect of transfer.

7. Concentrations of monomer and initiator have to be carefully selected. High $[M]_0$, is useful to increase ratio of propagation to transfer to counterion. However, high $[M]_0$, increases propagation rate in comparison with the rate of exchange reactions and may lead to higher polydispersities.

High initiator concentration is useful to reduce contribution of transfer, but high $[I]_0$ results in shorter polymer chains which will have less exchange steps and may increase polydispersities.

Thus, the ratio $[M]_0/[I]_0$, which defines polymerization degree, has to be carefully chosen and should be limited to values at which chain breaking reactions are insignificant, e.g. in the case of monomer transfer, the following dependence should be fulfilled to assure preservation of >90% chain end functionalities:

$$DP \approx \Delta[M]/[I]_0 \leq 0.1 \, k_p/k_{trM} \qquad (1)$$

At the same time, degrees of polymerization should be large enough to form polymers with sufficiently low polydispersities. Conversion, p, the ratio of propagation and deactivation rate constants (k_p/k_{deact}), concentration of the initiator and deactivator ($[D]_0$) affect the polydispersities:

$$M_w/M_n = 1 + (k_p/k_{deact}) \cdot ([I]_0/[D]_0) \cdot (2/p-1) \qquad (2)$$

For the synthesis of block copolymers it is also very important to preserve activities of all chains until the second monomer is added. Thus, shelf-time is important, since a system can become totally inactive in spite of final low polydispersities, if it is kept for a too long time.

7. Controlled Radical Systems [3]

Controlled radical polymerization can be approached in a way similar to the controlled carbocationic process. Dormant covalent species (C-X) can be activated spontaneously (thermally) to provide free radicals (C·) and stable persistent radical (X·). The same process can be also accomplished catalytically in the presence of redox active transition metal (Mt^n). The resulting free radicals can propagate with the rate constant of propagation (k_p), terminate (with k_t) and being deactivated by the persistent radical (X·), redox conjugate ($X-Mt^{n+1}$·) and also by the species with even number of electrons, Y, via reversible formation of hypervalent persistent radicals, CY·. None of the dormant species can propagate. The catalytic activation of the dormant species corresponds to the cationic process in the presence of Lewis acids, which however is not accompanied by the redox chemistry. The deactivation with Y formally resembles deactivation of carbocations with nucleophiles via onium ions. The third route in the cationic process involving salts has no formal correspondence to the controlled radical reactions. The other major difference between Schemes 1 and 2 is the bimolecular termination, which may occur via coupling or disproportionation (not shown in Scheme 2).

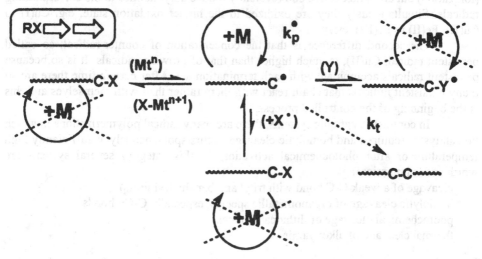

Scheme 2. Approaches towards controlled radical polymerization.

Below the role of various reagents is discussed analogously to cationic systems.

1. Initiators, RX, should cleave homolytically at least as efficiently as the corresponding dormant species, ...-C-X. Thus, it is useful to use groups R, which stabilize radicals such as those with benzyl or ester substituents and tertiary rather than secondary species. In addition, a leaving group, X, in the initiator should be at least as labile as that in dormant species. Since the leaving groups at the chain end and those attached to the catalyst may exchange, it is useful to use bromo rather than chloro derivatives with transition metal chlorides as catalysts, to accelerate the initiation process.

2. Catalyst, Mt^n should activate both the RX and ...-C-X species reversibly and dynamically. Neither too strong nor too weak catalysts are efficient. Ligands very strongly adjust the catalyst's strength.

3. The deactivators with even number of electrons, Y, have not yet been as successfully used as nucleophiles in cationic systems.

4. Solvents play less important role in radical reactions. They may, however, affect catalyst structure and may participate in transfer.

5. Temperature has a strong effect on controlled radical polymerization. Temperature should be high enough to assure faster propagation than termination but also low enough to avoid transfer.

6. Rules for selecting concentrations of monomer and initiator are similar to cationic systems but, since Trommsdorf effect is not significant in controlled systems, many reactions are carried out in bulk to accelerate propagation vs. termination.

7.1. REVERSIBLE HOMOLYTIC CLEAVAGE OF COVALENT SPECIES.

The closest system to the controlled carbocationic polymerization is the transition metal catalyzed process also known as atom transfer radical polymerization (ATRP). The main difference between carbocationic and radical process is that ATRP utilizes weakly acidic transition metal complexes which do not cause heterolytic cleavage of C-X bond (ionization) but are redox active and reversibly reduce alkyl halides to the corresponding radicals. Simultaneously they are oxidized to the higher oxidation state, e.g. Cu(I) to Cu(II), Fe(II) to Fe(III), etc.

The second difference is that the concentration of counter-radical, so called persistent radical, Cu(II), is much higher than that of growing radicals. It is so because persistent radicals accumulate with each termination act. After certain time there are so many persistent radicals that chain react with them rather than with themselves and this is the beginning of the controlled process.

In contrast to cationic systems, there are many radical polymerizations in which no catalyst is required and homolytic cleavage occurs spontaneously at sufficiently high temperature or after photochemical activation. In this category several systems are worth special mentioning:

- cleavage of a weak C-C bond with trityl and benzhydryl group
- homolytic cleavage of organometallic species, especially C-Co bonds
- photochemical cleavage of dithiocarbamates
- thermal cleavage of alkoxyamines

Although similar reactions have been observed in cationic systems (e.g. C-I for vinyl ethers), they are very rare.

$$\sim\!\!P_n\text{-}X \quad (+ Y) \rightleftharpoons \sim\!\!P_n^{\,\circ} + X \bullet (Y)$$

$$10^{-2}\text{-}10^{-1} \qquad 0\text{-}10^{-1} \qquad\qquad 10^{-9}\text{-}10^{-7} \qquad 10^{-5}\text{-}10^{-2}$$

-X = -CAr₂L; -Cr^{III}L$_n$; -Co^{II}(TMP); -Co^{II}(DMG); -S-C(S)-NR₂; -O-NR₂

-X = -Cl; -Br; -I; -SCN   Y=Cu^{I}·L$_n$; Ru^{II}·L$_n$,; Fe^{II}·L$_n$; Ni^{II}·L$_n$ (3)

7.2. ATOM TRANSFER RADICAL POLYMERIZATION [3, 1 1-13]

In the last few years many efforts have been devoted to developments of Controlled/"living" radical polymerization. Stable free radicals such as TEMPO, various organometallic species, degenerative transfer and atom transfer radical polymerization (ATRP) were developed. ATRP seems to be the most robust of these systems.

The extension of atom transfer radical addition (ATRA) to atom transfer radical polymerization (ATRP) provides an efficient way to conduct a controlled/"living" radical polymerization. With a variety of alkyl halides as the initiators and a transition metal species complexed by suitable ligand(s) as the catalyst, ATRP of vinyl monomers such as styrene, acrylates, methyl methacrylates, and acrylonitrile proceeds in a "living" fashion.

The resulting polymers have a degree of polymerization predetermined by $\Delta[M]/[I]_0$ up to $M_n \approx 10^5$ low polydispersity, $1.1 < M_w/M_n <1.5$. For example, when an alkyl chloride, l-PhEtBr is used as initiator, and a CuBr/4,4'-diheptyl-2,2'-bipyridine of 4,4'-di(5-nonyl)-2,2'-bipyridine complex is used as catalyst, styrene polymerizes by repetitive atom transfer radical additions to yield a well-defined, high molecular weight polymer, with narrow molecular weight distribution (M_w/M_n = 1.05) [13]. Under appropriate conditions, the contribution of termination becomes insignificant and polymerization occurs in a controlled way.

ATRP is a versatile tool for preparation of random, block, alternating, and gradient copolymers with controlled molecular weights, narrow polydispersities and desired architectures [14]. There is a multitude of initiators for ATRP. Any alkyl halide with activated substituents on α-carbon such as an aryl, carbonyl, or allyl can be used, in addition to polyhalogenated compound (CCl₄, HCCl₃) and others with weak halogen bonding. This includes not only low molar mass compounds but macromolecular species as well, used to form the corresponding block/graft copolymers. Moreover, it is possible to use multifunctional initiators to generate growth of a macromolecule in more than one direction and form stars and other novel architectures [15]. The functional initiators provide end functionalities for polymer chains. Additionally, halogen atoms from the other chain end can be quantitatively replaced by many useful functional groups, including azide, amine, hydroxy, etc. [16].

268

8. Conclusions

The above discussion demonstrates similarities and differences between conventional and new controlled/"living" carbocationic and radical systems. Conventional cationic and radical polymerizations show more differences than similarities. They are carried out under very different conditions, employ different monomers, initiators and reaction conditions. They are typically run at very low concentrations of active sites and provide polymers with undefined molecular weights and high polydispersities. On the other hand, recently developed controlled/"living" carbocationic and radical polymerizations use the same mechanistic principle based on dynamic equilibration between dormant covalent species and growing species. In both systems the control of molecular weights, polydispersities and functionalities is satisfactory for relatively short chains. To prepare these polymers while maintaining a low concentration of active species, it is necessary to establish a dynamic equilibrium between growing species and dormant species. Most successful approaches toward the controlled/"living" carbocationic and radical polymerizations involve reversible activation of alkyl halides as dormant species. They are similar in some aspects but also different since they include heterolytic vs. homolytic cleavage, different concentrations of counteranions and counterradicals and different rules for the stabilization of growing species.

Acknowledgements. Support from the Petroleum Research Fund administered by the American Chemical Society, the Office of Naval Research, and industrial sponsors of Atom Transfer Radical polymerization Consortium at Carnegie Mellon University Acknowledged.

9. References

1. Matyjaszewski, K. Ed., (1996) *Cationic Polymerizations: Mechanisms Synthesis and Applications*; Marcel Dekker, New York.
2. Moad, G. and Solomon, D.H. (1995) *The Chemistry of Free Radical Polymerization*; Pergamon, Oxford.
3. Matyjaszewski, K. Ed.; *Controlled Radical Polymerization*; ACS: Washington, D.C., 1998; Vol.685.
4. Matyjaszewski, K., Gaynor, S., Greszta, D., Mardare, D., and Shigemoto, T. (1995) *J. Phys. Org Chem.*, **8**, 306.
5. Matyjaszewski, K. (1993) *Macromolecules*, **26**, 1787.
6. Matyjaszewski, K. and Sigwalt, P. (1994) *Polymer Int.*, **35**, 1.
7. Greszta, D., Mardare, D., and Matyjaszewski, K. (1994) *Macromolecules*, **27**, 638.
8. Szwarc, M. (1968) *Carbanions, Living Polymers and Electron Transfer Processes*; Interscience Publishers, New York.
9. Kennedy, J.P. and Ivan, B. (1992) *Designed Polymers by Carbocationic Macromolecular Engineering. Theory and Practice*; Hanser, Munich.
10. Matyjaszewski, K. (1996) *Macromol. Symp.*, **107**, 53.
11. Kato, M., Kamigaito, M., Sawamoto, M. and Higashimura, T. (1995) *Macromolecules*, **28**, 1721.
12. Wang, J.S. and Matyjaszewski, K. (1995) *J. Am. Chem. Soc.*, **117**, 5614.
13. Patten, T.E., Xia, J., Abernathy, T., and Matyjaszewski, K. (1996) *Science*, **272**, 866.
14. Patten, T.E. and Matyjaszewski, K. (1998) *Adv. Materials*, **10**, 901.
15. Matyjaszewski, K. and Gaynor, S.G. (1998) *ACS Symp. Series*, **685**, 396.
16. Matyjaszewski, K., Coessens, V., Nakagawa, Y., Xia, J., Qiu, J., Gaynor, S., Coca, S., and Jasieczek, C. (1998) *ACS Symp. Series*, **704**, 16.

RECENT DEVELOPMENTS IN ANIONIC SYNTHESIS OF MODEL GRAFT COPOLYMERS

J. W. MAYS
Department of Chemistry
University of Alabama at Birmingham
Birmingham, AL 35294 USA

Abstract. Recent advances in anionic polymerization methodology have led to the synthesis of graft copolymers of precisely defined architecture. It is now possible to produce singly-grafted species, having very narrow molecular weight distribution backbones and branches, with complete control over branch point placement (symmetric simple graft, asymmetric simple graft). Species with two regularly spaced branched points (II and H architectures) have also been produced and thoroughly characterized. Most recently we have achieved the synthesis of "regular comb grafts" and "co-centipedes". These materials have 1, 4-polyisoprene backbone with multiple, regularly spaced branch points; at each branch point one or two polystyrene branches are connected, respectively for the comb and centipede. We review the chemistry that has been used to produce these controlled architecture non-linear block copolymers. The chemistry is useful in making polymers with virtually any desired molecular weight, i.e. it is not restricted to low molecular weight materials as is often the case with other chemistries. The basic strategy is living anionic polymerization using monofunctional and difunctional initiators in hydrocarbon solvents. Controlled chlorosilane linking chemistry, taking advantage of stoichiometry, chlorosilane volatility, and steric effects, is then used to assemble the final products. Characterization methods which provide unassailable proof of architecture will also be discussed. These methods include sampling and absolute molecular weight determinations for all homopolymer segments, as well as absolute molecular weight characterization and compositional analysis of the final products.

1. Introduction

Graft copolymers are polymers composed of a main chain of a certain type of polymer with one or more chemically different side chains connected to the backbone through covalent bonds. Unlike block copolymers, which are liner chains, graft copolymers are a class of branched polymers, i.e. graft copolymers have more than two chain ends per molecule. Graft copolymers are important from a scientific perspective because they exhibit properties that reflect the combined effects of thermodynamic incompatibility of the polymer segments and the branching architecture. The introduction of a grafted

269

J.E. Puskas et al. (eds.), Ionic Polymerizations and Related Processes, 269–281.

architecture into block copolymers is known to influence flow properties [1], bulk morphologies (2-10), and extent of long range order [4, 9-11]. The properties of graft copolymers are exploited commercially as adhesives, emulsifiers and compatibilizing agents, and as tough plastics, e.g. high-impact polystyrene (HIPS), impact modified acrylics (butadiene-styrene-methacrylate) and ABS resins (acrylonitrile-butadiene-styrene graft copolymers) [12, 13].

In the development of structure property relationships for graft copolymers, progress has been hindered due to synthetic difficulties in producing precisely tailored grafted architectures. Ideally one would like to be able to produce materials that have monodisperse backbones and monodisperse side chains, combined with precise control over the number of branches per molecule and the location(s) of branch point placement. All molecules in a synthesized batch of such polymers would be identical. Unfortunately, such molecules have never been made by synthetic polymer chemists! However, living polymerization processes (especially living anionic polymerization) have been extensively used to produce graft copolymers with narrow molecular weight distribution (MWD) backbones and branches, and with some control over branch number (for a recent review, see ref. 14). In Figures 1-3, respectively, typical anionic-based schemes for "grafting from", "grafting onto", and the "macromonomer approach" are shown.

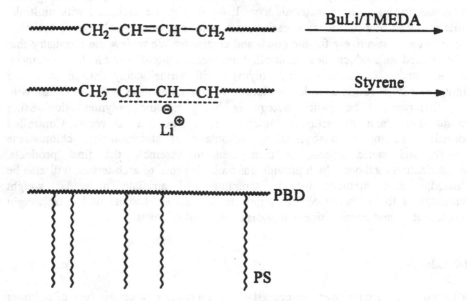

Figure 1. A "grafting form" strategy. The polydiene backbone is lithiated, followed by growth of polystyrene branches from the anionic sites along the backbone.

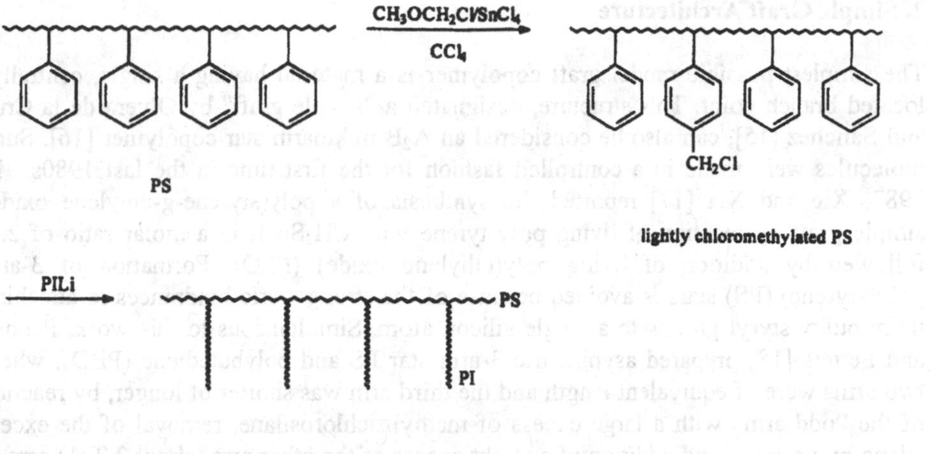

Figure 2. An example of "grafting onto". The polyisoprenyllithium anions react with chloromethyl groups on the polystyrene chain.

Figure 3. A polyisoprene macromonomer is copolymerized with methylmethacrylate to generate poly(methylmetacrylate-g-isoprene).

While all three of these methods can control backbone and side chain molecular weight and MWD to produce a wide range of graft copolymers (14), none of these methods allow tight control over the number of branches and their placement along the backbone.

In this paper, we will briefly review some recent advances in anionic polymerization methodology that has made possible more precise control over branch point number and placement. An emphasis will be placed on chlorosilane-linking strategies because these methods readily facilitate sampling and characterization of the components used to construct the model graft copolymers.

2. Simple Graft Architecture

The simplest possible model graft copolymer is a material having a single, centrally-located branch point. This structure, designated a "simple graft" by Olvera de la Cruz and Sanchez [15], can also be considered an A_2B miktoarm star copolymer [16]. Such molecules were made in a controlled fashion for the first time in the last 1980s. In 1987, Xie and Xia [17] reported the synthesis of a poly(styrene-g-ethylene oxide) simple graft by reaction of living polystyrene with CH_3SiCl_3 in a molar ratio of 2:1, followed by addition of living poly(ethylene oxide) (PEO). Formation of 3-arm poly(styrene) (PS) stars is avoided because of the strong steric hindrances to attaching three bulky styryl groups to a single silicon atom. Simultaneous to this work, Pennisi and Fetters [18] prepared asymmetric 3-arm star PS and polybutadiene (PBD), where two arms were of equivalent length and the third arm was shorter of longer, by reaction of the "odd arm" with a large excess of methyltrichlorosilane, removal of the excess silane by vacuum, and addition of a slight excess of the other arm (about 2.2 :1) arm to dichlorosilane-endcapped polymer). After linking is completed, the excess arm is removed by fractionation. Mays [19] employed this approach to prepare a narrow MWD poly(isoprene-g-styrene) simple graft (Figure 4). This method is more versatile than the approach of Xie and Xia [17] which fails if anions less bulky than PSL; are reacted with CH_3SiCl_3. (for example, a reaction of 2:1 PBD or polyisoprene (PI) anions with CH_3SiCl_3 would lead to a mixture of products including some 3-arm star). Mays, Hadjichristidis, and their collaborators [4, 7, 20] subsequently used this method to prepare a wide range of narrow MWD (polydispersities <1.05) simple graft copolymers of different compositions and molecular weights (up to well over 100,000).

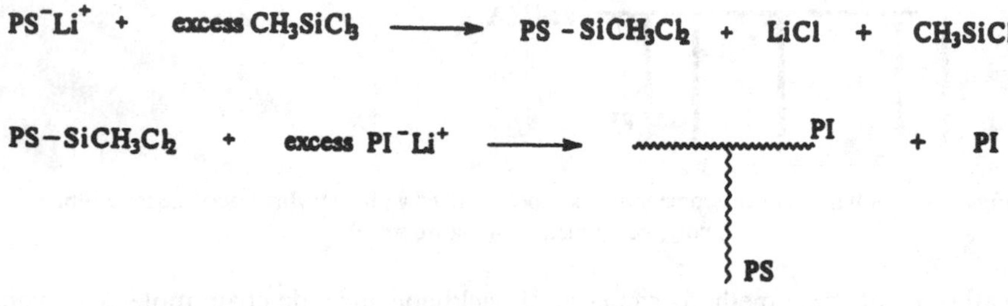

$$PS^-Li^+ \; + \; \text{excess } CH_3SiCl_3 \longrightarrow PS-SiCH_3Cl_2 \; + \; LiCl \; + \; CH_3SiCl_3\uparrow$$

$$PS-SiCH_3Cl_2 \; + \; \text{excess } PI^-Li^+ \longrightarrow \begin{array}{c} PI \\ | \\ PS \end{array} \; + \; PI$$

Figure 4. Synthesis of simple graft, poly(isoprene-g-styrene).

Other approaches have also been employed to make simple graft structures by anionic polymerization. Kahn *et al.* [21] reacted two moles of PSL: with CH_3SiCl_2H to produce a PS backbone with a Si-H bond at the middle of the chain. The vinyl-terminated poly(2-vinylpyridine) branch was attached to the backbone by hydrosilation addition chemistry. The products exhibited broader than expected MWDs [1.33 to 1.50]. Other approaches have been reported by Teyssie *et al.* [22], using naphthalene terminated polymers, and by Naka *et al.*, by reacting complexes of Ru [III] with PEO followed by addition of polyoxazoline [23]. The products of the former work [22] exhibited polydispersities of 1.2-1.3; no characterization data were given in reference 23. There has also been interest in recent years in utilizing 1,1-diphenylethylene (DPE)- based living linking reactions to make "heteroarm star" copolymer architectures [24-27]. These approaches, which can be separated into DPE- macromonomer methods [24,26,27] and double - DPE methods [25] offer the potential for making A₂B ABC or A₂B₂ [28] miktoarm stars, although it appears that the A₂B simple graft architecture has not yet been made using this methodology [29] [Figure 5].

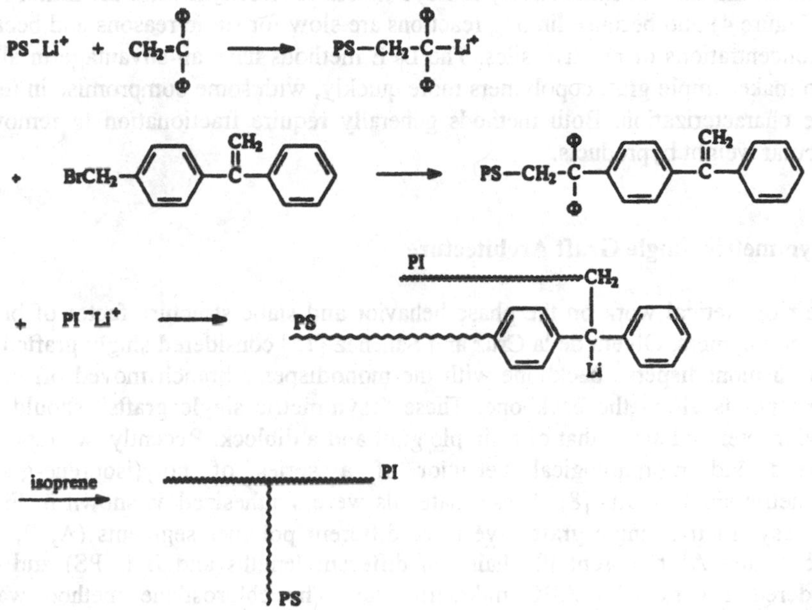

Figure 5. A DPE macromonomer strategy for simple graft synthesis.

On comparing the chlorosilane method with DPE approaches, it is clear that only the chlorosilane methods affords the opportunity to sample all three polymer segments that make up the simple graft. The determination of absolute molecular weights of each of the segments and the final graft, combined with compositional analysis via spectroscopy, provides rigorous proof of structure [19,20]. With the DPE methods the last polymer segment is made by polymerization from a living backbone (Figure 5), so this segment is not available for independent characterization. Depending on the molecular weight of the final segment, it may be difficult to detect the presence of non-grafted material that could be produced by termination of some active sites during the final monomer addition. In characterizing branched copolymers, SEC cannot be depended on for reliable molecular weights although it does provide some insight into the MWD of the products. Membrane osmometry (MO) or matrix-assisted laser desorption/ionization time-of-flight mass spectrometry (MALDI/TOF/MS) are the preferred molecular weight methods for use because they can yield absolute average masses of copolymers with branching. Although light scattering (LS) from block copolymers gives only an apparent weight-average molecular weight [30], it has been our experience that LS gives very reasonable values as compared with MO data, on compositionally homogeneous, narrow MWD block copolymers provided the refractive index increments of both components are large (ca. 0.1 mL/g or higher) [31], in accord with theoretical work by Bushuk and Benoit [32].

While the chlorosilane technique is extremely powerful for producing well-defined, rigorously characterized, simple grafts (and a wide range of more complex architectures discussed below), it is time-consuming (many days to make a polymer) because of the need to exhaustively remove excess chlorosilane after the initial capping step (Figure 4) and because linking reactions are slow for steric reasons and because of low concentrations of reactive sites. The DPE methods have an advantage in allowing one to make simple graft copolymers more quickly, with some compromise in the rigor of the characterization. Both methods generally require fractionation to remove low molecular weight byproducts.

3. Asymmetric Single Graft Architecture

In their theoretical work on the phase behavior and static structure factor of branched block copolymers, Olvera de la Cruz and Sanchez [15] considered singly grafted chains having a monodisperse backbone with the monodisperse branch moved off-center to various points along the backbone. These "asymmetric single grafts" should exhibit behavior intermediate to that of a simple graft and a diblock. Recently, we reported the synthesis and morphological behavior of a series of poly(isoprene-g-styrene) asymmetric single grafts [8]. These materials were synthesized as shown in Figure 6. These asymmetric single graft have three different polymer segments (A, B, and A' where A and A' represent PI chains of different lengths and B is PS) and can be considered a variety of ABC miktoarm star. The chlorosilane method was first employed for the synthesis of ABC miktoarm star terpolymers by Iatrou and Hadjichristidis [33], who made stars having a PS, and PI, and PBD arm tethered to one another via a silicon (from CH_3SiCl_3). The approach utilized in Figure 6 is the same one employed by Iatrou and Hadjichristidis. The key step is the "vacuum titration" of the chlorosilane-endcapped long PI segment with polystyryllithium. This process involves the drop-wise addition of orange PSLi solution to the endcapped PI under vacuum. The color disappears as the reaction to form the diblock with a chlorosilane group at the junction point takes place, providing a visual means for monitoring the progress of the reaction. The "endpoint" can be achieved while forming negligible amounts of PS_2PI stars because steric effects caused by the bulky nature of the styryl anion [17, 34, 35] prevent reaction with the third Cl on each Si atom. The short PI anion, which is less bulky, reacts with these last chlorines to produce the asymmetric single graft (Figure 6).

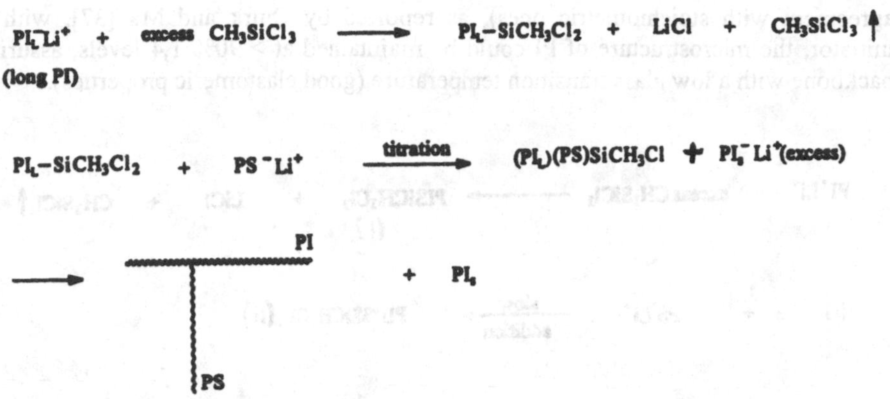

$$PI_L^-Li^+ + \text{excess } CH_3SiCl_3 \longrightarrow PI_L-SiCH_3Cl_2 + LiCl + CH_3SiCl_3 \uparrow$$
$$\text{(long PI)}$$

$$PI_L-SiCH_3Cl_2 + PS^-Li^+ \xrightarrow{\text{titration}} (PI_L)(PS)SiCH_3Cl + PI_L^-Li^+\text{(excess)}$$

Figure 6. A chlorosilane strategy for asymmetric single graft synthesis.

DPE-based methods, discussed above, have also been used to make ABC miktoarm star terpolymers [24-29], and could thus be adapted for the synthesis of asymmetric single graft copolymers. As with simple graft synthesis, the chlorosilane approach to asymmetric grafts gives a very well-defined product and allows rigorous characterization [8]. Polymers produced by the DPE methods tend to be somewhat less well-defined (broader in MWD and not so readily characterized with rigor), but the DPE method allows for much more rapid synthesis.

4. The π Architecture

π graft copolymers have been synthesized [7,10] with two PS branches regularly placed along a PI backbone. These model graft materials were produced using the chlorosilane linking scheme shown Figure 7. A diblock copolymer of PS and PI containing a chlorosilane group at the junction point is made in the same manner described above for asymmetric single graft synthesis. Two moles of this reactive diblock are reacted with one mole of living dianionic poly(isoprenyllithium) to produced the π (PIPS) PI (PIPS) graft. Notice that the length of the central PI connector may be the same length or different than that of the end PI segments. This allows flexibility in branch placement along the backbone. This synthetic approach, as with other chlorosilane-based methods, allows for thorough molecular characterization of the final product and all its segments [7,10]. It also allows the synthesis of very high molecular weight products, if desired. π grafts with molecular weights as high as 2×10^5 have recently been reported [10].

A key factor in the successful synthesis of our π copolymers has been the selection of an appropriate difunctional initiator. We have utilized (1,3- phenylene) bis (3 - methyl-1-phenylpen-tylidine) dilithium (PLi) [36] in our work. The use of LioBu as a polar modifier for polymerizations in benzene gave reasonably efficient initiation rates for both reactive sites of DLi (MWD of 1.2; measured molecular weights in agreement with stoichiometric ones), as reported by Quirk and Ma [37], with this initiator, the microstructure of PI could be maintained at > 90% 1,4 levels, assuring a backbone with a low glass transition temperature (good elastomeric properties).

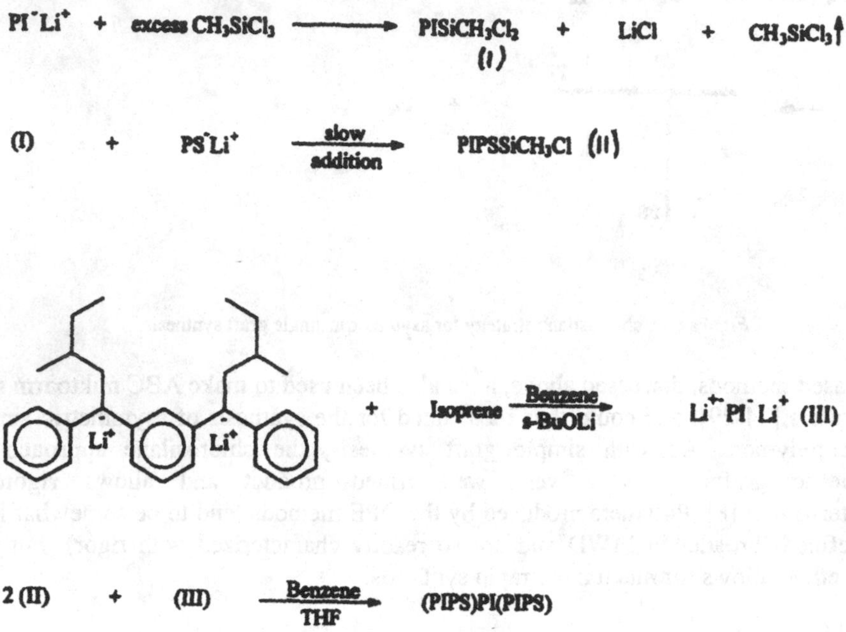
PI⁻Li⁺ + excess CH3SiCl3 → PISiCH3Cl2 + LiCl + CH3SiCl3↑
(I)
(I) + PS⁻Li⁺
slow
addition
PIPSSiCH3Cl (II)
Li⁺
Li⁺
+ Isoprene
Benzene
s-BuOLi
Li⁺PI⁻Li⁺ (III)
2 (II) + (III)
Benzene
THF
(PIPS)PI(PIPS)

Figure 7. Synthesis of double graft architecture.

Methods other than the chlorosilane method have not, to our knowledge, been used to make π - shaped graft copolymers. However, Hsieh and Quirk [29] have outlined the strategy for such a synthesis. Namely, a living polymer (backbone segment) is reacted with a DPE macromonomer (branch segment), followed by addition of monomer to grow additional backbone from the DPE anions, and the cycle is repeated. In practice, it might be difficult to repeat these cycles without mistakes in adding the correct amount of macromonomer or without deactivation of some anions.

5. H - Shaped Graft Architecture

The H - graft is an unusual graft copolymer in that it has two branches attached at each end of the backbone (connector), which is chemically different from the branches. The strategy followed to produce H- graft copolymers with a PI backbone and PS branches [7,9] was similar to that employed by Roovers and Toporowski [38] in their initial synthesis of H- shaped PS homopolymers. Two moles of PS were reacted with one mole of methyltrichlorosilane (using vacuum titration). Two moles of this PS, which contained a chlorosilane group at its midpoint, was reacted with one mole of living dianionic PI to produce the H- copolymer (in a manner similar to that shown in Figure 7) [7,9]. Various H - graft copolymers with molecular weights as high as 3 x 10^5 as polydispersities < 1.1 have been made and rigorously characterized.

6. Graft Copolymers with Multiple Regularly Spaced Branched Points

As mentioned previously, Hsieh and Quirk [29] have proposed a general method, based on DPE macromonomers, that in principle could lead to precisely tailored multigrafts. However, this strategy remains untested (to our knowledge) and has some potential shortcomings, discussed above, if large numbers of side chains are desired. An alternate method for synthesizing such regular multigrafts is by step-growth polymerization of two difunctional monomers, at least one of which is a step-growth macromonomer. Rempp and co-workers [39] first demonstrated the "polycondensation" of polymers containing reactive anions at both ends with difunctional electrophiles, e.g. dibromobutane, to generate a polymer of increased molecular weight. Of course, the statistical nature of the step-growth polymerization results in a polymer, assuming high degrees of polymerizations with a MWD of about two. Strazielle and Herz [40] utilized this step-growth concept to make regular comb PS by the scheme shown in Figure 8. Triallyloxytriazine (a trifunctional reactant) was copolymerized with dianionic PS prepared using a dipotassium initiator based an oligomeric α - methyl - styrene. Only two of the allyloxy groups reacted with the lining dipotassium PS, leaving the remaining allyloxy groups reacted for subsequent reactions with living PS bearing a single reactive end [40]. However, star PS made by the same workers [40] using a tetraallyloxy linking agent yielded broader than expected MWSs, suggesting that possibility of side reactions. Recently, a group at the Dow Chemical Company [41] prepared branched polystyrenes of known architecture by a similar step-growth process involving reaction of living disodium PS with a mixture of α, α' - dichloro-p-xylene and α, α', α" - trichloromesitylene.

Very recently we have developed a chlorosilane linking route to regular multigrafts which employs this combination of living anionic polymerization and step-growth polymerization [42]. The chemistry is outlined in Figure 9 for the synthesis of "co-centipedes". This novel architecture has a backbone of PI with double branches of PS at each branch point along the PI backbone. The number-average degree of polymerization is controlled by the stoichiometry and the extent of conversion (Carothars equation). Raw products had polydispersities of 2.1 to 2.5, with , 5% unreacted species ("monomer"). Fractionation was used to obtain the highest molecular weight components from four separate runs. Their molecular characteristics and compositions are presented in Tables 1 and 2, respectively. The polydispersities are in the range of 1.3 for the four fractions; narrower fractions could be obtained by more careful fractionation procedures. Degrees of polymerization of 9-13 were measured for these materials, corresponding to weight -average molecular weights measured by LS as high as 1.26 x 10^6 for the co-centipede products. Compositions, measured by NMR, and refractive index increments (dn/de) measured by differential refractometry, were in good agreement with values expected based on absolute molecular weights and dn/dc of

278

the segments (dn/dc graft = x dn/dc (PS) + (1-x) dn/dc (PI) where x is the weight fraction of PS)

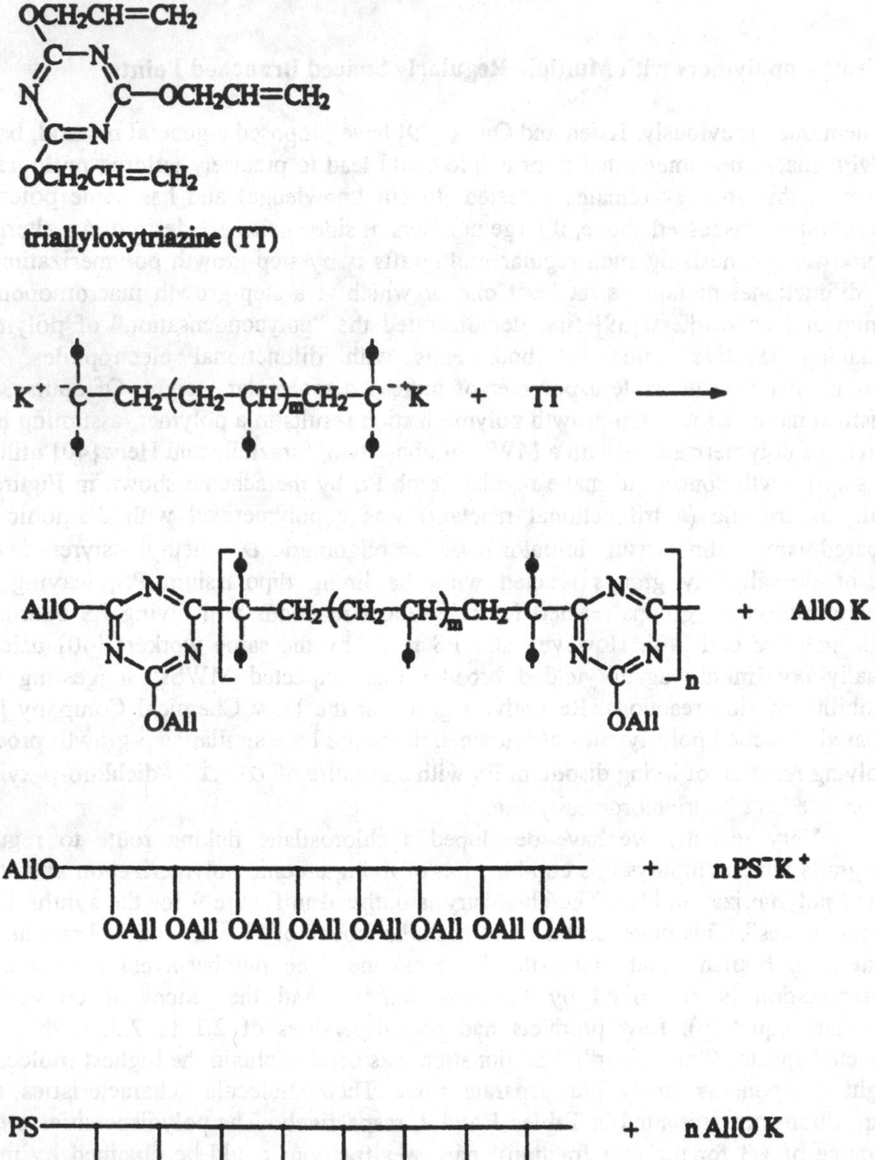

OCH₂CH=CH₂

triallyloxytriazine (TT)

K⁺ ⁻C—CH₂(CH₂—CH)ₘCH₂—C⁻ ⁺K + TT ⟶

AllO—C ... C—CH₂(CH₂—CH)ₘCH₂—C ... C + AllO K

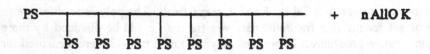

Figure 8. Synthesis of regular comb-branched polystyrene by the method of Strazielle and Herz [40].

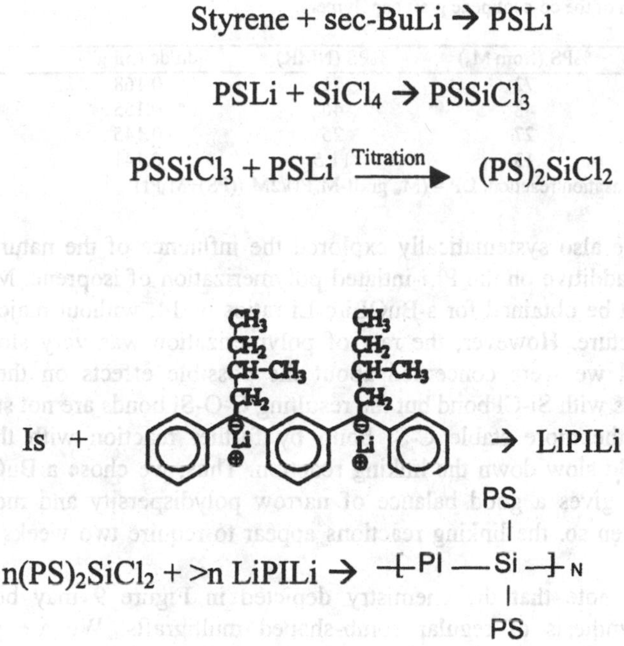

$$Styrene + sec\text{-}BuLi \rightarrow PSLi$$

$$PSLi + SiCl_4 \rightarrow PSSiCl_3$$

$$PSSiCl_3 + PSLi \xrightarrow{\text{Titration}} (PS)_2SiCl_2$$

$$n(PS)_2SiCl_2 + >n\ LiPILi \rightarrow \quad \text{--}[PI\text{---}Si\text{---}]_N$$

Figure 9. Synthesis of co-centipedes having a PI backbone and PS branches.

TABLE 1. Molecular characteristics of the Co-centipedes.

sample	M_nPS arm[a] x10^{-4}	M_nPS con[a] x10^{-4}	M_w/M_n[b]	M_w x 10^{-6c}
GSI 25-35	3.44	2.74	1.30	1.22
GSI 60-25	2.88	6.18	1.29	1.26
GSI 70-15	1.35[d]	7.27	1.30	1.17
GSI 90-7	0.67[d]	9.16	1.32	1.02

[a] Membrane osmometry in toluene at 37 °C. [b] Polydispersity index obtained by SEC in THF at 25 °C. [c] Obtained by MALDI/TOF/MS measurements.

TABLE 2. comparison of the co-centipede graft copolymers.

sample	%PS (from M_n)	%PS (NMR)	dn/dc (mLg^{-1})	DP[a]
GSI 25-35	72	71	0.168	13
GSI 60-25	48	46	0.155	10
GSI 70-15	27	25	0.145	11
GSI 90-7	13	11.5	0.141	9

[a] Degree of polycondensation reaction. DP = $(M_w$ graft-$M_nPI)/2M_n((PS)+M_nPI)$

In this work [42], we also systematically explored the influence of the nature of and amount of the polar additive on the PLi-intiated polymerization of isoprene. MWDs as narrow as 1.03 could be obtained for s-BuOLi/c-Li ratios of 14, without major impact on the PI microstructure. However, the rate of polymerization was very slow under these conditions and we were concerned about the possible effects on the linking raction. BuOLi reacts with Si-Cl bond but the resulting C-O-Si bonds are not stable and are transformed to the more stable C-Si bond by further reaction with the living connector. This would slow down the linking reaction. Thus, we chose a BuOLi/C-Li ratio of four, which gives a good balance of narrow polydispersity and more rapid linking reactions. Even so, the linking reactions appear to require two weeks to reach completion.

Finally, we note that the chemistry depicted in Figure 9 may be readily modified for the synthesis of regular comb-shaped multigrafts. We are presently pursuing this work, along with the synthesis of regularly branched homopolymers, in our laboratories.

Acknowledgment. This work was supported by the U.S. Army Research Office (Grants DAAH04-94-G-0245, DAAH04-95-1-0306, and DAAG55-98-1-0005).

7. References

1. Falk, J.C., Schlott, R., Hoeg, D.F., and Pendleton, J.F. (1973) *Rubber Chem. Technol.* **46**, 1044.
2. Hadjichristidis, N., Iatrou, H., Behal, S.K., Chludzinski, J.J., Disko, M.M., Garner, R.T., Liang, K.S., Lohse, D.J., and Milner, S.T. (1993) *Macromolecules* **26**, 5812.
3. Milner, S.T. (1994) *Macromolecules* **27**, 2333.
4. Pochan, D.J., Gido, S.P., Pispas, S., Mays, J.W., Ryan, A.J., Fairelough, P.A., Hanley, I.W., and Terrill, N.J. (1996) *Macromolecules* **29**, 5091.
5. Pochan, D.J., Gido, S.P., Pispas, S., and Mays, J.W. (1996) *Macromolecules* **29**, 5099.
6. Tselikas, Y., Hadjichristidis, N., Iatrou, H., Liang, K.S., and Lohse, D.J. (1996) *J. Chem. Phys.* **105**, 2456.
7. Gido, S.P., Lee, C., Pochan, D.J., Pispas, S., Mays, J.W., and Hadjichristoidis, N. (1996) *Macromolecules* **29**, 7022.
8. Lee, C., Gido, S.P., Pitsikalis, M., Mays, J.W., Beck Tan, N., Trevino, S.F., and Hadjichristidis, N. (1997) *Macromolecules* **30**, 3732.
9. Lee, C., Gido, S.P., Poulos, Y., Hadjichristidis, N., Beck Tan, N., Trevino, S.F., and Mays, J.W. (1997) *J. Chem. Phys.* **107**, 6460.
10. Lee, C., Gido, S.P., Poulos, Y., Hadjichristidis, N., Beck Tan, N., Trevino, S.F., and Mays, J.W. (1998) *Polymer* **39**, 4631.
11. Xenidou, M., Beyer, F.L., Hadjichristidis, N., Gido, S.P., and Beck Tan, N. (1998) *Macromolecules*, submitted.

12. Buchnall, C.B. (1977) *Toughened Plastics*, Applied Science Publishers, London.
13. Rempp, P.F and Lutz, P.J. (1989) in Eastmond, G.C., Ledwith, A., Russo, S., and Sigwlt, P. (eds.), *Comprehensive Polymer Science* **6**, 403.
14. Pitsikalis, M., Pispas, S., Mays, J.W., and Hadjichristidis, N. (1998), *Adv. Polym Sci.* **135**, 1-137.
15. Olvera de la Cruz, M. and Sanchez, I. S. (1986) *Macromolecules* **19**, 2501.
16. Iatrou, H., Tselikas, Y., Hadjichristidis, N., and Mays, J.W. (1996) in Salamone, J.C. (ed.), *Polymeric Materials Encyclopedis*, CRC Press, Boca Raton, FL 6, 4398-4406.
17. Xie, H. and Xia, J. (1987) *Makromol. Chem.* **188**, 2543.
18. Pennisi, R.W. and Fetters, L.J. (1988) *Macromolecules* **21**, 1094.
19. Mays, J.W. (1990) *Polym. Bull.* **23**, 247.
20. Iatrou, H., Siakali-Kioulafa, E., Hadjichristidis, N., Roovers, J., and Mays, J.W. (1995) *J. Polym. Sci., Phys. Ed.* **33**, 1925.
21. Khan, I.M., Gao, Z., Khougaz, K., and Eisenberg, A. (1992) *Macromolecules* **25**, 3002.
22. Ba-Gia, H., Jerome, R., and Teyssie, P. (1980) *J. Polym. Sci., Polym. Chem. Ed.* **18**, 3483.
23. Naka, A., Sada, K., Chujo, Y., and Saegusa, T. (1991) *Polym. Prepr. Jp.* **40(2)**, E116.
24. Fujimoto, T., Zhang, H., Kazama, T., Isono, Y., Hasegawa, H., and Hashimoto, T. (1992) *Polymer* **33**, 2208.
25. Quirk, R.P. and Yoo, T. (1993) *Polym. Preprints* **34(2)**, 578.
26. Quirk, R.P. and Kim, Y.J. (1996) *Polym. Preprints* **37(2)**, 643.
27. Huckstadt, H., Abetz, V., and Stradler, R. (1996) *Mactomol. Rapid Commun.* **17**, 599.
28. Quirk, R.P., Yoo, T., and Lee, B. (1994) *J.M.S. - Pure Appl. Chem* **A31**, 911.
29. Hsieh, H.L. and Quirk, R.P. (1996) *Anianic Polymerization, Principles and Practical Applications*, Marcel Dekker, New York, pp. 386-388.
30. Huglin, M.B. (ed) (1972) *Light Scattering from Polymer Solutions,* Academic Press, London.
31. Mays, J.W. and Hadjichristidis, N. (1998) unpublished observations.
32. Bushuk, W. and Benoit, H. (1958) *Can. J. Chem.* **36**, 1616.
33. Iatrou, H. and Hadjichristidis, N. (1992) *Macromolecules* **25**, 4649.
34. Morton, M., Helminiak, T.E., Gadkary, S.D., and Bueche, F. (1962) *J. Polym Sci.* **57**, 471.
35. Roovers, J.E.L. and Bywater, S. (1972) *Macromolecules* **5**, 385.
36. Tung, L.H. and Lo, G.Y-S. (1994) *Macromolecules* **27**, 2219.
37. Quirk, R.P. and Ma, J.-J. (1991) *Polym. Int.* **24**, 197.
38. Roovers, J. and Toporowski, P.M. (1981) *Macromolecules* **14**, 1174.
39. Finaz, G., Gallot, Y., Parrod, J., and Rempp, P. (1962) *J. Polym Sci.* **58**, 1363.
40. Strazielle, C. and Herz, J. (1977) *Eur. Polym. J.* **13**, 223.
41. Hahnfeld, J.L., Pike, W.C., Kirkpatrick, D.E., and Bee, T.G. (1996) *Polym. Preprints* **37(2)**, 733.
42. Iatrou, H., Mays, J.W., and Hadjichristidis, N. (1998) *Macromolecules*, submitted.

KINETICS OF ELEMENTARY REACTIONS IN CYCLIC ESTER POLYMERIZATION

S. PENCZEK*, A. DUDA, R. SZYMANSKI, J. BARAN,
J. LIBISZOWSKI, AND A. KOWALSKI
*Department of Polymer Chemistry, Center of Molecular
and Macromolecular Studies, Polish Academy of Sciences,
Sienkiewicza 112, PL-90-363 Lodz, Poland*

Abstract. Kinetics of the polymerization of L-lactide (**LA**) and ε-caprolactone (**CL**) initiated with dialkylaluminum alkoxide, aluminum trialkoxide and related metal alkoxides is reviewed. The former give aggregating active species, and a method is described allowing simultaneous determination of the rate constants of propagation and aggregation equilibrium constants. It has been shown that tin^{II} dicarboxylate (dioctoate) requires a coinitiator in order to start polymerization. Both kinetic and spectroscopic evidence indicate that propagation proceeds on the –Sn-OR bonds. The major reaction accompanying propagation is the chain transfer-to polymer; either inter or intramolecular. Methods are described giving access to the rate constants of both transfers. Finally, correlation is given between the atomic number of metal involved in the active species and the determined rate constants: the larger the atomic number, the higher is the rate constant of propagation and the less selective is the polymerization process.

1. Introduction

Ring-Opening Polymerization (ROP) of cyclic esters has become an efficient tool in studies of the mechanism of anionic and pseudoanionic (covalent) ROP. This is because in many cyclic ester/initiator systems termination does not take place. There are, however, two well documented chain transfer reactions. Both are based on transesterification, known also in polycondensation: back- and/or end-to-end-biting and chain transfer to foreign macromolecules followed by chain rupture.

In this paper we review major kinetic phenomena related to the polymerization of cyclic esters, mostly L-lactide (**LA**) and ε-caprolactone (**CL**), studied during the last few years in our laboratory with initiators belonging to three groups of compounds: $(R)_n(R'O)_m Mt$, $(RO)_p Mt$, and $(RCOO)_q Mt$. In this review we describe polymerizations conducted with dialkylaluminum alkoxide, aluminum trialkoxide, and tin^{II} dicarboxylate.

J.E. Puskas et al. (eds.), Ionic Polymerizations and Related Processes, 283–299.
© 1999 *Kluwer Academic Publishers. Printed in the Netherlands.*

An expanded version of this review will appear in the ACS Monograph Series. **LA** and **CL** are sufficiently strained to polymerize efficiently: the equilibrium monomer concentration is low enough at 25°C (0.1 wt% for **LA** and 0.6 wt% for **CL**) but increases with temperature and may reach a few percent at higher temperatures, when polymerization is conducted in the polymer melt (e.g. in the range 150-200°C). The corresponding equilibrium concentrations at the required temperature can be calculated from ΔH_p and $\Delta S^°_p$, being equal to: (**LA**): -22.9, -25.0 and (**CL**)-28.8, -53.8, respectively. ΔH_p is given in kJ/mol; ΔS in J/mol·K.

2. Initiators

2.1. ALKYLMETAL ALKOXIDES.

In this group of initiators the most extensively studied were dialkylaluminum alkoxides (R_2AlOR') introduced over 30 years ago [1, 2] but studied quantitatively and understood only recently [3 -7].

Depending on the size of substituents [R and R') these initiators are known to exist in solution mostly as dimeric or trimeric species [8-10]. More complicated structures have also been proposed. Aggregates are usually in equilibria with unimeric structures. The rate of interconversion depends on a number of factors (e.g. solvent, temperature, R and R' substituents). Diethylaluminum ethoxide (Et_2AlOEt) used in our studies gives, however, simple ^1H NMR spectrum, indicating that either one type of species (unimers or aggregates) are present or there is a fast exchange between these two forms [11]. The same dialkylalkoxy compound may assume in one solvent the aggregated structure and deaggregated in another one. Deaggregation proceeds easier in polar and nucleophilic solvents, able to interact with Al atoms. If solvent is too strong as a complexing agent, stronger than monomer itself, the "initiator" may be exclusively unimeric, but initiation may not take place at all or become very slow. This kind of behavior was observed in our attempts of initiating the **CL** polymerization by Et_2AlOEt in hexamethylphosphorous triamide (HMPT) [12].

The ^1H NMR spectrum of a low molecular weight oligomer, formed when a high enough ratio $[R_2AlOR']_0/[monomer]_0$ is used, shows exclusively signals expected for the stoichometric reaction product, responsible for further polymerization. The alkyl groups stay intact on Al and only the alkoxy group is involved in initiation [4,11]:

$$\begin{array}{c} R \\ \diagup \\ R \end{array} Al-OR' + (n+1)\; O \overset{O}{\underset{\|}{=}} C \longrightarrow \begin{array}{c} R \\ \diagup \\ R \end{array} Al-O \overset{O}{\underset{\|}{C}} - \left(O \overset{O}{\underset{\|}{C}} \right)_{\overline{n}} OR' \qquad (1)$$

2.2. METAL ALKOXIDES.

Metal alkoxides give stronger aggregates in comparison to alkylmetal alkoxides. This results from the presence of a larger proportion of oxygen atoms, forming bridges between the metal atoms. Therefore, whenever steric factors permit, relatively stable aggregates are formed.

Aluminum *tris*-isopropoxide has been the most often used in polymerization. For a long time it was not clear, why various authors observed different numbers of chains supposedly growing from one aluminum atom. Then, it was found that the two known aggregates, namely a trimer (A_3) and a tetramer (A_4), do not only exchange slowly, in comparison with the rate of polymerization (eq 2):

$$4\ (A_3) \underset{K_{34},\ slow}{\rightleftharpoons} 3\ (A_4) \qquad (2)$$

(where K_{34} denotes the equilibrium constant)

but react with monomers with rates, that differ from 10^2 times (**LA**) to 10^5 times (**CL**), at least at moderate temperatures [13-15]. Similar difference of reactivities between A_3 and A_4 was already reported for the Meerwein-Ponndorf-Verley reaction [16]. However, these differences of reactivities are not observed when Al(O^{i}Pr)$_3$ is used in the presence of an alcohol, since the exchange between A_3 and A_4 is then much faster [17].

Similar behavior of the Al(O^{i}Pr)$_3$ initiator, we observed also in the polymerization of **LA**- the much less reactive monomer [15]. In Table I kinetic parameters are given, revealing difference of A_3 and A_4 reactivities in their reactions with **CL** and **LA** monomers.

Fortunately, A_3 and A_4 can be prepared as individual compounds, or separated from their mixtures, since A_3 is soluble in some solvents in which A_4 is practically insoluble at all (e.g. in pyridine). Therefore, almost pure A_3 and A_4 were obtained and applied then in polymerizations. Moreover, as mentioned above, the rate of interconversion is relatively low: time required for conversion of a few percent of A_4 (unreactive) into A_3 (reactive) is, at room temperature, over 10^3 times longer than time needed for complete (over 90%) polymerization of **CL** or **LA**. Reactivities of both aggregates tend to converge with increasing temperature [15].

TABLE I. Comparison of the Rate Constants of Propagation (k_p) and Initiation (k_i) for Polymerization of CL and LA Initiated with A_3 and A_4[a] [15]

Monomer	ε-caprolactone	L,L-dilactide
$\dfrac{k_p}{mol^{-1} \cdot L \cdot s^{-1}}$	0.5	$7.5H10^{-5}$
$\dfrac{k_i(A_4)}{mol^{-1} \cdot L \cdot s^{-1}}$	$5H10^{-6}$	$2.7H10^{-8}$
$k_i(A_3)/ k_i(A_4)$	10^5	$4.1H10^3$

[a] Conditions of polymerization (concentrations given in $mol \cong L^{-1}$): $[CL]_0 = 2.0$, $[LA]_0 = 1.0$, $3[A_3]_0 = 4[A_4]_0 = [Al(O^iPr)_3]_0 = 0.01$; THF as a solvent, 20°C.

Fortunately, A_3 and A_4 can be prepared as individual compounds, or separated from their mixtures, since A_3 is soluble in some solvents in which A_4 is practically insoluble at all (e.g. in pyridine). Therefore, almost pure A_3 and A_4 were obtained and applied then in polymerizations. Moreover, as mentioned above, the rate of interconversion is relatively low: time required for conversion of a few percent of A_4 (unreactive) into A_3 (reactive) is, at room temperature, over 10^3 times longer than time needed for complete (over 90%) polymerization of CL or LA. Reactivities of both aggregates tend to converge with increasing temperature [15].

A_4 is thermodynamically more favorable at room and lower temperatures, whereas at higher temperatures the equilibrium shifts towards of A_3. Thus, some energy is needed to break A_4 and to produce the less stable (less favorable) A_3 form.
Polymerization with some other initiators, like μ-oxoalkoxides of lanthanium, samarium or yttrium also look today bizzare, and perhaps their closer studies will reveal that their strange behavior (not covered in this review) has its origin in aggregation.

3. Kinetics of Propagation.

It has conclusively been shown in a few papers of our own, that for initiators having the alkoxide groups, all of these groups start the polymer chains. Similar observations were made by some other authors. This relationship has been quantitatively studied for $(C_2H_5)_2AlOR$, $(^iC_4H_9)_2AlOR$, $Al(O^iC_3H_7)_3$ (A_3 and A_4 forms), $Fe(OC_2H_5)_3$, $Ti(O^iC_3H_7)_4$, and $Sn(OC_4H_9)_2$.

3.1. DIALKYLALUMINUM ALKOXIDES

Structures and behavior of active species in propagation reflect the properties of initiators. The "only" difference, when an initiator like dialkylaluminum alkoxide is used, is related to the size of the alkoxy substituent, becoming during propagation a polymer chain. Indeed, propagation initiated with R_2AlOR' proceeds exclusively on the aluminum-oxygen bond, therefore during propagation the size of the alkoxy substituent increases.

It was assumed in further analysis that the total concentration of active species ($[P_i^*]$) is equal to the starting concentration of alkoxide groups in the initiator. Thus, in the present case $[P_i^*]$ is equal to the initiator concentration in the feed ($[I]_0$).
It was observed that the rate of CL polymerization initiated with R_2AlOR' is directly proportional to the first power of concentration of the active species $[P_i^*]$ only at low enough $[P_i^*]$. At higher concentration the rate is either proportional to $[P_i^*]^{1/2}$ when R = ^iBu, or to $[P_i^*]^{1/3}$ when R = Et [6].

It follows from these changes of the kinetic propagation order in initiator (total active species), that aggregation of active species takes place into dimers: ($[P_i^*]^{1/2}$) or trimers ($[P_i^*]^{1/3}$) (where $[P_i^*] = [I]_0$) and that only the deaggregated species propagate, whereas the aggregated ones are dormant. At these conditions only a certain fraction of chains are instantaneously active; they become dormant upon aggregation and restore their reactivity when deaggregate. The corresponding kinetic scheme reads:

$$x \ldots\text{-}P_n^* \overset{k_{ag}}{\underset{k_{dag}}{\rightleftharpoons}} (\ldots\text{-}P_n^*)_x \qquad (K_{ag} = k_{ag}/k_{dag}) \tag{3}$$

$$\ldots\text{-}P_n^* + M \overset{k_p}{\underset{k_d}{\rightleftharpoons}} \ldots\text{-}P_{n+1}^* \quad ; \quad (\ldots\text{-}P_n^*) + M \longrightarrow\!\!\!\times\!\!\!\longrightarrow$$

where k_p, k_d, k_{ag}, and k_{dag} denote the rate constants of propagation, depropagation, aggregation, and deaggregation, respectively; M the monomer molecule; $\ldots\text{-}P_n^*$ the growing macromolecule: $R'O[C(O)(CH_2)_5O]_n\text{-}AlR_2$ in the CL polymerization ; x = 2 for R = ^iBu and x = 3 for R = Et.

Using the analytical method elaborated by us, solution of the kinetic scheme 3 and thus determination of k_p and the k_{ag}/k_{dag} ratio was eventually possible [5-7]. The pertinent data are collected in Table II.

Both k_p and K_{ag} decrease with increasing solvating power of the solvent; this is in agreement with the proposed coordinaton-insertion mechanism of propagation. Moreover, in spite of the fact that degree of aggregation (x) depends on the size of the alkyl substituent at the Al atoms, k_p remains practically the same (at least in THF as a solvent).

Table II. Propagation Rate Constants (k_p) and the Aggregation Equilibrium Constants (K_{ag}) for Polymerization of ε-Caprolactone Initiated with Dialkylaluminum Alkoxides (R$_2$AlOR')[a] [6,7]

Solvent	ε[b]	R$_2$AlOR'	x[c]	$\dfrac{k_p}{mol^{-1} \cdot L \cdot s^{-1}}$	$\dfrac{K_{ag}}{(mol \cdot L^{-1})^{1-x}}$
CH$_3$CN	37	Et$_2$AlOEt	3	$7.5*10^{-3}$	77
THF	7.3	Et$_2$AlOEt	3	$3.9*10^{-2}$	$4*10^4$
C$_6$H$_6$	2.3	Et$_2$AlOEt	3	$8.6*10^{-2}$	$2.4*10^5$
THF	7.3	^{i}Bu$_2$AlOMe	2	$3.9*10^{-2}$	77

[a] Conditions: [CL]$_0$ = 2.0 mol≅L^{-1}, 25°C. [b] Solvent dielectric constant [c] Aggregation degree (scheme 3) (77 appearing two times namely for CH$_3$CN and THF in the table above is an unusual coincidence).

3.2. ALUMINUM TRIS-ALKOXIDE.

Propagation on aluminum tris-alkoxides active species, was a matter of controversy, as discussed in the section describing initiation. Eventually, it has been established that, for both CL and LA three polyester chains grow from one Al atom.

Polymerization of CL exhibited approximately first order of propagation both in monomer (internally) and in active species [14].

Thus, determination of the absolute rate constant of propagation k_p seems to be straightforward. The average k_p value is equal to 0.62 mol^{-1}≅L≅s^{-1} (25°C, THF solvent) [14], thus approximately ten times higher than with dialkylaluminum alkoxides. On the other hand, the ^{27}Al NMR spectrum of the model system reveals that [Al(6)]/[Al(5)] . 3:1, therefore the real k_p could be even higher than that reported above, if polymerization proceeds exclusively on the pentacoordinated (Al.(5)) aluminum atom.

In poly(LA), according to the ^{27}Al NMR spectrum, hexacoordinated species are not present. This may result from the steric hindrance caused by the presence of the methyl groups. Moreover, the fractional orders in active species, observed by us in the polymerization of LA initiated with A$_3$ [15] suggest that an aggregation - deaggregation equilibrium: 2 Al(4)1 ≈ Al(4)2 takes place and propagation proceeds on the non-aggregated species Al(4)1. Therefore the kinetics of LA polymerization was analyzed in terms of propagation with aggregation, like polymerization initiated by dialkylalkoxy aluminum. K_{ag} = 92 mol^{-1}≅L and k_p = 8.2≅10^{-3} mol^{-1}≅L≅s^{-1} (80°C, THF solvent) were determined this way.

3.3. METALCARBOXYLATES – TIN OCTOATES

Tin octoate ($Sn(OC(O)CH(C_2H_5)C_4H_9)_2$), (tin(II) 2-ethylhexanoate), denoted further as $Sn(Oct)_2$), is probably the most often used initiating compound in the polymerization of cyclic esters [18,19,20-34]. However, it is also the least understood, and several "mechanisms" of initiation with $Sn(Oct)_2$ were proposed [e.g., ref. 21,24,30,32].

One of the most developed ideas is based on the assumption that $Sn(Oct)_2$ is not bound to the polyester chain during the chain growth, and that $Sn(Oct)_2$ merely acts as a catalyst, by activating the monomer before the actual propagation step. We shall call this mechanism the "activated monomer mechanism" (AMM) by analogy with a formally similar concept developed earlier: this expression was coined by Szwarc for anionic polymerization of N-carboxy anhydrides [35]. We discovered cationic AMM and elaborated the kinetics analysis of this process [36]. The AMM can be presented schematically for the polymerization of a cyclic esters, according to the papers cited above, in the following way:

(4)

(where K_{ac} denotes the equilibrium constant of activation, k_p the rate constant of propagation; ROH stands for water or alcohol molecule, either present as adventitious impurities or intentionally added).

This mechanism would lead to the simple kinetics, and the rate of polymerization (propagation) (R_p) would be given by the following equation:

$$R_p = k_p K_{ac} \cong [\ldots\text{-OH}]_0 \cong [Sn(Oct)_2]_0 \cong [M]$$

(5)

(where $[\ldots\text{-OH}]_0$ denotes starting concentration of compounds containing the hydroxy group, equal to invariant concentration of hydroxyl groups, [M] is the instantaneous monomer concentration).

The work of our own, to be discussed below, shows that AMM does not operate [18,19] in the $Sn(Oct)_2$/cyclic ester systems. We also proposed a general mechanism of initiation with $Sn(Oct)_2$ (and presumably all of the metal carboxylates) in the polymerization of cyclic esters.

There was no systematic kinetic work before we started our measurements, although a work of Zhang et al. [21,22] has to be mentioned, since these authors came earlier to some conclusions we describe in the present paper. However, our conclusion is based on direct evidence: observation of the growing species, backed by kinetic measurements.

Two series of kinetic experiments were performed with butyl alcohol (BuOH) as a coinitiator. In Figure 1 dependencies of the relative rate of polymerization ($r_p = R_p/[M]$ on $[BuOH]_0$ (for constant $[Sn(Oct)_2]_0$) and on $[Sn(Oct)_2]_0$ (for constant $[BuOH]_0$) are given.

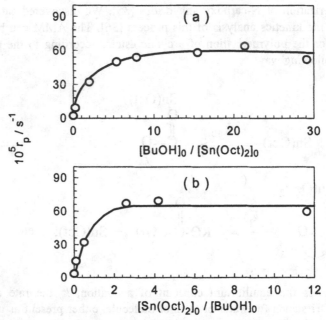

Figure 1. Dependencies of the rate of polymerization (r_p = -d[CL]/[CL]dt = $(1/t)\cong\ln([CL]_0/[CL])$) of (CL) initiated with tin octoate ($Sn(Oct)_2$) on: (a) the ratio of starting concentrations of butyl alcohol (BuOH) to $Sn(Oct)_2$ ($[BuOH]_0/[Sn(Oct)_2]_0$) with $[Sn(Oct)_2]_0 = 0.05$ mol$\cong$L^{-1}, and kept constant through this series of experiments; (b) the ratio of starting concentrations of BuOH to $Sn(Oct)_2$ ($[Sn(Oct)_2]_0/[BuOH]_0$) with $[BuOH]_0 = 0.10$ mol$\cong$L^{-1}, and constant through this series of experiments (THF solvent, 80°C) [19].

These results show that, in both cases, the rate first increases and then levels off. Such a kinetic behavior suggests strongly that $Sn(Oct)_2$ is not a catalyst or initiator exclusively by itself. On the contrary, these results indicate that $Sn(Oct)_2$ has to react with the alcohol added in order to produce the true initiator. The rate of polymerization increases with added alcohol as long as there is enough $Sn(Oct)_2$ in the system to give the initiator (Figure 1 (a)). The same argument holds when $[BuOH]_0$ is kept constant and $Sn(Oct)_2$ is being added (Figure 1 (b)). When all of the BuOH is already used in reaction with

Sn(Oct)$_2$ its further addition does not increase the rate. Thus, the only possibility is that the following reactions take place:

$$Sn(Oct)_2 + ROH \rightleftharpoons OctSn\text{-}OR + OctH$$

$$OctSn\text{-}OR + nM \longrightarrow OctSn\text{-}O\text{-}(m)_n\text{-}R \tag{6}$$

(where R denotes H or alkyl group, OctH the 2-ethylhexanoic acid, M the CL or LA molecules, and m the polyester repeating unit).

Thus, the general mechanism involving Sn(Oct)$_2$ and **CL**, used as an example, can be presented by scheme 7 given below. In our studies by MALDI-TOF-MS we observed all of the populations of macromolecules shown in this scheme, including macromolecules containing Sn atoms, linear and cyclic ones [37].

The actual monomer addition proceeds most probably as a multicenter concerted reaction, similarly to the propagation step we have illustrated for Al-alkoxides [38].

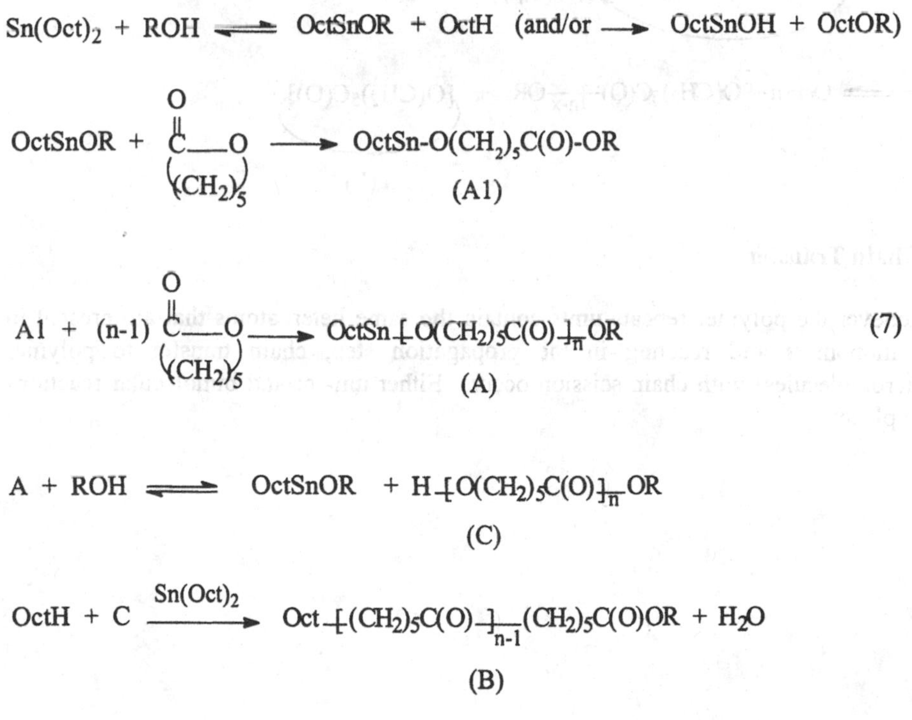

$$Sn(Oct)_2 + ROH \rightleftharpoons OctSnOR + OctH \quad (and/or \longrightarrow OctSnOH + OctOR)$$

$$OctSnOR + \underset{(CH_2)_5}{\overset{O}{\overset{\|}{C}}\!\!-\!\!O} \longrightarrow OctSn\text{-}O(CH_2)_5C(O)\text{-}OR \tag{A1}$$

$$A1 + (n\text{-}1)\,\underset{(CH_2)_5}{\overset{O}{\overset{\|}{C}}\!\!-\!\!O} \longrightarrow OctSn\text{-}[O(CH_2)_5C(O)]_n\text{-}OR \tag{7}$$
$$(A)$$

$$A + ROH \rightleftharpoons OctSnOR + H\text{-}[O(CH_2)_5C(O)]_n\text{-}OR$$
$$(C)$$

$$OctH + C \xrightarrow{Sn(Oct)_2} Oct\text{-}[(CH_2)_5C(O)]_{n\text{-}1}(CH_2)_5C(O)OR + H_2O$$
$$(B)$$

$$Sn(Oct)_2 + H_2O \rightleftharpoons OctSnOH + OctH$$

$$OctSnOH + n \underset{(CH_2)_5}{\overset{\overset{\displaystyle O}{\|}}{C-O}} \longrightarrow OctSn\text{-}[O(CH_2)_5C(O)]_n\text{-}OH$$

$$C + OctSn\text{-}[O(CH_2)_5C(O)]_n\text{-}OH \rightleftharpoons A + H\text{-}[O(CH_2)_5C(O)]_n\text{-}OH$$

(E)

$$OctH + E \xrightarrow{Sn(Oct)_2} Oct\text{-}[(CH_2)_5C(O)]_{n-1}(CH_2)_5C(O)OH + H_2O$$

(D)

$$OctSnO(CH_2)_5C(O)\text{-}[O(CH_2)_5C(O)]_{n-1}\text{-}OR \rightleftharpoons$$

$$\rightleftharpoons OctSn\text{-}[O(CH_2)_5C(O)]_{n-x}\text{-}OR + [O(CH_2)_5C(O)]_x$$

(F)

4. Chain Transfer

Whenever the polymer repeats units contain the same heteroatoms that are present in the monomers and reacting in the propagation step, chain transfer to polymer (macromolecules) with chain scission occurs. Either uni- or/and bimolecular reactions take place:

$$\cdots (m)_n\, m^* + M \quad \underset{k_d}{\overset{k_p}{\rightleftharpoons}} \quad \cdots (m)_{n+1}\, m^* \qquad \text{(propagation)} \qquad \text{(8 (a))}$$

$$\cdots (m)_n\, m^* \quad \underset{k_{p(x)}}{\overset{k_{tr(1)}}{\rightleftharpoons}} \quad \cdots (m)_{n-x}\, m^* + (m)_x \qquad \begin{array}{l}\text{(unimolecular} \\ \text{chain transfer)}\end{array} \qquad \text{(8(b))}$$

$$\cdots (m)_n\, m^* + \cdots (m)_p\, m^* \quad \underset{k_{tr(2)}}{\overset{k_{tr(2)}}{\rightleftharpoons}} \quad \cdots (m)_{n+q}\, m^* + \cdots (m)_{p-q}\, m^* \qquad \text{(8 (c))}$$

$$\text{(bimolecular chain transfer)}$$

(where m denotes the polymer repeating unit, m* - the involved active species, M - the monomer molecule; $k_{tr(1)}$, $k_{tr(2)}$ - the rate constants of the uni- and bimolecular chain transfer, respectively; k_p and $k_{p(x)}$ - the rate constants of propagation of monomer and cyclic x-mer, respectively; k_d - the rate constant of depropagation)

These reactions of transfer are of various intensities and have been observed in polymerizations of cyclic ethers, acetals, esters, amides, and sulfides [39-42]

4.1. INTRAMOLECULAR CHAIN TRANSFER

Initially, we elaborated methods of measuring the $k_p/k_{tr(1)}$ ratio in the polymerization of CL and we correlated this ratio with k_p, determined from the polymerization kinetics. The $k_{tr(1)}$ was determined from the rates of appearance of cyclics measured by SEC [3,11,43,44]. On the basis of eq 8 (a) and 8 (b), assuming $k_p \gg k_d$ and introducing the selectivity parameter $\beta = k_p/k_{tr(1)}$ we have:

$$\beta = \frac{k_p}{k_{tr(1)}} = \frac{\ln([M]_0/[M])}{[M(x)]_{eq} \cdot \ln\{[M(x)]_{eq}/([M(x)]_{eq} - [M(x)])\}} \qquad (9)$$

(where [M] denotes concentration of monomer, $[M]_0$ the starting monomer concentration, [M(x)] the concentration of a cyclic x-mer, $[M(x)]_{eq}$ the equilibrium concentration of a cyclic x-mer)

The corresponding chemistry is visualized in scheme 10:

$$
\cdots \!\!-\!\!(\,)_{\overline{n}}\!\!\overset{O}{\overset{\|}{C}}(CH_2)_5\dot{O}Mt \;+\; \text{[cyclic monomer]} \xrightarrow{\;k_p\;} \cdots\!\!-\!\!(\,)_{\overline{n+1}}\overset{O}{\overset{\|}{C}}(CH_2)_5OMt
$$

$$
k_{tr(1)} \Big\Downarrow\; k_{p(x)}
$$

$$
\cdots\!\!-\!\!(\,)_{\overline{n-x}}\overset{O}{\overset{\|}{C}}(CH_2)_5OMt \;+\; [\overset{O}{\overset{\|}{C}}(CH_2)_5O]_{\overline{x}}
$$

(10)

(where Mt: (e.g.) Na, Sm<, Al<)

The parameters β (for $x = 2$), determined in the polymerization of **CL** conducted with ionic and covalent active species, are compared with the respective propagation rate constants in Table III.

Analysis of these data shows that there are two factors influencing the $k_p/k_{tr(1)}$ ratio, namely the intrinsic reactivity of the growing species and the steric hindrance. The Reactivity-Selectivity Principle has already been discussed in our previous work in terms of the early and late transition states formation [44]. In the most simplified way it could be said, that the less reactive species are more discriminating; therefore with increase of k_p the $k_p/k_{tr(1)}$ ratio is decreasing. The second factor is steric hindrance influencing transfer but not changing the propagation rate constant.

There is only one chain growing from the Al atom for dialkyloalkoxy aluminum and three chains grow from one Al atom when trialkoxy derivative is used. Reaction with a monomer molecule is not very much perturbed by steric hindrance, because of the small size of the monomer. Reaction with a polymer chain requires, however, penetration of the bulky structure through a polymer coil in order to find a convenient conformation for the transfer reaction to proceed. Therefore, for the more bulky growing species k_p may not be (cf. Table III, entries:-AlEt$_2$ and -Al^{i}Bu$_2$) affected, although k_{tr} is decreasing, as discussed above.

TABLE III . Propagation rate constants (k_p) and the selectivity parameters (ß = $k_p/k_{tr(1)}$) for the polymerization of CL (THF, 20°C) [44]

Active species	$\dfrac{k_p}{mol^{-1} \cdot L \cdot s^{-1}}$	$\beta = \dfrac{k_p / k_{tr(1)}}{mol^{-1} \cdot L}$
...-$(CH_2)_5O^{\ominus}Na^{\oplus}$	31.70	$1.6*10^3$
...-$(CH_2)_5O-Sm[O(CH_2)_5-...]_2$	2.00	$2.0*10^3$
...-$(CH_2)_5O-Al(C_2H_5)_2$	0.03	$4.6*10^4$
...-$(CH_2)_5O-Al[CH_2CH(CH_3)_2]_2$	0.03	$7.7*10^4$
...-$(CH_2)_5O-Al[O(CH_2)_5-...]_2$	0.50	$3.0*10^5$

4.2. INTERMOLECULAR CHAIN TRANSFER.

As we already mentioned in the previous paragraphs, both intramolecular and intermolecular chain transfer usually coexist. In the polymerization of CL we could measure *exclusively* the $k_{tr(1)}$ (intramolecular) by observing the rate of cyclics formation, whereas for the polymerization of LA intramolecular transfer seems to be less important and therefore we were able to determine $k_{tr(2)}$ (intermolecular). In the systems studied we secured quantitative and fast initiation (in comparison with propagation). Thus, we had to exclusively consider propagation and the intermolecular transfer. In the absence of any transfer, Poisson distribution should result; when transfer to polymer proceeds, the molar mass distribution (MWD) broaden with conversion. As a first approximation we assumed, that this change of MWD is exclusively due to intermolecular transfer. Further work will introduce a correction, taking into account the intramolecular transfer that can also influence the M_w/M_n ratio.

In the respective kinetic scheme (eq 11) propagation is accompanied by intermolecular chain transfer (bimolecular segmental exchange). Computation was made for various arbitrarily assumed $k_p/k_{tr(2)}$ and monomer to initiator starting concentration ([M]$_0$/[I]$_0$) ratios. Then, the experimentally determined plots of M_w/M_n = f(conversion) were compared with a set of computed dependencies and the best fit gave the $k_p/k_{tr(2)}$ ratio. An example of such computed dependencies is given in Figure 2.

296

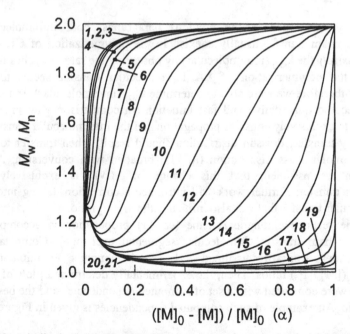

$$(11)$$

(where Mt studied were: Al<, Fe<, La<, Sm<, Sn<, -Ti<)

Figure 2. Numerically simulated plots of M_w/M_n versus monomer conversion degree (α). Assumptions: $[M]_0 = 1.0$ mol≅L^{-1}, $[I]_0 = 10^{-2}$ mol≅L^{-1}; $k_p/k_{tr(2)} = 10^{-3}$ (*1*), 10^{-2} (*2*), $2\cdot10^{-2}$ (*3*), $5\cdot10^{-2}$ (*4*), 10^{-1} (*5*), $2\cdot10^{-1}$ (*6*), $5\cdot10^{-1}$ (*7*), 10^0 (*8*), $2\cdot10^0$ (*9*), $5\cdot10^0$ (*10*), 10^1 (*11*), $2\cdot10^1$ (*12*), $5\cdot10^1$ (*13*), 10^2 (*14*), $2\cdot10^2$ (*15*), $5\cdot10^2$ (*16*), 10^3 (*17*), $2\cdot10^3$ (*18*), $5\cdot10^3$ (*19*), 10^4 (*20*), 10^5 (*21*); $k_p>>k_d$. [45].

Thus, as expected, M_w/M_n increases with monomer conversion reaching eventually the value predicted by the theory of the segmental exchange [46]. This is of $M_w/M_n = 2$, characteristic for the most probable distribution. Moreover, for a given ∀ the higher M_w/M_n are obtained for the lower $k_p/k_{tr(2)}$ ratios.

Two examples are given: the results obtained for the polymerization of **LA** initiated with Al(O^{i}Pr)$_3$ trimer and Fe(OEt)$_3$ are shown below in Figure 3.

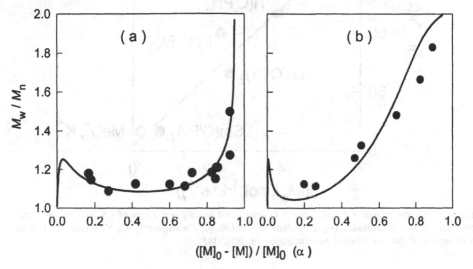

Figure 3. Dependencies of M_w/M_n on degree of monomer conversion (∀) determined for LA polymerization initiated with: (a) {Al(O^{i}Pr)$_3$}$_3$ and (b) Fe(OEt)$_3$. Conditions (concentrations given in mol≡L^{-1}): [LA]$_0$ = 1.0, [Al(O^{i}Pr)$_3$]$_0$ = 10^{-2}, [Fe(OEt)$_3$]$_0$ = 1.4·10^{-3}; THF solvent, 80°C. Points experimental, lines computed assuming $k_p/k_{tr(2)}$ = 100 (a) and 60 (b) [44,45].

Data collected in Figure 4 summarize results obtained in this way for polymerization of **LA** with a series of initiators. The selectivity parameters ($\gamma = k_p/k_{tr(2)}$) for the intermolecular transfer are compared with the corresponding propagation rate constants (k_p). It is remarkable that, at least for metal alkoxides used as initiators (viz. the resulting active species), this dependence conforms to the rules of the Reactivity-Selectivity Principle, i.e. the high selectivities are linked to the low reactivities.

The highest selectivities are provided by tin(II) derivatives (e.g., tin octoate (Sn(Oct)$_2$) or tin dibutoxide (Sn(OBu)$_2$), for which $k_p/k_{tr(2)}$. 200 was detrmined. Unfortunately, for the **LA**/Sn(Oct)$_2$ and **LA**/Sn(OBu)$_2$ polymerizing systems concentrations of the actually (momentary) propagating active species are not yet known and the absolute rate constants of propagation (k_p) could not be determined. Therefore those points are not introduced to the Figure 4.

298

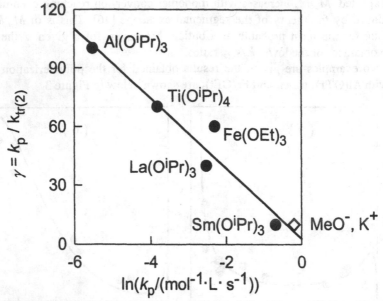

Figure 4. Dependence of $\gamma = k_p/k_{tr(2)}$ on ln k_p determined in polymerizations of LA initiated by metal alkoxides. Covalent alkoxides: (•); ionic alkoxide - MeOK: (◊). Conditions: $[LA]_0 = 1.0$ mol$\cong$L^{-1}, THF as a solvent, 80°C (MeOK initiated polymerization - at 20°C) [45].

5. Conclusion

Polymerization of **LA** and **CL** proceeds on all of the alkoxide groups attached to the metal atom. In the case of metal carboxylates (like $Sn(Oct)_2$) first a -Sn-OR (alkoxide) bond is formed and then polymerization proceeds on this bond. Rate constants of propagation were determined. They increase with increasing atomic metal number in – Mt-OR active species. Chain transfer with chain rupture was quantitatively studied. Finally, correlation between the atomic number of metal involved in the initiator and the rate constants were determined: the larger the atomic number, the higher is the rate constants of propagation and less selective is the polymerization process.

Acknowledgement. This work was supported financially by the Polish State Committee for Scientific Research (KBN) grant 3 T09B 105 11.

6. References

1. Cherdron, H., Ohse, H., and Korte, F. (1962) *Makromol. Chem.* **56**, 187.
2. Hsieh, H.L. and Wang, I.W. (1985) *Am. Chem. Soc., Symp. Ser.* **286**, 161.
3. Hofman, A., Slomkowski, S., and Penczek, S. (1987) *Makromol. Chem., Rapid Commun.* **8**, 387.
4. Duda, A., Florjanczyk, Z., Hofman, A., Slomkowski, S., and Penczek, S. (1990) *Macromolecules* **23**, 1640.

5. Penczek, S. and Duda, A. (1991) *Makromol. Chem., Macromol. Symp.* **47**, 127.
6. Duda, A. and Penczek, S. (1994) *Macromol. Rapid Commun.* **15**, 559.
7. Biela, T. and Duda, A. (1996) *J. Polym. Sci., Part A: Polym.Chem.* **34**, 1807.
8. Mole, T. and Jeffery, E.A. (1972) *Organoaluminium Compounds*, Elsevier Publishing Company, Amsterdam, London, New York.
9. Bradley, D.C., Mehrothra, R.C., and Gaur, D.P. (1978) *Metal Alkoxides*, Academic Press, London, pp. 74, 122.
10. Eisch, J.J. (1982) *"Aluminium"*, in *Comprehensive Organometallic Chemistry*, ed. G.Wilkinson et al., Pergamon Press, Oxford U.K., vol.1, p.583.
11. Penczek, S. and Duda, A. (1996) *Macromol. Symp.* **107**, 1.
12. Duda, A. and Penczek, S. unpublished data.
13. Duda, A. and Penczek, S. (1995) *Macromol. Rapid Commun.* **196**, 67.
14. Duda, A. and Penczek, S. (1995) *Macromolecules* **28**, 5981.
15. Kowalski, A., Duda, A., and Penczek, S. (1998) *Macromolecules* **31**, 2114.
16. Shiner, V. J. and Whittaker, D. (1969) *J. Am. Chem. Soc.* **91**, 394.
17. Duda, A. (1994) *Macromolecules* **27**, 577; (1996), *Macromolecules* **29**, 1399.
18. Kowalski, A., Libiszowski, J., Duda, A., and Penczek, S. (1998) *Polym. Prepr. (Am. Chem. Soc., Div. Polym. Chem.)* **39(2)**, 74.
19. Kowalski, A., Duda, A., and Penczek, S. (1998) *Macromol. Rapid Commun.* **19**, in press.
20. Kleine, J. and Kleine, H.H. (1959) *Makromol. Chem.* **30**, 23.
21. Leenslag, J.W. and Pennings, A.J. (1987) *Makromol. Chem.* **188**, 1809.
22. Jamshidi, K., Eberhard, R.C., Hyon, S.-H., and Ikada, Y. (1987) *Polym. Prepr. (Am. Chem. Soc.,Div. Polym. Chem.)* **28(1)**, 236.
23. Nijenhuis, A.J., Grijpma, D.W., and Pennings, A.J. (1992) *Macromolecules* **25**, 6419.
24. Doi, Y.J., Lemstra, P.J., Nijenhuis, A.J., van Aert, H.A.M., and Bastiaansen, C. (1995) *Macromolecules* **28**, 2124.
25. Dahlman, J., Rafler, G., Fechner, G., and Meklis, B. (1990) *Brit. Polym. J.* **23**, 235.
26. Rafler, G. and Dahlman, J. (1992) *Acta Polym.* **43**, 91.
27. Dahlman, J. and Rafler, G. (1993) *Acta Polym.* **44**, 103.
28. Zhang, X., Wyss, U.P., Pichora, D., and Goosen, M.F.A. (1992) *Polym. Bull. (Berlin)* **27**, 623.
29. Zhang, X., MacDonald, D.A., Goosen, M.F.A., and McCauley, K.B. (1994) *J. Polym. Sci., PartA: Polymer. Chem.* **32**, 2965.
30. Kricheldorf, H.R., Kreiser-Saunders, I., and Boettcher, C. (1995) *Polymer* **36**, 1253.
31. In't Veld, P.J.A., Velner, E.M., van de Witte, P., Hamhuis, J., Dijkstra, P.J., and Feijen, J. (1997) *J. Polym. Sci., Part A: Polym. Chem.* **35**, 219.
32. Schwach, G., Coudane, J., Engel, R., and Vert, M. (1997) *J. Polym. Chem., Part A: Polym. Chem.* **35**, 3431.
33. Witzke, D.R., Narayan R., and Kolstad, J.J. (1997) *Macromolecules* **30**, 7075.
34. Storey, R.F. and Taylor, A.E. (1998) *J. Macromol. Sci.-Pure Appl. Chem.* **A35**, 723.
35. Szwarc, M. (1966) *Pure Appl. Chem.* **12**, 127.
36. Penczek, S., Sekiguchi, H., and Kubisa, P. (1997) Activated monomer polymerization of cyclic monomers, in Hatada, K., Kitayama, T., and Vogl, O. (eds), *Macromolecular Design of Polymeric Materials*, Marcel Dekker, Inc, pp 199-221.
37. Kowalski, A., Duda, A., and Penczek, S., to be published.
38. Penczek, S., Duda, A., and Libiszowski, J. (1998) *Macromol. Symp.* **128**, pp 241-254.
39. Rozenberg, B.A., Irzhak, V.I., and Enikolopyan, N.S. (1975) *Interchain Exchange in Polymers* (in Russian), Khimiya, Moscow.
40. Sosnowski, S., Slomkowski, S., Penczek, S., and Reibel, L. (1983) *Makromol. Chem.* **184**, 2159.
41. Ito, K. and Yamashita, Y. (1978) *Macromolecules* **11**, 68.
42. Goethals, E.J., Simonds, R., Spassky, N., and Momtaz, A. (1980) *Makromol. Chem.* **181**, 2481.
43. Penczek, S., Duda, A., and Slomkowski, S. (1992) *Makromol. Chem., Macromol. Symp.* **54/55**, 31.
44. Baran, J., Duda, A., Kowalski, A., Szymanski, R., and Penczek, S. (1998) *Macromol. Symp.* **128**, 241.
45. Penczek, S., Duda, A., and Szymanski, R. (1998) *Macromol. Symp.* **132**, 441.
46. Flory, P.J. (1942) *J. Am. Chem. Soc.* **64**, 2205.

SYNTHESIS AND PROPERTIES OF AMPHIPHILIC AND FUNCTIONAL COPOLYMERS AND NETWORKS

R. VELICHKOVA, D. CHRISTOVA, I. PANCHEV, V. GANCHEVA

Institute of Polymers, Bulgarian Academy of Sciences
Acad. G. Bonchev St. 103A, 1113 Sofia, Bulgaria

Abstract. Well-defined amphiphilic and functional copolymers and networks based on cyclic ethers, oxazolines and aziridines were prepared. The synthesis of copolymers with controlled hydrophilic-hydrophobic balance, composition and structure was achieved using sequential living cationic polymerization, copolymerization of macromonomers, functionalization of conventional polymers and following grafting or crosslinking by telechelic polymers. Properties, displaying the amphiphilic nature of the copolymers were examined: solubility, swelling in water and hydrocarbons. Anion-exchange and sorption capacity towards metal ions of the copolymers bearing quaternary ammonium groups were evaluated. Environmentally responsive conformational changes, aggregation and microphase separation were observed for some of the products.

1. Introduction

The development of advanced materials as specialized polymers able to compete efficiently with traditional materials and to extend their application, requires well-defined and specially shaped macromolecules. Amphiphilic and functional copolymers are among the numerous structures currently receiving increasing interest as high performance materials. Amphiphilic copolymers are macromolecules consisting of covalently bonded blocks or segments of different chemical nature and opposite philicity, usually hydrophilic and hydrophobic [1-3]. The amphiphilic nature of these copolymers containing incompatible parts gives rise to their unique properties in selective solvents, at surfaces and interfaces as well as in bulk. Their characteristic self-organization in the presence of selective media (solvents or surface) often results in formation of aggregates such as micelles, microemulsions and adsorbed polymer layers.

Functional polymers are macromolecules bearing specific functions or functional groups on the backbone or side chains. Functional groups, besides imparting specific properties, are very efficient in promoting and stabilizing particular self-organization of the blocks by complexation, co-ordination or H-bonding.

J.E. Puskas et al. (eds.), Ionic Polymerizations and Related Processes, 301–323.
© *1999 Kluwer Academic Publishers. Printed in the Netherlands.*

The most attractive features of the amphiphilic block copolymers, based on their inherent self-assembling ability, include micellization in solvents and microphase separation in bulk. The driving force for the self-organization into micelles or microdomains is the incompatibility of the blocks, characterized by Flory-Huggins interaction parameter, χ. Similarly to the classical surface active agents, amphiphilic copolymers form micelles in selective solvents with a core consisting of the insoluble block and a shell of the soluble one. Changing the polarity of the solvent, inverse micelles could be obtained. Among the important parameters for the applications based on the micellar aggregates are the critical micelle concentration (cmc), the aggregation number, Z (the number of macromolecules in the micelle) and the densities of the core and the shell. By the analogy with low-molecular weight surfactants the aggregation number, Z could be determined from Equation (1), where N_A and N_B are the degrees of polymerization of the insoluble and the soluble block, respectively; and Z_o is the packing parameter of the copolymer [4].

$$Z = Z_o . N_A^2 . N_B^{-0.8} \tag{1}$$

The micelles are able to solubilize organic compounds and inorganic materials within the core. The solubilization efficiency is determined mainly by the difference in the Hildebrand solubility parameters of the core-forming blocks and the solubilizate and depends on the molecular volume of the solubilizate and its interfacial tension against the solvent. The solubilization increases the micellar core radius and the aggregation number thus decreasing the cmc.

Amphiphilic copolymers are interfacially active and self-organize at interfaces to form various thin films assemblies of highly ordered structures. This feature is a base for their application as stabilizers for colloidal dispersions and emulsions. In bulk, the repulsive interactions between the unlike blocks or segments, which are the thermodynamic driving force for demixing, induce microphase separation into domains. The microdomain formation and the morphology of the domain structures are a result of the balance between the repulsion of the unlike blocks which causes the domains growth and reduces the surface/volume ratio and the opposing entropic forces.

The structure and the nature of the amphiphilic copolymers constituents determine their surface activity and morphology parameters. The substantial advantage of the polymeric amphiphiles is the possibility to adjust the level of interaction, the kinetic stability of the aggregates and the critical micelle concentration by changing the structure of the copolymer.

The most powerful modes for the synthesis of amphiphilic copolymers with predictable molecular architecture are the living polymerization processes. Under appropriate reaction conditions they proceed without transfer and termination thus providing opportunities to control the molecular weight parameters, sequence of blocks, structure and end-functionality of the copolymers. Recently, in addition to the anionic [5, 6] living cationic [7, 8], cationic ring-opening [9], group-transfer [10] and even radical [11] polymerizations have been achieved.

This paper summarizes our results in the synthesis of well-defined amphiphilic copolymers and polymer networks based on O- and N-containing cyclic monomers and investigation of their properties. Cyclic ethers, 2-oxazolines, *tert*-butyl aziridine, isoprene and styrene were chosen as monomers for the synthesis of well-defined building blocks due of the living nature of their polymerization and their useful properties.

2. Block Copolymers

Amphiphilic copolymers of AB and ABA type containing blocks of opposite philicity and/or flexibility were prepared by the well-known sequential living polymerization - anionic and cationic - and by coupling reaction between living propagating species and prepolymers with appropriate end-functions. Cationic ring-opening polymerization of N-tert-butylaziridine (TBA) showing living nature [12] provides possibility for precise control of molecular weight characteristics and subsequent functionalization followed by chain extension. Diblock copolymers were obtained by end-capping the living polyTBA species with defined prepolymers functionalized with carboxylic or alcoholate end-groups. Thus, the hydrophobic, rigid and crystallizable polyTBA chains were coupled with blocks of the highly hydrophilic, flexible, crystallizable poly(ethylene oxide) (polyEO); highly hydrophilic, rigid poly(N-vinyl pyrrolidone) (polyNVP) or hydrophobic elastomeric polyisoprene (polyIs) [13]. In consequence of the amphiphilic nature of polyTBA-*b*-polyEO, the $\overline{M}_n$ values obtained by GPC in THF were essentially underestimated in comparison with $\overline{M}_n$, calculated from the ^{1}H NMR spectra. This phenomena could be explained with intramolecular hydrophobic interaction of polyTBA blocks which causes shrinkage of the molecules and the small hydrodynamic volumes lead to apparently low molecular weight (Table 1).

TABLE 1. Block copolymers obtained by end-capping living polyTBA with polymeric nucleophiles.

Living polyTBA	Polymeric nucleophile			Block copolymer		
$\overline{M}_n$ (NMR)	Structure	$\overline{M}_n$ (NMR)	$\overline{M}_n$ (GPC)	$\overline{M}_n$ (NMR)	T_g, °C (DSC)	T_m, °C (DSC)
3 900	polyIs-COOH	3 100	7 800	6 200	24	130
4 600	polyEO-ONa	2 000	1 100	6 500	20	37; 128
2 800	polyNVP-COOH	900	3 000	3 600	-	-

The expected phase separation of these block copolymers was proved by DSC measurements. The thermograms of polyTBA-*b*-polyEO, shown in Figure 1, illustrate the influence of the solvent/precipitant system used for isolation on the morphology of polyTBA-*b*-polyEO copolymer consisting of two crystallizable blocks. The product is precipitated from THF in water and then dried. The tendency for hydrophobic association in solution is more pronounced and the formation of the ordered phase starts from polyTBA. During the first heating, intensive melting endotherm at 141 °C for polyTBA is observed whereas the melting of polyEO is almost undetectable. Evidently, cooling at 10 °C/min rate does not allow polyTBA crystallization and favors ordering of polyEO chains. Thus, during the second heating two melting endotherms are present: strong at 44 °C for polyEO and week at 132 °C for polyTBA.

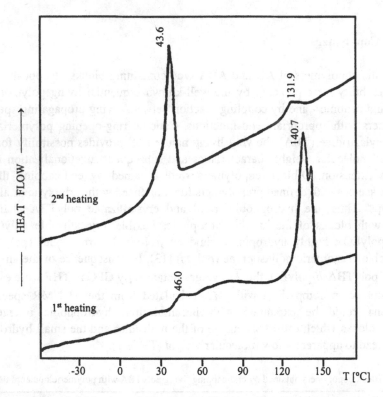

Figure 1. DSC-curves of first and second heating of polyTBA-*b*-polyEO.

Another promising route to realize the advantages of living cationic ring-opening polymerization is the sequential block growth. It was applied for the synthesis of poly(THF-*b*-MeOx) diblock copolymers which possess emulsifying properties [14]. In cooperation with Goethals *et al.* [15] this approach was used to prepare poly(MeOx-*b*-THF-*b*-MeOx) ABA triblock copolymers. A series of well-defined linear triblock copolymers were synthesized according to Scheme 1. The composition of the block copolymers was varied by changing the molecular weight of the central polyTHF block from 4000 to 19000 g/mol keeping while the molecular weight of the external blocks constant at a value of 1500 g/mol.

On the basis of the large differences in solubility parameters of polyTHF ($\delta=17,3$ MPa$^{1/2}$) and polyMeOx ($\delta=25$ MPa$^{1/2}$), microphase separation in solution and in bulk was expected. In addition, both constituents strongly differ in flexibility, crystallinity and transition temperatures. PolyTHF is a flexible, semi-crystalline polymer with a T_g of about -65 °C and melting point ranging from 20 °C to 40 °C depending on the molecular weight, while the polyMeOx is rigid and amorphous with a T_g ca. 80 °C. The phase separation was established by DSC and DMA analysis in all studied compositions.

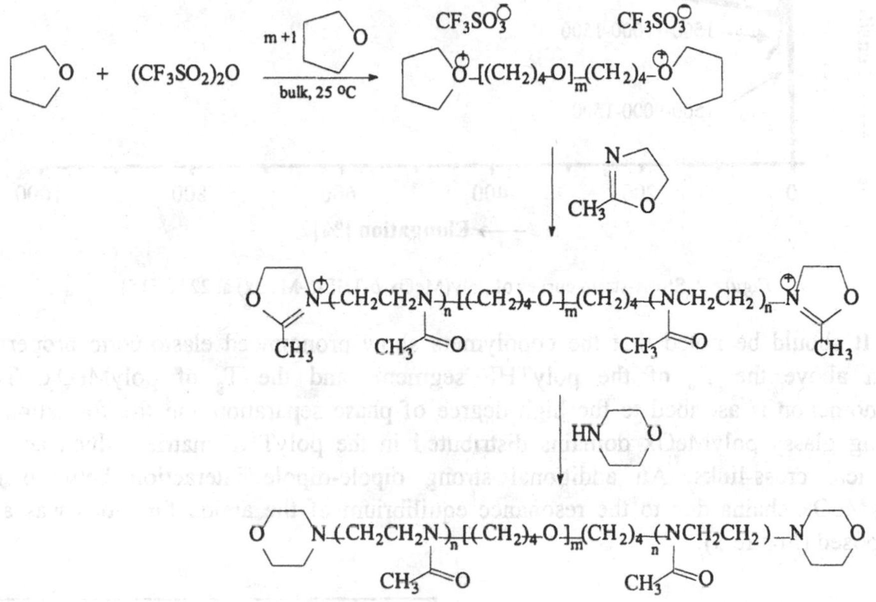

Scheme 1.

The stress-strain curves of the copolymers of several different composition are presented in Figure 2. The samples with a central polyTHF block of $\overline{M}_n$ ca. 13000 g/mol and higher show high elasticity moduli at small tensile stress. These excellent elastomeric properties point to stress-induced crystallization in the polyTHF continuous phase. A lower limit of molecular weight of polyTHF blocks exists, below which the segments are not able to crystallize.

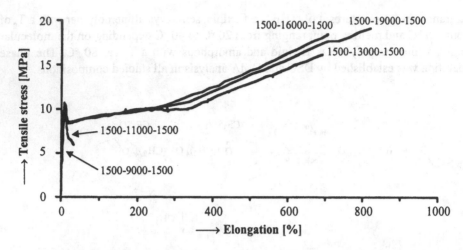

Figure 2. Stress-strain curves of poly(MeOx-*b*-THF-*b*-MeOx) at 22 ºC [15].

It should be noted that the copolymers show pronounced elastomeric properties even above the T_m of the polyTHF segments and the T_g of polyMeOx. This phenomenon is ascribed to the high degree of phase separation and the formation of strong glassy polyMeOx domains distributed in the polyTHF matrix, which acts as physical cross-links. An additional strong dipole-dipole interaction between the polyMeOx chains due to the resonance equilibrium of the amide functions was also proposed (Figure 3).

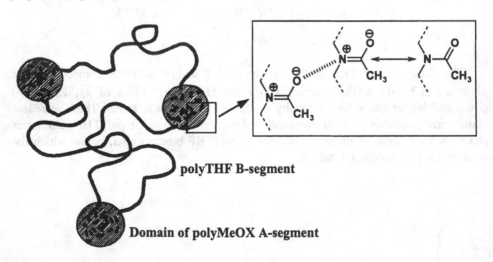

Figure 3. Proposed model of phase separation and dipole-dipole interaction in poly(MeOx$_{1500}$-THF$_{19000}$-MeOx$_{1500}$) [15].

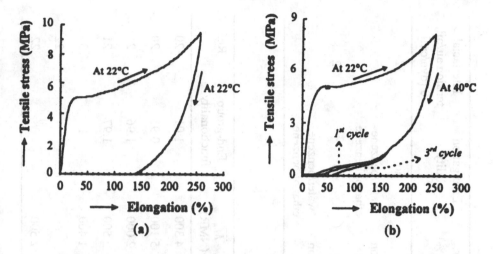

Figure 4. Cyclic tension behavior of poly(MeOx-THF-MeOx) block copolymers a) loading and unloading at 22 °C, and b) loading at 22 °C and unloading at 40 °C [15].

Very attractive in view of potential applications is the observed pronounced shape memory effect in the copolymers with molecular weight of the polyTHF blocks exceeding 13 000 g/mol (Figure 4). When samples are elongated for 250% at 22 °C the residual deformation is almost permanent due to the crystallization of the polyTHF segments. Heating of the stretched samples to a temperature above the melting point of the polyTHF crystals (40 °C) causes abrupt shrinkage and almost complete deformation recovery (Figure 4, b). This stress-relaxation behavior is reversible. The observed shape memory effect is the consequence of the macromolecular organization of the amphiphilic block copolymer which results in response to the external stimuli.

Sequential living anionic copolymerization of styrene and isoprene is a very convenient method for the preparation of thermoplastic elastomers with definite molecular weight and microstructure [5, 16]. Chain extension of living polySt-polyIs copolymers with polyEO blocks was performed by two methods: 1) sequential polymerization of ethylene oxide after transformation of the active species (carbanions) into potassium alkoxides; 2) coupling reactions between the alkoxide-ended copolymers and α-methoxy-ω-chloro-polyEO (Table 2) [17]. Water absorption properties are strongly influenced by the length of the polyEO segments. Films of polymers with polyEO blocks of mid-range molecular weight swell in water (water uptake ca. 50-150%). Copolymers with relatively long polyEO blocks stabilize water/benzene and ethanol/benzene emulsions for weeks. Their amphiphilic nature is also illustrated by the foam-forming ability. A stable equilibrium surface tension of 32 dyn/cm is found in oil-air system at 1 wt.% concentration of copolymers.

TABLE 2. Tri- and pentablock copolymers by sequential living anionic polymerization [17].

Structure	$\overline{M}_n$ of the respective block (NMR)	$\overline{M}_n$ (GPC)	Water uptake (%)	Emulgation ability	Interfacial tension air/oil, θ [dyn/cm]
pSt-b-pIs-b-pEO	1100-2900-500	39 000	59	no	32
pEO-b-Is-b-pSt-b-pIs-b-pEO	5500-8400-44000-8400-5500	56 000	emulsion	water/benzene ethanol/benzene	
pIs-b-pSt-b-pEO	11000-18500-700	29 000	150	no	32
pIs-b-pSt-b-pEO	16800-13500-2000	35 600	emulsion	water/benzene ethanol/benzene	

TABLE 3. Macromonomers by living cationic and anionic polymerization.

Structure	$\overline{M}_n$ (GPC)	MWD (GPC)	$\overline{M}_n$ (NMR)	End-group functionality	Ref.
polyTBA-NHCH$_2$CH=CH$_2$	4 500	1.25	4 300	1.01	20
polyTBA-OCH$_2$CH$_2$OCOC(CH$_3$)=CH$_2$	6 100	1.19	5 100	0.95	20
CH$_2$=CHCOO-polyMeOx-OCOCH=CH$_2$	1 800	1.11	2 000	1.96	21
CH$_2$=CHCOO-polyEtOx-OCOCH=CH$_2$	2 100	1.14	2 500	1.97	21
polyMeOx-N(CH$_2$CH$_2$OH)$_2$	1 600	1.07	1 400	1	33
polyGly^{protected}-CH$_2$C$_6$H$_5$CH=CH$_2$	2 600	1.19			22
polyGly-CH$_2$C$_6$H$_5$CH=CH$_2$	2 800	1.25	2 500		22

3. Graft Copolymers and Networks

Recently, an aspect of considerable interest in the polymer synthesis is related to graft copolymers constituted of main chains and grafts quite different in hydrophilicity and/or rigidity [4, 18, 19]. Thus, keeping the inherent properties of the backbone, it is possible to "add" the properties of the side chains. The co-existence of hydrophilic and hydrophobic segments in the macromolecule and their proper orientation favors the self-assembling and microphase separation of these copolymers by the analogy with the linear counterparts. However, because of their specific structure, they considerably differ in some features.

In our investigations different synthetic approaches were applied for preparation of graft copolymers, namely: copolymerization of macromonomers; "grafting from" method and grafting telechelics onto conventional functionalized polymers.

3.1. MACROMONOMER METHOD

The copolymerization of macromonomers, hydrophilic or hydrophobic, with comonomers of opposite philicity offers precise control over the graft length, however the number of grafts per polymer chain and their distribution along the backbone could be controlled only in certain cases. In order to prepare amphiphilic graft copolymers, two types of macromonomers were synthesized (Table 3) - hydrophobic, based on TBA, and hydrophilic, based on MeOx and glycidol (Gly). Hydrophobic polyTBA macromonomers and hydrophilic polyoxazoline macromonomers were prepared by reacting the propagating cationic species with nucleophiles bearing polymerizable groups.

Hydrophilic polyglycidol (polyGly) macromonomers were synthesized by living anionic polymerization [22]. To obtain linear well defined polymers, the hydroxyl group of the monomer was protected according to a described method [23]. Macromonomers were prepared by end-capping the living anionic species with p-chloromethyl styrene (Scheme 2). A successful deprotection with formic acid was carried out without any degradation of the polyether chains.

PolyTBA macromonomers were radically copolymerized with N-vinyl-2-pyrrolidone (NVP), HEMA and Is (Table 4). Two types of products were obtained: 1) Amphiphilic copolymers with hydrophilic polyNVP or polyHEMA main chains and hydrophobic uniform polyTBA grafts; 2) Hydrophobic copolymers of elastic polyIs main chains and stiff crystallizable polyTBA side chains (with controlled length), which become amphiphilic after quaternization of the polyaziridine grafts. The amphiphilic behavior of the copolymers was manifested by changed solubility, e.g. polyNVP-g-polyTBA dissolves in ethanol (precipitant for polyTBA) and in CHCl$_4$ (precipitant for polyNVP). Depending on the composition, the quaternized polyIs-polyTBA graft copolymers swell up to 300 % or form emulsions in water.

It is known that GPC analysis, when calibrated with linear standard polymers gives underestimates of the molecular weights for branched polymers [24]. This discrepancy is more pronounced for amphiphilic graft copolymers, as has been established for similar copolymers [25]. As seen from Table 4, the $\overline{M}_n$ values of the graft copolymers measured by GPC are several times lower than expected from the NMR analysis and even lower than that of the initial macromonomers.

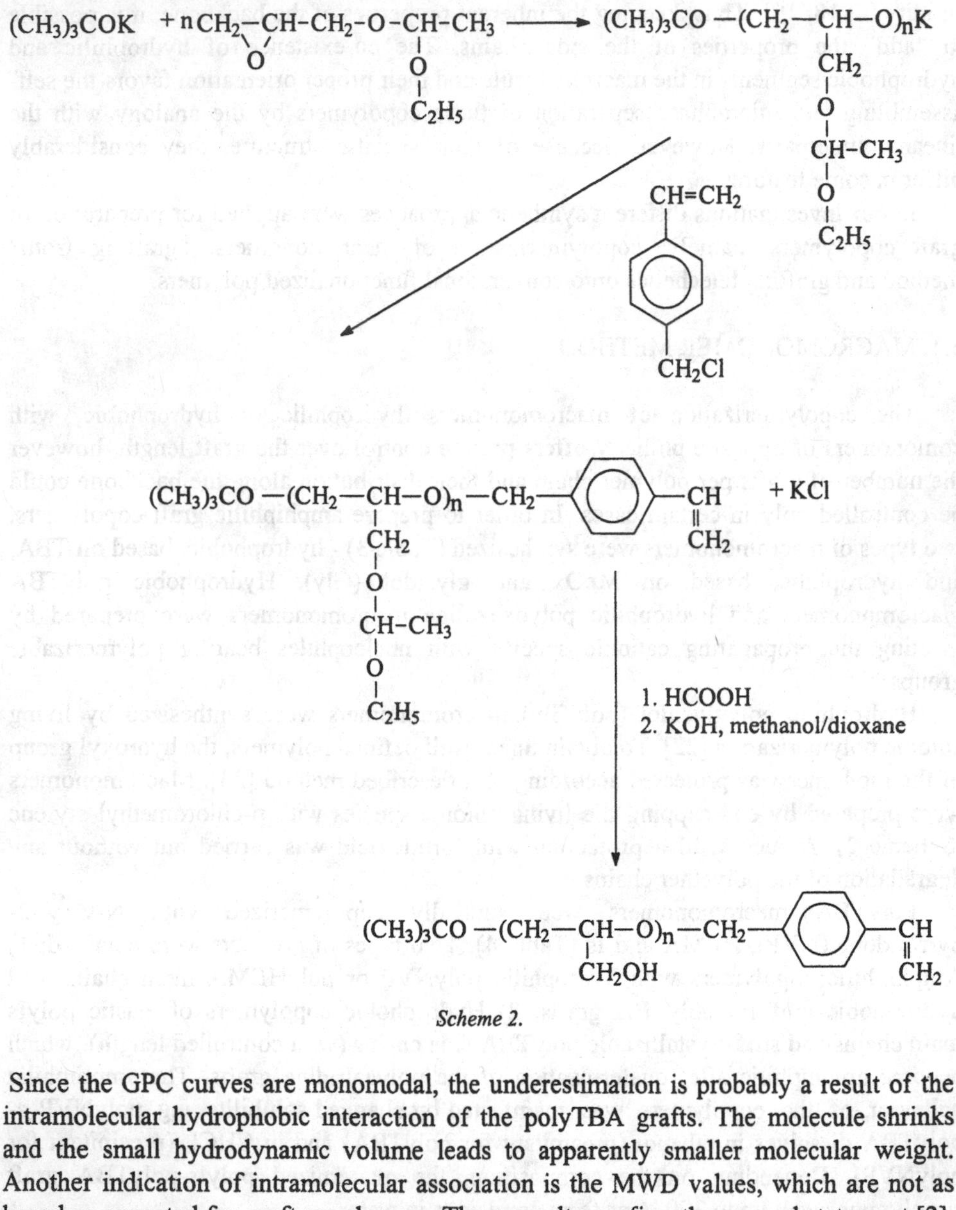

$(CH_3)_3COK$ $+ n\,CH_2{-}CH{-}CH_2{-}O{-}CH{-}CH_3$ $\longrightarrow$ $(CH_3)_3CO{-}(CH_2{-}CH{-}O)_nK$

Scheme 2.

Since the GPC curves are monomodal, the underestimation is probably a result of the intramolecular hydrophobic interaction of the polyTBA grafts. The molecule shrinks and the small hydrodynamic volume leads to apparently smaller molecular weight. Another indication of intramolecular association is the MWD values, which are not as broad as expected for graft copolymers. These results confirm the general statement [2], that mainly intramolecular micelle-like association occurs for copolymers with long pendant hydrophobic grafts on flexible hydrophilic chains, depending on the concentration, molar volume and the relative placement of the hydrophobe.

TABLE 4. Graft copolymers from polyTBA macromonomers [20].

Macromonomer			Comonomer	Copolymer					
Structure	$\overline{M}_n$ (GPC)	MWD (GPC)		$\overline{M}_n$ (GPC)	MWD (GPC)	Mole fraction of pTBA (NMR)	Content of pTBA in wt.% (NMR)	T_g, °C (DSC)	T_m, °C (DSC)
$CH_3(NCH_2CH_2)n\text{-}NHCH_2CH=CH_2$ $C(CH_3)_3$	4 500	1.19	NVP	3 200	3.01	0.26	23.9	21	130
$CH_3(NCH_2CH_2)n\text{-}OCH_2CH_2OCC=CH_2$ $\overset{\|}{O}$ CH_3 $C(CH_3)_3$	6 100	1.20	HEMA			0.11	8.7	96	-
$CH_3(NCH_2CH_2)n\text{-}OCH_2CH=CH_2$ $C(CH_3)_3$	5 800	1.25	Is	4 200	1.84	0.32	40.7	15	133
$CH_3(NCH_2CH_2)n\text{-}OCH_2CH_2OCC=CH_2$ $\overset{\|}{O}$ CH_3 $C(CH_3)_3$	5 200	1.21	Is	14 100	2.47	0.06	8.8	-56	136

In bulk, the strong tendency of TBA side chains to phase separate was also observed (Table 4). For polyNVP-g-polyTBA, both T_m and T_g are observed for TBA phase, while for the copolymer with HEMA the rigid backbone and the low polyaziridine content hamper the polyTBA segregation. Thermal behavior of polyIs copolymers is influenced mainly by the length of the flexible segments. Long polyIs backbone enable the polyTBA grafts to reach a high degree of organization and crystalline phase with T_m = 136°C is observed. In the copolymer with short polyIs segments, but with high polyTBA contents, amorphous polyaziridine phase is detected along with the crystalline one.

The Polyglycidol macromonomer was expected to be a non-ionic amphiphilic macromonomer, as it combines a hydrophobic polymerizable styryl group with a strongly hydrophilic polyether chain containing a hydroxyl group in each repeat unit. The macromonomer of polyglycidol with vinylbenzyl end-groups was copolymerized with styrene in benzene solution and in water emulsion (Table 5). In the solution copolymerization, the acetal form of the macromonomer (I) was used. The amphilphilic copolymer was obtained after the removal of the protective acetal groups. In the emulsion polymerization the polyglycidol macromonomer (II) acts as a polymer surfactant, as well as a comonomer.

TABLE 5 Copolymerization of protected and deprotected polyGly macromonomer with styrene [22].

Macromonomer (M₁)	Type of polymerization	Graft copolymer		
		Conversion of M₁, %	$\overline{M}_n$ (GPC)	MWD (GPC)
PolyGly^{protected}-CH₂C₆H₅CH=CH₂	solution	62	59 000	1.7
polyGly-CH₂C₆H₅CH=CH₂	dispersion	81	81 000	3.7

3. 2. GRAFTING FROM POLYMERS WITH REACTIVE SITES

This method was applied for the synthesis of amphiphilic graft copolymers from poly(vinyl acetate) (polyVAc) and poly(ethylene-co-vinyl acetate) (EVA), which are commercially available in a variety of compositions and molecular weights. The starting polymers polyVAc and EVA were subjected to complete alkaline or partial acidic alcoholysis. Two functionalization reactions - phosgenation and tosylation - were applied to convert the hydroxyl groups into reactive sites able to initiate cationic ring-opening polymerization of oxazolines [26-29].

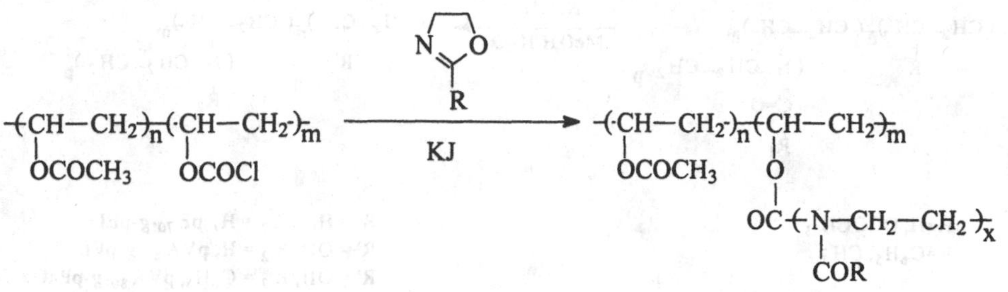

$$\begin{array}{c} \overset{\displaystyle N\!\!-\!\!O}{\underset{\displaystyle R}{\diagdown\!\!\diagup}} \\ \end{array}$$

$+\!\!\left(\!CH\!-\!CH_2\!\right)_{\!n}\!\left(\!CH\!-\!CH_2\!\right)_{\!m}\quad\xrightarrow{\quad KJ \quad}\quad +\!\!\left(\!CH\!-\!CH_2\!\right)_{\!n}\!\left(\!CH\!-\!CH_2\!\right)_{\!m}$
$\quad\;\; \underset{\displaystyle OCOCH_3}{|}\quad \underset{\displaystyle OCOCl}{|} \qquad\qquad\qquad\qquad\qquad \underset{\displaystyle OCOCH_3}{|}\quad \underset{\displaystyle O}{|}$

$$OC\!\!\left(\!N\!-\!CH_2\!-\!CH_2\!\right)_{\!x}$$
$$\underset{\displaystyle COR}{|}$$

Mn = 28 000;	Mn = 43 000; x = 10
n = 270, m = 48	R = -CH₃, -C₆H₅

Scheme 3.

The chloroformates are efficient polyfunctional macroinitiators for grafting of hydrophilic MeOx and hydrophobic PhOx (Scheme 3). The disadvantage of the method is that the polyoxazoline grafts are bound to the main chain by means of hydrolizable urethane bonds, which limits their potential applications. When the macromolecular chloroformates were reacted with α-hydroxy-ω-methoxy polyEO, graft copolymers with more stable carbonate bond were obtained (Scheme 4) [28].

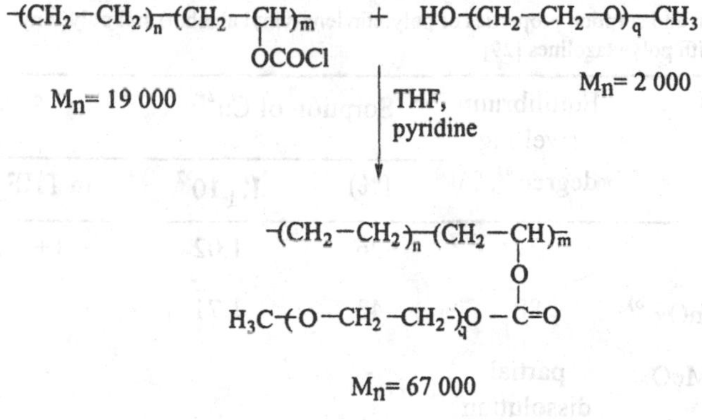

$$+\!\!\left(\!CH_2\!-\!CH_2\!\right)_{\!n}\!\!\left(\!CH_2\!-\!\underset{\underset{\displaystyle OCOCl}{|}}{CH}\!\right)_{\!m} \quad + \quad HO\!\!\left(\!CH_2\!-\!CH_2\!-\!O\!\right)_{\!q}\!\!CH_3$$

$$M_n = 19\ 000 \qquad\qquad\qquad\qquad\qquad M_n = 2\ 000$$

$$\Big\downarrow \begin{array}{l} THF, \\ pyridine \end{array}$$

$$+\!\!\left(\!CH_2\!-\!CH_2\!\right)_{\!n}\!\!\left(\!CH_2\!-\!\underset{\underset{\displaystyle O}{|}}{CH}\!\right)_{\!m}$$

$$H_3C\!\!\left(\!O\!-\!CH_2\!-\!CH_2\!-\!\right)_{\!q}\!\!O-\overset{\displaystyle |}{C}\!\!=\!\!O$$

$$M_n = 67\ 000$$

Scheme 4.

In contrast to polychloroformate initiators, tosylate derivatives of poly(ethylene-vinyl alcohol) and poly(vinyl acetate-vinyl alcohol) form unhydrolizable bond between backbone and side chains [29]. The advantage of such stable bond is that amphiphilic copolymers with different structure could be prepared by selective hydrolysis of both ester and amide functions (Scheme 5).

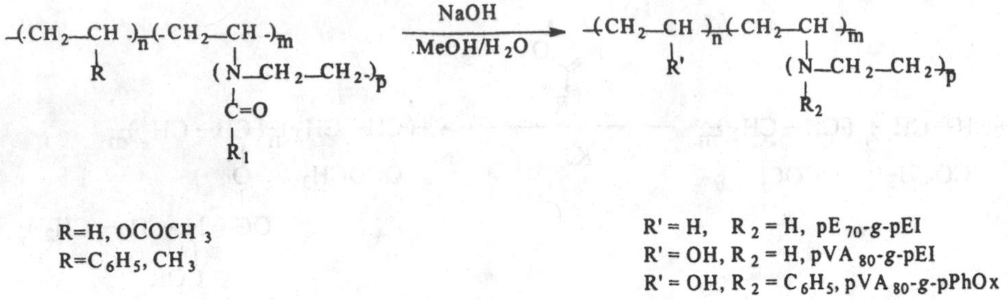

$$-(CH_2-CH)_n-(CH_2-CH)_m \quad \xrightarrow[\text{MeOH/H}_2O]{\text{NaOH}} \quad -(CH_2-CH)_n-(CH_2-CH)_m$$

(Scheme structures showing R, R' substituents with N—CH₂—CH₂ and C=O, R₁ groups)

R=H, OCOCH₃
R=C₆H₅, CH₃

R' = H, R₂ = H, pE₇₀-g-pEI
R' = OH, R₂ = H, pVA₈₀-g-pEI
R' = OH, R₂ = C₆H₅, pVA₈₀-g-pPhOx

Scheme 5.

Thus the method permits to control the hydrophobic-hydrophilic properties of the graft copolymers ranging from the water repellent polyE-g-polyPhOx to the water soluble poly(vinyl alcohol-g-ethylenimine) polymers (polyVA-g-polyEI) (Table 6).

Depending on the composition, the graft copolymers show quite different equilibrium swelling degree in water and different solubility (Table 6). Thus polyE₂₀-g-polyMeOx forms a stable emulsion in water, while polyE₇₀-g-polyMeOx, with nearly the same content of MeOx units only swell. This deference could be assigned to the different macromolecular architecture: many short methyl oxazoline chains being solvated by water are sufficient to hold the chain in solution, whereas a small amount of long grafts has no such effect.

TABLE 6. Solution and sorption properties of poly(ethylene-vinyl alcohol) and poly(vinyl acetate-vinyl alcohol) grafted with polyoxazolines [29].

Polymer	Equilibrium swelling degree [a], %	Sorption of Cu^{2+}		Solubility [c]	
		(%)	$K_d \cdot 10^{-2}$	in THF	in CHCl₃
polyVAc	27	26	1.02	++	+
pVAc₈₀-g-pPhOx [b]	88	43	1.71	+	++
pVAc₈₀-g-pMeOx	partial dissolution	-	-	-	+
pVA-g-pPhOx	200	72	2.87	+	++
pVA-g-pEI	soluble	-	-	-	-
pE₇₀-g-pMeOx	135	31	1.24	-	+
pE₂₀-g-pMeOx	emulsion	33	1.32	-	++
pE₇₀-g-pEI	135	94	3.75	-	-

[a] in distilled water; [b] the subscripts indicate wt.% of the units;
[c] ++ soluble, + soluble under heating, - insoluble.

The sorption ability of the swollen graft copolymers towards Cu^{2+} was checked and the distribution coefficients K_d were estimated (Table 6). The copolymers polyE$_{70}$-g-polyEI and polyVA-g-polyPhOx show high complexation ability. The high chelating efficiency of polyEI is well known [30]. We found that polyE$_{20}$-g-polyMeOx take up Cu^{2+} from the solution. Contrary to some literature data [31], this sorption could only be ascribed to the polyMeOx segments. Thus playing with the nature of the segments, distribution of grafting sites and hydrophilic/hydrophobic balance it is possible to obtain products with controlled swelling and chelating properties.

3. 3. GRAFTING OF TELECHELICS ONTO POLYMERS

The "grafting onto" method is a synthetic route to amphiphilic graft copolymers and networks which enables control over the side-chains length and the grafting density. Using appropriate combination of prepolymers it is possible to prepare a number of new materials with exceptional properties. Telechelic prepolymers of different structure and bearing different reactive sites were synthesized by terminating the corresponding living polymers with end-capping agents [32, 33]. Telechelics of polyEO were prepared by transformation of the hydroxyl end-groups (Table 7).

TABLE 7. Characterization of telechelic polymers.

Structure	$\overline{M}_n$ (GPC)	MWD	$\overline{M}_n$ (NMR)	F [a]	Ref.
polyTBA-OH	3300	1.22	3 900	0.96	32
polyTBA-NH$_2$	3600	136	3 600	0.94	32
polyMeOx-OH	3000	1.09	3 400	1	33
HO-polyMeOx-OH	2000	1.30	2 100	2	33
polyMeOx-NH$_2$	2200	1.09	2 400	0.95	
HO-polyTHF-OH	9600	1.27	13 800	2	
polyEO-NH$_2$	2020	1.16		0.97	33
J-polyPhOx-J	1600	1.21	1 500	1.93	37
J-polyMeOx-J		1.40	1 900	1.94	37

[a] end-groups functionality, determined from ^{1}H NMR spectra.

Copolymers with elastic main chain and polyamine segments in the side chain were obtained by free-radical addition of amino-telechelics of polyTBA and polyTBA-b-polyTHF to the polybutadiene (polyBd) double bonds (Scheme 6) [13, 34]. The reaction proceeds as alkylation induced by free radicals and is considered to be a chain reaction. The graft copolymers were found to have phase-separated morphologies manifested by the presence of melting endotherms for each phase: in the

316

polyBd-g-polyTBA thermogram at -7°C for polyBd and at 128°C for polyTBA; in the polyBd-g-(polyTBA-b-polyTHF) curve at –8 °C for polyBd, at 19 °C for polyTHF and at 101 °C for polyTBA. The lower melting temperature of the polyaziridine phase could be explained by the location of the polyTBA segments, restricted between the polyBd backbone and polyTHF side blocks.

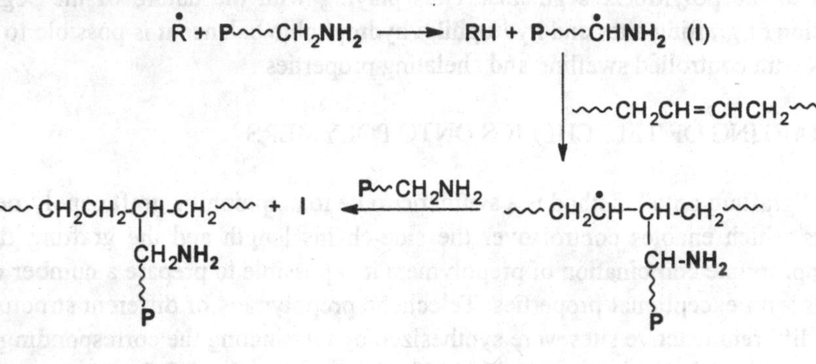

$$\dot{R} + P\sim\sim CH_2NH_2 \longrightarrow RH + P\sim\sim \dot{C}HNH_2 \quad (I)$$

$$\sim\sim CH_2CH=CHCH_2\sim\sim$$

$$\sim\sim CH_2CH_2-CH-CH_2\sim\sim + I \xleftarrow{P\sim CH_2NH_2} \sim\sim CH_2\dot{C}H-CH-CH_2\sim\sim$$

$$\begin{array}{c} CH_2NH_2 \\ | \\ P \end{array} \qquad\qquad \begin{array}{c} CH-NH_2 \\ | \\ P \end{array}$$

P= polyTBA; polyTHF-b-polyTBA

Scheme 6.

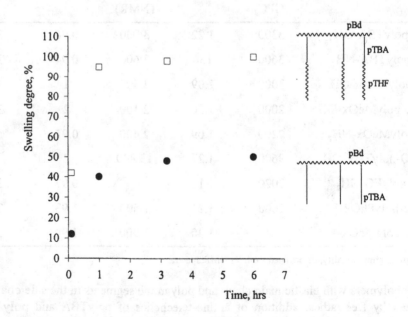

Figure 5. Swelling in water at pH=2 of (polyBd)$_{80}$-g-(polyTBA)$_{20}$ (●) and (polyBd)$_{67}$-g-{ (polyTBA)$_{22}$-b-(polyTHF)$_{11}$} (□).

After protonation of the polyamine segments the graft copolymers become amphiphilic and swell in water and in organics. The extent of the water uptake depends on the chemical nature and structure of the copolymer (Figure 5). The equilibrium swelling degree of the polyBd grafted with polyTBA-*b*-polyTHF is higher due to the lower compactness of the structure.

An alternative approach to amphiphilic graft copolymers was applied for hydrophilization of styrene-isoprene block copolymers [35, 36]. The isoprene block was functionalized with maleic anhydride under mild conditions. The reaction proceeds as a substitutive addition to the methylene groups in the allylic position. Thus, prepolymers functionalized up to 15-20% with single succinic anhydride rings along the backbone were formed. By opening the succinic rings using α-hydroxy-ω-methoxy polyEO, side chains of definite length were introduced (Table 8).

TABLE 8. Amphiphilic properties of grafted St-Is block copolymers [36].

Sample	Water uptake	Contact angle, θ ($^\circ$)	Interfacial tension, σ (dyn/cm)	Recovery of Cu^{2+} (% per hr)
PS-*b*-PI	5	102	—	—
PS—*b*—PI ├—COOH └—COOH	4	68	—	17
PS—*b*—PI HOOC C=O PEG$_{55}$	10	6	47.2	25
PS—b—PI HOOC C=O PEG$_{200}$	32	totally wetted	32.0	—

These copolymers could be used as non-aqueous surfactants since they form stable emulsions in methanol, heptane, hexane and ether. Their foam-forming properties are illustrated by the stable equilibrium values of the surface tension (σ) in a model oil/air system containing 1wt.% copolymer. The contact angle (ϑ) markedly decreases after introducing the polyEO side chains. The use of these copolymers as cation-exchange membranes is of potential interest. The dialysis of Cu^{2+} proceeds better through membranes combining the carboxyl group action with the solvating effect of adjacent polyEO chain.

The reaction of polyfunctional tertiary amines with mono- or dihalides was applied as an another approach in the synthesis of amphiphilic copolymers and networks [37].

Living mono- or bifunctional polyoxazoline iodides were terminated by the water-soluble poly(dimethylaminoethyl methacrylate) (polyDMAEM).

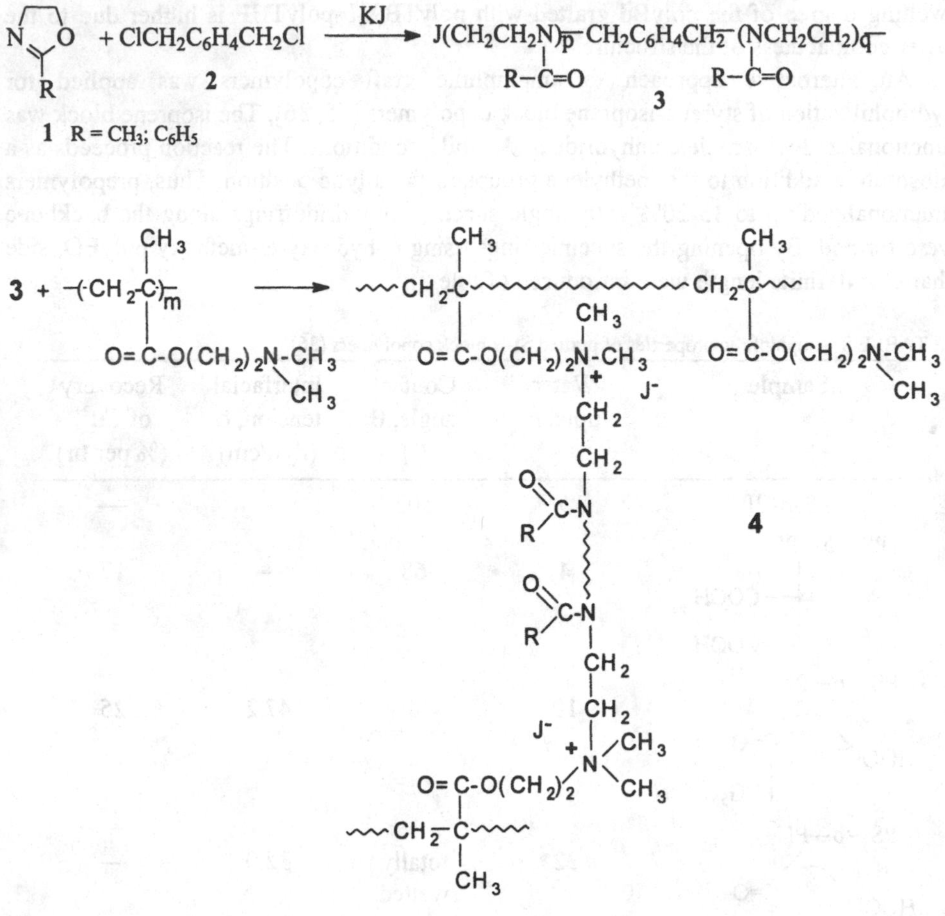

$$1 \quad R = CH_3; \; C_6H_5$$

Scheme 7.

At first, the efficiency of the reaction was examined in the grafting of α-benzyl-ω-iodide polyROx. Using the specified conditions, polymer networks were prepared according to Scheme 7. Characteristic feature of the copolymers and networks is that bonding proceeds through quaternary ammonium groups. The high reaction yields allowed the conclusion that the average density of the network could be controlled by the amine/iodide ratio and the length of the bridges. The graft copolymers and networks contain both strong-base (quaternary ammonium) and weak-base (tertiary amine) groups. They behave as hydrogels and amphigels, since the backbone is hydrophilic and the grafts or the bridges could be hydrophilic or hydrophobic, depending on the substituent (methyl or phenyl) in the oxazoline unit. Further transformation of the free tertiary amine groups into quaternary ammonium is possible as well. Both networks with polyMeOx and polyPhOx crosslinkers, swell in water and in chloroform.

As expected, the degree of swelling depends on the nature and the length of the crosslinkers and on the average molecular weight of polyDMAEM segments between the nearest crosslinking points (Figure 6).

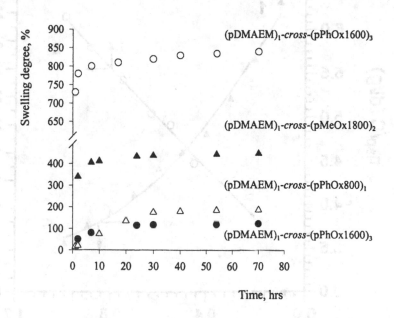

Figure 6. Swelling degree of polyDMAEM-polyoxazoline networks. Full symbols present the swelling in water; open symbols - in CHCl₃.

The presence of ionic groups is expected to modify the solution properties of the copolymer. Typical polyelectrolyte behavior was established for the water soluble polyDMAEM-g-polyMeOx. The reduced viscosity - concentration dependence does not obey the Huggins relation. However, it fits very well the Fuoss-Strauss equation (Eq. 2) for the reduced viscosity of solutions containing flexible polyelectrolytes, dissociated in H_2O (Figure 7). Constants A and B are characteristic of the copolymer. The estimated A value of 10.63 dL/g suggests high extension of the polymer chains in water.

$$\frac{\eta_{sp}}{c} = A\left(1 + B.c^{1/2}\right) \qquad (2)$$

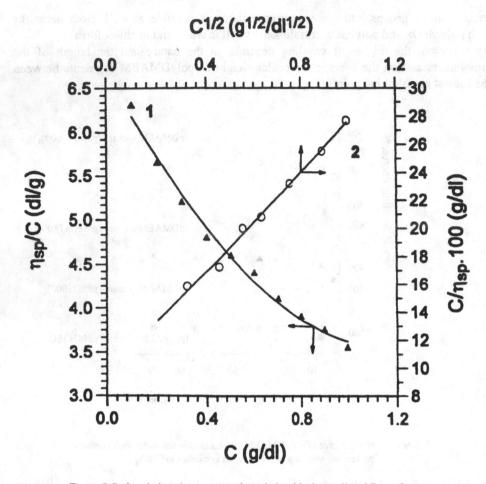

Figure 7. Reduced viscosity-concentration relationship (curve 1) and Fuoss-Strauss
plot (curve 2) of polyDMAEM-*g*-polyMeOx [37].

The potentiometric titration curves (Figure 8) in salt-free water and in solution with
constant ionic strength of $\mu=0,1$ mol/L (KCl) are characteristic for intermediate
polybase electrolytes. The effect of the neighboring ammonium groups (*m*) and the
dissociation constant, pKa were determined using a modified Henderson-Hasselbach
equation (Eq. 3) [38]. The value of $pK_a=5.21$ suggests that there is a polycation in the
solution and the low *m* value of 0.54 shows week interaction of the groups.

$$pH = pK_a - m.\log \frac{1-a}{a} \qquad (3)$$

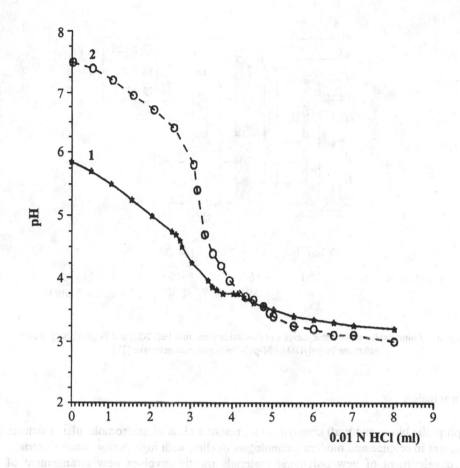

Figure 8. Potentiometric titration courses of polyDMAEM-*g*-polyMeOx [2g/L] with HCl [0.01N]: 1 - in distilled water, 2 - in KCl solution (μ = 0.1 mol/L) [37].

The amphiphilic polyelectrolyte networks extract over 60% of the ferrocyanide and rhodanide ions from their aqueous solutions due to the high charge density of the ammonium groups (Figure 9).

The anion-exchange capacity could be controlled during synthesis by changing the length of the grafts or crosslinking blocks, the charge density of ammonium groups and the density of the networks.

322

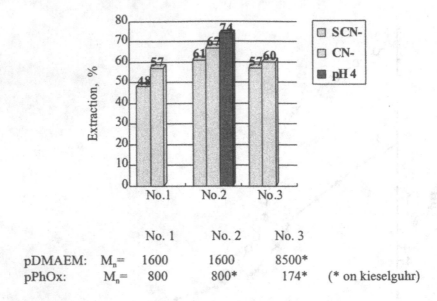

		No. 1	No. 2	No. 3	
pDMAEM:	$M_n=$	1600	1600	8500*	
pPhOx:	$M_n=$	800	800*	174*	(* on kieselguhr)

Figure 9. Anion-exchange of ferrocyanide and rhodanide ions from $Fe(CNS)_3$ and $Fe_4[Fe(CN)_6]_3$ water solutions by polyDMAEM-polyPhOx polymer networks [37].

4. Conclusion

Amphiphilic block and graft copolymers represent a class of macromolecular structures important in science and modern technologies dealing with high performance materials. The development of new polymeric materials mostly involves new arrangement of already known segments and blocks differing in properties. The increasing number of living systems provides the possibility to prepare new surface active polymers with particular features in aggregation phenomena, phase behavior and interface interactions, especially those adapting themselves to external stimuli and environmental changes.

Acknowledgments. Financial support from the Bulgarian National Science Fund is gratefully acknowledged. The authors express their appreciations to Prof. E. Goethals, Polymer Chemistry Division, University of Gent and Dr. A. Dworak, Institute of Coal Chemistry, Polish Academy of Sciences for their fruitful cooperation.

5. References

1. Riess, G., Hurtrez, G., and Bahadur, P. (1985) Block copolymers, in Mark, H.F., Bikales, N.M., Overberger, C.G., and Menges, G. (eds.), *Encyclopedia of Polymer Science and Engineering*, Wiley, New York, Vol. 2, pp. 324-434.
2. Mc Cormick, C.L., Bock, J., and Schulz, D.N. (1989) Water-soluble polymers, in Mark, H.F., Bikales, N.M., Overberger, C.G., and Menges, G. (eds.), *Encyclopedia of Polymer Science and Engineering*, Wiley, New York, Vol. 17, pp. 730-784.
3. Velichkova, R. and Christova, D. (1995) *Prog. Polym. Sci.* **20**, 819-997.
4. Dan, N. and Tirrel, M. (1993) *Macromolecules* **26**, 637-642.
5. Rempp, P. and Merrill, E.W. (1985) *Polymer Synthesis*, Hüting & Wepf Verlag, Heidelberg.
6. Hsieh, H. and Quirk, R. (1996) *Anionic Polymerization. Principles and Practical Application*, Marcel Dekker Inc., New York.
7. Kennedy, J. and Ivan, B. (1991) *Designed Polymers by Carbocationic Macromolecular Engineering: Theory and Practice*, Hanser Publishers, Munich, New York.
8. Fukui, H., Sawamoto, M., and Higashimura, T. (1993) *Macromolecules* **26**, 7315-7321.
9. Penczek, S., Kubisa, P., and Matyjaszewski, K. (1985) *Adv. Polym. Sci.* **68/69**, 1-317.
10. Sogah, D.J., Hertler, W.R., Webster, O.W., and Cohen, G.M. (1987) *Macromolecules* **20**, 1473-1488.
11. Hawker, C.J. (1996) *Trends in Polymer Science* **4**, 183-188.
12. Munir, A. and Goethals, E.J. (1980) Synthesis of "living" linear poly(N-tert. butyl imino ethylene) with predictable molecular weight, in Goethals, E.J. (ed.), *Polymeric Amines and Ammonium Salts*, Pergamon Press, New York, pp. 19-24.
13. Velichkova, R., Christova, D., and Gancheva, V. (1994) *Macromol. Symp.* **85**, 143-165.
14. Kobayashi, S., Hyama, H., Ihara, E., and Sacgusa, T. (1990) *Macromolecules* **23**, 1585-1589.
15. Van Caeter, P., Goethals, E.J., Gancheva, V., and Velichkova, R. (1997) *Polym. Bull.* **29**, 589-596.
16. Velichkova, R., Toncheva, V., Getova, C., Pavlova, S., Dubrovina, L., Gladkova, E., and Ponomareva, M. (1991) *J. Polym. Sci., A: Polym. Chem.* **29**, 1107-1112.
17. Mitov, Z. and Velichkova, R. (1992) *Eur. Polym. J.* **28**, 771-775.
18. Tezuka, Y., Okabayashi, A., and Imai, K. (1991) *J. Colloid Interface Sci.* **141**, 586-588.
19. Nuyken, O., Sanchez, J.R., and Voit, B. (1997) *Macromol. Rapid Commun.* **18**, 9125-131.
20. Christova, D. and Velichkova, R. (1995) *Macromol. Chem. Phys.* **196**, 3253-3266.
21. Christova, D., Velichkova, R., and Goethals, E.J. (1997) *Macromol. Rapid Commun.* **18**, 1067-1073.
22. Dworak, A., Panchev, I., Trzebicka B., and Walnch, W. (1998) *Polym. Bull.* **40**, 461-468.
23. Taton, D., le Borgne, A., Sepulchre, M., and Spasski, N. (1994) *Makromol. Chem. Phys.* **195**, 139-148.
24. Dreyfuss, P. and Quirk, R.P. (1987) Graft copolymers, in Kroschwitz, J.I. (ed.), *Encyclopedia of Polymer Science and Engineering*, Wiley, New York, Vol. 7, pp. 551-579.
25. Wesslen, B. and Wesslen, K.B. (1992) *J. Polym. Sci.,A: Polym. Chem.* **30**, 355-362.
26. Aoi, K. and Okada, M. (1996) *Prog. Polym. Sci.* **21**, 151-208.
27. Dworak, A., Schulz, R.C., Panchev, I., Trzebicka, B., and Velichkova, R. (1994) *Polym. Int.* **34**, 157-161.
28. Pantchev, I., Velichkova, R., Dworak, A., and Schulz, R.C. (1995) *Reactive and Functional Polymers* **27**, 53-59.
29. Pantchev, I., Velichkova, R., and Dworak, A. (1997) *Reactive Polymers* **32**, 241-248.
30. Kobayashi, S. (1990) *Prog. Polym. Sci.* **15**, 751-823.
31. Rivas, B. and Geckeler, K. (1992) *Adv. Polym. Sci.* **102**, 171-188.
32. Christova, D., Velichkova, R., and Panayotov, I.M. (1993) *Makromol. Chem.* **194**, 2975-2983.
33. Angelova, N. and Velichkova, R., unpublished results.
34. Christova, D. and Velichkova, R., to be published.
35. Mitov, Z. and Velichkova, R. (1993) *Eur. Polym. J.* **29**, 597-601.
36. Mitov, Z. and Velichkova, R. (1993) *Eur. Polym. J.* **29**, 1129-1136.
37. Panchev, I., Velichkova, R., Lakov, L., Peshev, O., and Goethals, E.J. (1998) *Polymer*, in press.
38. Morawetz, H. (1980) *Macromolecules in solution*, Wiley Interscience, New York.

MODELING, SIMULATION AND CONTROL OF POLYMERIZATION PROCESSES: SOME ASPECTS FOR TAILORED SYNTHESIS

A. F. JOHNSON, R. G. GOSDEN, Z. G. MESZENA*
IRC in Polymer Science and Technology, School of Chemistry, University of Leeds, Leeds, West Yorkshire LS2 9JT, England
Department of Chemical Information Technology, Technical University of Budapest, Budapest, 1521, Hungary

Abstract. Polymerization reactor control can sometimes be used to augment chemistry for the tailored synthesis of polymers. This theme is explored giving attention to generic methods for the control of molecular weight distribution for living polymerization processes when carried out in flow reactors. The role of modeling and simulation for achieving targets are considered.

1. Introduction

Although significant progress has been made towards the control of polymerization processes in recent years, much still remains to be done because of limitations in (i) quantitative understanding of the precise mechanism of many polymerization processes, particularly when taken to very high conversions of monomer or co-reactants where there is not enough information required to formulate mathematical models of processes; (ii) our ability to numerically simulate the molecular features of polymers which are of particular interest, notably molecular weight distribution, composition distribution, branching and crosslinking, particularly in the context of spatial variations which occur within polymerization reactors, and (iii) our capacity to measure rapidly, accurately and directly in real-time many of the molecular features of interest in the polymeric products as they are being produced so that one has the appropriate feedback to enact good control.

Since the theme of the conference is ionic polymerization reactions, the aim is to explore aspects of the modeling, simulation and control of polymerization processes with a view to assessing our capability of 'tailoring' polymer structures. In keeping with the 'tutorial' element of the NATO ASI, the content is aimed at those not experienced in the modeling, simulation and control fields.

325

J.E. Puskas et al. (eds.), Ionic Polymerizations and Related Processes, 325–349.
© 1999 *Kluwer Academic Publishers. Printed in the Netherlands.*

2. Control Action

2.1. TARGETS (SET-POINTS)

In order to take control action in any process it is necessary to identify the target (or set-point). Some *physical* control targets are obvious. For example, it is easy to define what is required if isothermal or isobaric conditions are needed as a specific temperature or pressure can be defined. There are also many different ways of measuring in real-time both temperature and pressure so that it is easy to see whether these features of a process are off-target. Often the target is *chemical* i.e. it is necessary to control the pH of a solution, comonomer concentration or mean molecular weight. It is a relatively easy task to directly measure pH in real-time as devices are readily available to continuously monitor pH. However, it may not be easy to measure monomer concentration or initiator/catalyst concentration so readily and perhaps impossible to measure the mean molecular weight of a polymer during polymerization. It can be seen that, even when the target can be readily defined, if appropriate measurements cannot be made of the target property of interest, there is an immediate problem of assessing whether a process is off-target. Often, indirect measurements are made to monitor or assess the set point. For example, viscometric measurements can be carried out in real-time and can give information about the mean molecular weight of the polymer as it is produced in a polymerization process. However, viscosity is not only a function of molecular mass but solution concentration, temperature, molecular mass distribution and much else. Despite such limitations, viscometric or rheological devices as commonly used because of the lack of appropriate real-time measurement devices for the specific property of interest.

2.2. FEEDBACK CONTROL

Control engineering is a mature field of study [1,2]. Feedback control action should be familiar to everyone.

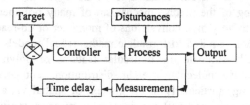

Figure 1. Feedback control loop

The rudiments of feedback control can be described pictorially as shown in Figure 1. For the sake of simplicity, the *process* in question can be considered to be the heating of a water bath to some predetermined temperature and the *product* no more than getting the water to the desired temperature. The *target* temperature can be selected and interpreted in terms of the voltage generated by a thermocouple if a thermocouple is the chosen *measurement device*. The temperature in the water bath is subject to various *disturbances* such as the loss of heat by cooling or a gain in heat because of stirring which will cause the measured temperature (voltage) to differ from that of the target. The difference in the target voltage and the measured voltage can be readily detected electronically by a device know as a *comparator* and the positive or negative error measured. Given that the size and direction of the error is known, *control action* can be taken to correct for the error i.e. a heater or cooling system can be switched on to compensate for the measured error. If the measurement takes time to make, a *time delay* is introduced into the feedback loop and this introduces a problem as the control action will be out of phase with the actual conditions in the process. It is not easy to correct for the time delay hence it is very desirable to have measurement devices which have a fast dynamic response in relation to the dynamics of the changes which might take place in the process.

2.3. FEEDFORWARD CONTROL

It is immediately obvious that there are problems if one wishes to apply feedback methods for the control of molecular weight distribution or copolymer composition distribution. Neither of these variables can be described by a single target parameter or set-point i.e. they are both distribution functions and fast measurement devices are not available. The basic concept of feedforward control is that the disturbances are detected before they enter the process and it is these that are controlled to ensure that the process delivers what has been targeted (Figure 2).

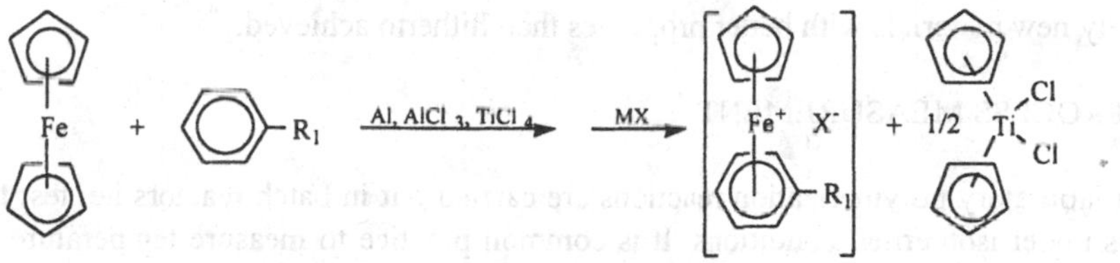

Figure 2. Feedforward control

2.4. CONTROL ACTION

The control action in response to the perceived error might be taken in various ways. In the case of a water bath, when the error is such that the temperature is too high, the heater may be switched off and switched on again when the error changes sign i.e. *bang-bang* control. More subtle controllers might alter the voltage of the heater/cooler input to the process so that it is either *proportional* to the magnitude of the measured error, or related to the *integral* of the error over a period of time, or to the rate at which the magnitude of the error is seen to change (a *differential* term which cannot be used alone).

Proportional control action (P-control) can be taken alone but leads to an *offset*. For this reason, it is commonly used in conjunction with integral control action (PI -control). Differential control action cannot be used alone but when used in conjunction with the other control modes (PID-control) speeds up the rate at which the target or set-point is achieved.

2.5. TARGETS IN POLYMERIZATION PROCESSES

In attempting the tailored synthesis of a polymer, the target(s) are self evident to the chemist and these might include molecular weight, molecular weight distribution, copolymer composition, copolymer composition distribution, branching, stereochemistry and much else.

Others might see the targets in quite different terms. The chemical engineer faced with the task of scaling up a laboratory polymerization reaction has a number of constraints. Notably, he or she only has access to well-defined reactor configurations (batch, semi-batch, tubular or CSTR) with fixed process measurement capabilities, finite stirring and mixing capabilities, fixed heat exchange capabilities and the capacity to measure temperatures, pressures, torques and flow rates only. The targets for process control might be temperature, pressure, monomer conversion or viscosity. Some of these parameters will have a relevance to molecular structure and composition, but the relationship is often indirect and not always easy to quantify. The end user might only be interested in using a product and then the target becomes cost effectiveness or some desirable physical property. It is often the demands of the end-user that provides the stimulus for finding new ways forward with the controlled synthesis of known monomer systems or for the making of entirely new materials with better properties than hitherto achieved.

2.6. PROCESS MEASUREMENT

Most laboratory polymerization reactions are carried out in batch reactors i.e. test tubes or flasks under isothermal conditions. It is common practice to measure temperature since it controls the rate and perhaps the course of polymerization reactions. Temperature can readily be measured by thermometer, thermocouple or thermistor and any measurement device with an electrical output can be used for real-time control in a feedback loop as described previously. Almost all the other important measurements relevant to the process are made off-line at the leisure of the experimentalist. Most spectroscopic methods used to examine polymer structure and composition are used in this way e.g. FT-Raman, FTIR or NMR as is the measurement of molecular weight by GPC. Some of these techniques lend themselves to real-time measurements. It is not uncommon for relatively slow polymerizations to be followed in an NMR tube or an infrared cell, techniques which not only allow the rate of the polymerization reactions to be followed in many instances but also allow aspects of the polymerization reaction mechanism to be elucidated. Increasingly, spectroscopic methods are being adapted for real-time measurements and infrared or Raman spectrometers are being attached to large-scale polymerization reactors for monitoring and control purposes, often remotely for safety reasons using optical fibers to link the cells within the reactor to a purpose-built process spectrometer. Gas chromatography (GC), gel permeation chromatography (GPC) and liquid chromatography (HPLC) are all used to monitor reactions off-line and can, in some instances be used in-line with the process for control purposes. *Continuous* measurement methods are intrinsically simpler to use for control purposes than *cyclical* measurement devices and are essential for the control of very fast processes i.e. where cycle rates might be long compared with reaction rates. Process monitoring and control devices together with the attendant interfacing technology and computers, can often cost as much, if not more, than the polymerization reactor system itself. The reactive processing of polymers is of growing interest and the reaction injection molding of polymers requires very specific control action [3].

3. Modeling and Simulation

3.1. GENERAL COMMENTS

There has been a very rapid increase in the capacity to represent the dynamics of chemical processes. There are now many readily available low-cost software packages which allow the solutions of the ordinary differential heat and mass balance equations which mathematically describe the dynamic behavior of chemical reactors when temperatures are uniform within a reactor or when mixing is perfect. It is a computationally intensive task to describe all reacting species in a polymerization reaction in order to get a picture of molecular features such as molecular weight distribution. Mathematical procedures have been developed which make it possible to get some impression of the molecular weight distribution through the moments of the distribution. It is a more difficult task to describe the spatial distribution of species or temperature but significant progress has been made with the use of computational fluid dynamics methods (CFD) to describe gas phase reactions and the first attempts to apply such methods to the model equations related to polymerization reactions have been reported. In addition to the deterministic approach to modeling, it is possible to use statistical or probabilistic methods to predict molecular weight distributions and chain compositions. In general, these methods do not take into account the influence of time in a chemical process, which can be a drawback if it is the dynamics of the reactor, which are of interest. A simplistic overview of deterministic methods only will be presented in an attempt illustrate the power of modeling and simulation for the controlled synthesis of polymers.

3.2. KINETIC MODELING

The textbooks usually consider the same basic reactions i.e. first order, second order, third order, simple reversible processes and chain reactions (sometimes classical free radical polymerization). Reactions are almost always considered at constant temperature and constant volume. In those cases where chain polymerization reactions are described, reactions are always considered at low conversion of monomer only i.e. in situations where time tends to zero. The primary reason for this limited range of reactions being considered is very simple; it is necessary for them to be solved by integration. Only cases where there is a simple analytical integral form can be dealt with without having to revert to numerical procedure. In the real world the range of chemical reactions is greater and the temperature of the reaction mixture may not be constant. Furthermore, in the case of semi-batch and flow reactors, reagents enter the reactor while reaction proceeds and the volume of the reaction mixture is not necessarily constant. Engineers have had to cope with these problems routinely for many years [4] and they have been amongst the first to exploit numerical methods to solve the sets of equations which ensue from these real cases. Simple examples of heat and mass balance equations and numerical methods for their integration are outlined in the following sections. More sophisticated methods of modeling polymerization processes are available but will not be discussed here.

3.3. MASS AND HEAT BALANCE EQUATIONS

A typical set of equations to describe an exothermic first order reaction (monomer loss is usually a first order process) carried out in a single-stage continuous flow stirred tank reactor (CSTR) is shown below ($n = 1$ for a first order reaction).

$$\frac{dV}{dt} = F_0 - F \tag{1}$$

$$\frac{d(VC_A)}{dt} = F_0 C_{A0} - FC_A - VC_A{}^n \alpha e^{-E/RT} \tag{2}$$

$$\rho C_p \frac{d(VT)}{dt} = \rho C_p (F_0 T_0 - FT) - \lambda V C_A{}^n \alpha e^{-E/RT} - UA_H (T - T_j) \tag{3}$$

$$F = K_V(V - V_{min}) \qquad \text{if} \quad V > V_{min} \tag{4}$$
$$F = 0 \qquad\qquad\qquad \text{if} \quad V \le V_{min}$$

where V is the reactor volume, F_0 the flow rate of the input and F the output, C_A the concentration of A in the reactor with A_0 representing A in the feed, α is the pre-exponential term, λ the heat of reaction and C_p the heat capacity at constant temperature, U is a heat transfer coefficient, A_H the area of the reactor in contact with the reactants and T and T_j the temperature of the reactants and reactor jacket, respectively.

These equations represent the total reactor continuity, reactor component continuity, reactor energy, and a flow control equation which assumes that some sort of measurement level device (for V) changes the outflow of the reactor in direct proportion to the volume of reagents in the reactor, where K_V is the gain for the proportional control action.

Mass balances of the species involved in the reaction are used because this circumvents the problem of volume change that might occur when material flows in and out of the reactor. The mass balance for any given reactant is given as follows:

$$\begin{bmatrix} \text{Rate of change} \\ \text{of mass} \\ \text{in the reactor} \end{bmatrix} = \begin{bmatrix} \text{Rate of mass} \\ \text{flow into} \\ \text{the reactor} \end{bmatrix} - \begin{bmatrix} \text{Rate of mass} \\ \text{flow out of} \\ \text{the reactor} \end{bmatrix} +/- \begin{bmatrix} \text{Rate of change} \\ \text{of mass} \\ \text{by reaction} \end{bmatrix}$$

In model equations, a mass balance equation has to be constructed for each species in the kinetic mechanism. For a simple first order reversible process of the type

$$A \rightleftharpoons B$$

the mass balance equation for A will differ only in the sign of the term relating to the rate of change of mass due to reaction.

The heat balance equations can be constructed in a similar manner.

$$\begin{bmatrix} \text{Rate of} \\ \text{heat change} \\ \text{in the reactor} \end{bmatrix} = \begin{bmatrix} \text{Rate of heat flow} \\ \text{into the reactor} \\ \text{by flow} \end{bmatrix} - \begin{bmatrix} \text{Rate of heat flow} \\ \text{out of the reactor} \\ \text{by flow} \end{bmatrix} +/- \begin{bmatrix} \text{Heat production or} \\ \text{consumption by} \\ \text{reaction, transfer} \\ \text{and stirring} \end{bmatrix}$$

Once the heat balance expression has been established, it is a relatively simple step to construct an expression for the rate of change of temperature in the reactor provided the heat capacity of the system is known.

Numerical integration methods, used for solving the balance equations, are widely described in texts but few explain the methods more clearly and simply than Luyben [4] to which the reader is referred for further information. Modeling and simulation methods have been discussed in greater detail elsewhere in the NATO Program [5].

3.4. MATHEMATICAL AND COMPUTATIONAL METHODS: MOMENTS AND TRANSFORMATIONS

The equations which describe the concentration of a polymer of any given chain length can be transformed by a variety of methods into a smaller finite set of equations. These can then be solved to give the moments of the molecular weight or chain length distribution from which it is possible to also obtain the actual distribution of chain sizes. This approach is not too difficult for monomodal homopolymers and simple reactions but can become quite complex, particularly in the case of copolymers or multicomponent systems and branching reactions.

The moments of a distribution function can be defined as follows,

$$\lambda_k = \sum_{j=1}^{\infty} j^k \cdot [P_j] \qquad k = 0,1,2,\ldots$$

where λ_0, λ_1, and λ_2 are the zeroth, first and second moments, respectively, the first three moments in a series which can be extended further as necessary,

The number average and weight average molecular weights and chain lengths, when defined in terms of the moments, are given in Equations (5) and (6)

$$M_n = w \cdot \lambda_1/\lambda_0 = w \cdot \mu_n \tag{5}$$

$$M_w = w \cdot \lambda_2/\lambda_1 = w \cdot \mu_w \tag{6}$$

where w is the molecular weight of the monomer. The dispersity index, D_n, may then be conveniently defined as shown in Equation (7):

$$D_n = M_w/M_n = \mu_w/\mu_n = \lambda_2 \cdot \lambda_0/\lambda_1^2 \tag{7}$$

The magnitude of D_n must always be greater than 1.0 by an amount equal to the normalized variance of the distribution function.

The lower moments have an obvious physical significance. The 1st moment is the total mass of polymer, the 0th moment the total number of polymer chains comprising that mass and the second moment the weighted sum of the number of polymer chains (the number of any given mass times their mass). The number and weight average molecular weights are accessible experimentally by colligative (e.g. osmometry) and light scattering methods, respectively. Some higher moments are also accessible by sedimentation methods (ultracentrifugation) but will not be considered here. The dispersity index is a useful measure of the breadth of the molecular weight distribution of monomodal homopolymers and has to have a value of 1.0 or greater. The variance of a distribution is sometimes used in characterizing very narrow molecular weight distributions and this is related to the 2nd moment about the mean; this and other ways of representing the breadth of distributions has been described [6].

Bamford and Tompa suggested the use of moments to describe the molecular weight distributions stemming from kinetic models of free radically initiated polymerization reactions [7]. It has been shown that it is possible to construct the actual molecular weight distribution curve from a finite number of the low moments, using Laguerre polynomials and other methods [7-9], the actual number of moments required depending on the precision required and on the complexity of the polymerization mechanism. At the time there was no absolute molar mass measurement method available to obtain the higher moments and this presented a problem for the earlier workers in this area. Gel permeation chromatography now presents us with a reasonable means (but a much less than perfect means than is commonly thought for making correlation with kinetic predictions) of obtaining molecular weight distributions with the predictions from simulations of polymerization kinetic models.

The use of moments, transfer functions and other procedures to render the simulation of polymerization processes more tractable have been well described in many articles and reviews [8,10]. For those prepared to accept commercial software, the problems of handling the simulation of polymerization processes have been greatly simplified with the advent of packages such as PREDICI [11]. This package is specifically designed for the dynamic modeling of polymerization processes and deals very effectively with the problems of molecular weight distributions and other parameters. This package has many unique features and has the capacity to handle:

- Full molecular weight distributions and their moments
- Elemental reaction steps and kinetics
- Temperature, volume, pressure and mass reactor cascades
- Modeling of kinetic parameters in terms of reaction time, reactor variables or reactants
- Heterogeneous reactions including particle size distribution
- Copolymer composition and branching
- Input of experimental distributions data
- Parameter estimation

4. Control of Molecular Structure

4.1. CHEMISTRY AND CHEMICAL TECHNOLOGY

It is self evident that most of the important molecular features in a polymer are introduced through the specific chemistry of the polymerization reaction. Hence it is essential that the chemistry be appropriate for any target molecular structure otherwise no control of any sort can be exercised whether the reactions be carried out on the laboratory-scale or in large-scale processes. The inherent mechanism of a polymerization reaction cannot be changed using engineering methods except to the extent that the reactions might be influenced by temperature, pressure or reactor feed composition. For example, there is no way that an engineering procedure can alter the stereoregulating nature of a catalyst, the relative reactivity of monomers in a copolymerization process or the propensity of a propagating species to engage in chain transfer agent reactions.

What engineering techniques do allow is the ability to select appropriate reactor configurations to deal with the real physical problems of manipulating highly exothermic processes or the mixing of viscous fluids. Engineering methods can be used to change the concentrations of reagents with time in any given reactor configuration i.e. manipulation of some of the key variables which influence the molecular weight, molecular weight distribution, copolymer composition, or copolymer composition distribution. Engineering considerations will often dictate whether a reaction should be carried out in bulk, solution, suspension, emulsion, or dispersion. With heterophase reaction conditions such as bulk polymerizations carried out in water suspension, it is engineering considerations that might dominate particle size and particle size distribution of the products.

Of relevance to the theme of the NATO Advanced Study Institute (ASI) on Ionic Polymerization and Related Processes is the 'dial-a-polymer' program in our laboratories which devoted to the development of a fully automatic computer controlled reactor system for the production of well tailored polymers from living polymerization processes [12-15]. This program seeks to make use of well established and well behaved living polymerization reactions for which the kinetic mechanism and reaction constants are known, and to control the reactors in which the reactions are carried out in ways which make it possible to control molecular weight distribution or composition distribution. The approach is unorthodox in that disturbances are deliberately introduced into flow reactors (the conventional control target is to maintain steady-state conditions in such reactors). However, it does provide the unique prospect of controlling molecular features that are impossible to control if left to chemistry alone. The approach is generic. Although the focus of the experimental testing of the control strategies which have been developed has been on a well behaved living anionic polymerization reactions, the logic can be extended to many other chemistries which give rise to living polymers (but not all). The aim in briefly outlining the logic behind one aspect of the 'dial-a-polymer' program is to illustrate that modeling, simulation and control consideration are requirements for tailored synthesis.

As might be expected, there are many chemical and physical factors in real processes which conspire to cause deviations from the expected behavior. These factors will also be briefly explored.

4.2. LIVING POLYMERIZATIONS

There has been an enormous expansion of interest in living polymerization processes in recent times and the question "livingness" is topical and merits exploration before proceeding further with the use of chemical technology for the control of molecular weight distribution.

Whatever other molecular features might be of interest when carrying out a living polymerization, the control targets during synthesis will usually include the production of narrow molecular weight distribution (MWD) materials with predictable average molecular weights. The extent to which these targets appear to be achieved is often used to assess the so-called livingness of the polymerization process. The theoretically predicted values of the average molecular weight and dispersity of products and that obtained experimentally, do not always concur. It is important to understand the physical and chemical factors which might influence the observed molecular weight and molecular weight distribution prior to examining any reactor control strategies. Such considerations have a wider significance, particularly when exploring novel chemistries where one of the objectives might be to establish whether a reaction is living or not.

Some of the ground rules pertaining to the MWDs to be expected in the products from living processes are well established. However, these will be briefly re-examined before giving attention to the impact of reactor type, reactor instabilities (both inadvertent and deliberate), mixing processes and the instrumental methods of analysis on what is actually observed.

When chain polymerization reactions are simplified by excluding any form of termination reaction and transfer and initiation is instantaneous, the theoretical molecular weight distribution of the products from such reactions can be explicitly described by a Poisson distribution [16-21]. The main criteria established for the identification of livingness initially were (i) the production of "monodisperse" polymer, (ii) an increase in molecular weight with conversion of monomer and (iii) a final molecular weight defined by the initial molar concentrations of the monomer and initiator used in the reaction. For the systems studied at that time, the propagating species were long lived and the ability to produce block copolymers by the sequential addition of monomers was also evidence of the living character of the reaction.

The nature of the living character or "livingness" of polymerization processes has become less clear cut as the range of polymerization chemistries described as "living" has expanded to include cationic [21-23], metathesis [24,25], group transfer [26-28], radical systems

[29], and metallocene catalyzed [30-32] polymerizations. The ability to initiate living anionic polymerization reactions with a number of vinyl and diene monomers using simple organometallic initiators can be attributed to the intrinsic stability of the propagating species in the solvents in which the reactions are conducted (usually, anaerobic conditions are needed). Other chemistries do not always give rise to such stable propagating centers and living polymerization characteristics are not as easily sustained or consequently identified. The definition of what is essential in the reaction mechanism for a reaction to be described has been the subject of debate. The current understanding of livingness has been reviewed by Szwarc [33-35]. Of particular interest has been the growing number of polymerization reactions where the lifetime of the "living" propagation center has been found to be comparable or even shorter than the total time for complete conversion of monomer. Although propagating chains might be truly living on a time-scale comparable to that required for monomer consumption, the making of block copolymers by the sequential addition of monomers is difficult or impossible in such systems. When the dispersity index of the final product is close to unity, it is often suggested that the reactions are living in character.

When polymerization processes give rise to narrow MWD products, say dispersity values less than 2.0, particular care has to be taken when interpreting the significance of the experimental information for the reasons outlined in the following parts of the lecture. Rigorous analytical expressions are available to describe the molecular weight distribution for some simple reaction mechanisms, and, as described earlier, there are now numerical procedures available to help predict what MWD should be generated for more complex mechanisms. Simulation predictions are good when there are no physical factors which impinge on the chemistry which might alter the molecular weight distribution (see later).

4.2.1. Reaction Mechanism

The simplest possible description of an idealized non-equilibrium living polymerization reaction with a single propagating center per living molecular is shown in Equations 8-9:

$$I + M \rightarrow P_1^* \; k_i \text{ (instantaneous)} \tag{8}$$

$$P_j^* + M \rightarrow P_{j+1}^* \; k_p \tag{9}$$

where $(*)$ indicates a propagating center of some type, j the number of repeat units in the polymer chain, P the polymer chain, I the initiating species and M the monomer. In this scheme there is only one propagating center per molecule and the mean degree of polymerization, dp, is given by Equation 10 where x_M is the monomer conversion.

$$dp = x_M \cdot [M]_0/[I]_0 \tag{10}$$

For a reaction of the type described by Equations 8 and 9, the number chain length distribution (NCLD) of the products from reactions carried out in a perfectly mixed isothermal batch reactor can be explicitly defined by a Poisson distribution function [16-17]:

$$[P_j] = [I]_0 e^{-dp} \cdot \frac{dp^{j-1}}{(j-1)!} \qquad j = 1, 2, \ldots \qquad (11)$$

The dispersity index is related to the moments of the distribution envelope as shown earlier and its magnitude is close to 1.0 for such reactions. It is possible to predict the value of D_n for various situations using Equation 11. The values for D_n shown in Table 1 are the theoretical lower limit, which can be achieved for a reaction conforming to Poisson statistics.

TABLE 1. Dispersity index for Poisson distribution as a function of the average degree of polymerization

dp	D_n
10	1.09
100	1.0099
1000	1.0010
10 000	1.0001
100 000	1.00001

The fact that a truly monodisperse product where $D_n = 1$ can not be obtained from a reaction where initiation is instantaneous is solely the result of the finite speed of the propagation reaction. Since D_n is more readily predicted in modeling and simulation studies than the overall distribution of molecular masses, it might be appropriate to examine the factors that can cause deviation from the theoretically calculated values. These factors can be broadly classified as chemical or physical and some of these will be described in more detail. Some of these factors can be used to 'control' the molecular weight distribution - but only in a significant way and generally they have been used to get better monodisperse products (which is a form of control but not really control of MWD as an independent parameter in a reaction).

4.3. FACTORS INFLUENCING THE DISPERSITY INDEX

4.3.1. Chemical Factors
Relative Rate of Initiation and Propagation Reactions. If the rate of the initiation reaction is very slow relative to the propagation reaction in a simple living reaction of the type illustrated above, it is possible for the monomer to be consumed before all the initiator has had time to become involved in an initiation reaction. This results in i) the mean molecular weight of the product being larger than that predicted from the ratio of initial molar monomer and initiator concentrations and ii) a broadening of the distribution. It is

noteworthy that the magnitude of D_n never exceeds 1.33 whatever the combination of reaction parameters.

The impact of slow initiation on the dispersity index is small and intrinsic to the reaction. Other broadening effects are extrinsic. There are ways in which the breadth of the distribution can be controlled such as the "pre-seeding" methods. However, Szwarc *et al* [35] has emphasized that this procedure has no effect on the MWD if the propagation reaction is irreversible. Feng and Yan [36] have suggested that a bimodal MWD can be deliberately produced from reactions which are pre-seeded if an appropriate change is made in the solvent for the second stage of the polymerization process. This is not a particularly practical suggestion and it is also assumed that there is no killing of active centers when changing the solvent and adding the second monomer.

When initiation reaction rates are slow compared with propagation rates, the MWD of the product cannot be correctly described by a Poisson distribution function. Gold [37] derived analytical expressions, which describe the MWD for living polymerizations in the case of isothermal reactions where there is a disparity in the relative rates of the two reaction steps. Nanda and Jain [38] arrived at similar expressions to those of Gold but using more acceptable mathematical methods. The MWD function for living polymerizations initiated by difunctional initiators has been established [39].

Termination Reactions. Kinetic schemes involving monomer termination [40], disproportionation and combination, and a variety of other situations such as a combination of spontaneous and impurity termination, spontaneous and impurity transfer and chain transfer to monomer [41, 42] have been reported. In most cases, the formulae have been derived analytically for the MWD and several calculate the expected dispersity index for a given reaction type and it has been assumed that the initiation reaction rate is instantaneous. As might be expected, termination and transfer reactions have a significant effect on the MWD if the order of magnitude of their rate constant is close to the propagation rate constant. In this case the dispersity index may be close to the most probable value of 2.0 or even higher in some instances. Many other situations have also been considered including [43, 44] which will not be considered further here.

Equilibria Between Living and Dormant Species. There are many living polymerization processes which can be described as having two propagating species of different reactivity co-existing in equilibrium as shown below: One could be so inactive as to be regarded as simply dormant.

$$A \Leftrightarrow B$$

Typically, A might be a polymeric free ion-pair and B a less active aggregate of the polymeric ion-pairs. If, in competing for monomer in the propagation process, one of the species has a very small propagation constant, it might be considered to be dormant. The broadening of the molecular weight distribution and consequential increase in the dispersity index which might stem from the competing propagation processes has been considered by many [45-47], usually by making the simplifying assumption that the exchanges between species are of the same degree of polymerization. The broadening effect of competing reactions increases as the relative rate of the equilibration reaction to the propagation rate increases. The dispersity index can be increased even when the relative reactivity of the propagating centers are similar provided the lifetime of the active centers is very different. A situation of this sort has been described involving an equilibrium between ion pairs, free ions and dormant species [48, 49] where the ion pair collapses readily to the dormant moiety. The molecular weight distribution observed in products from reactions of this sort have been examined by Matyjaszewski [50, 51]. Pepper [52] has suggested that bimodal distributions can arise in situations where a fast chain transfer reaction occurs. It has been a long-held view that a bimodal distribution can be produced when the exchange of the type shown above is very slow, however, drawing a distinction between the mechanisms which might give rise to a bimodal distribution product is a matter of conjecture. The impact of equilibrium processes on MWD has been further investigated by Müller [53,54]. Some obvious noteworthy facts are (i) that the D_n will always increase when the molecular weight distribution broadens and (ii) D_n only has meaning for comparing the properties of polymers when the products have monodal distributions.

4.3.2. Physical Factors Influencing the MWD

Reactor Design Reactions of the type described by Equations 8 and 9 carried out in an ideal plug flow tubular reactor under steady-state flow conditions give rise to polymers with the same MWD characteristics as those produced in a batch reactor [55, 56]. The reason for this is evident if a small segment of the tube is regarded as a well mixed batch reactor which simply moves along the length of the tube. Real tubular reactors do not usually conform to this simplistic description as rapid viscosity changes during reaction cause a change from turbulent well-mixed flow behavior to that of laminar flow with some radial diffusion. However, for polymerizations carried out in relatively dilute solution, it is possible to approach experimental conditions where polymers with very low D_n values can be produced.

If the same ideal living polymerization described in Equations 8 and 9 are carried out in a single-stage continuous flow stirred tank reactor (CSTR) then a polymer with the "most probable" (also called the Schulz-Flory) MWD is produced when ideal conditions pertain for the reaction and reactor. This distribution has the form

$$[P_j] = q^{j-1}.(1-q).\tau.k_i.[M].[I] \qquad (12)$$

where the quotient, q, is given by

$$q = \tau.k_p.[M]/(1+\tau.k_p.[M]) \qquad (13)$$

and τ is the mean residence time.

The number- and weight average chain lengths and D_n are given by Equations (14), (15) and (16).

$$\mu_n = 1/(1-q) \tag{14}$$
$$\mu_w = (1+q)/(1-q) \tag{15}$$
$$D_n = 1+q \tag{16}$$

In most cases the value of the quotient, q, is very close to one, hence $D_n \approx 2.0$. It has been established that for polymerization reactions carried out on well behaved living polymerizations in dilute solution the theoretical predictions are substantiated by experiment.

If a number of continuous flow stirred tank reactors (CSTRs) are connected in series, the magnitude of D_n for the polymer decreases as the number of CSTRs is increased. In theory, if an infinite number of perfectly mixed CSTRs are used, the product from such a reactor at complete conversion of monomer D_n would approach a value of 1.0 (i.e., a train of CSTRs approximates to the behavior of an ideal plug flow tubular reactor [55].

Quality of Stirring and Mixing of Reactants. Small, unsuspected disturbances can influence the nature of the products from living polymerization reactions when using more conventional laboratory methods (i.e. flasks, test tubes or NMR tubes). For example, when the initiator is simply added to a monomer solution in a flask to initiate polymerization, if the rate of addition is slow compared to the rate of the slowest reaction step, this could have a significant effect on the MWD of the product. The same would be true if monomer were being added to a monomer solution. Even if the rate of addition is fast, slow mixing of the initiator into the reaction mixture as a result of inadequate stirring could have a significant effect.

Mixing effects are difficult to quantify [57-59]. However, it is possible to demonstrate the effect of inefficient mixing on the magnitude of D_n by making use of a simple example as described by Meszena et al [60]. Consider the addition of a monomer to a large-scale batch reactor where the following conditions hold:(a) The monomer is introduced from the top into a well mixed initiator solution and the monomer becomes rapidly distributed in some way along the horizontal axis of the reactor so that within any of the striations through the height of the reactor both reactants can be considered to be perfectly mixed. (b) The added monomer is only transported in an axial direction by slow diffusion processes and the axial diffusion of polymer is infinitely slow. (c) The polymerization process can be described by Equations 8 and 9 and termination reactions are unimportant.

The purpose of the example is only to estimate the order of magnitude D_n. Qualitative results are obtained by assuming some approximate profile for the degree of polymerization, *dp* with respect to height. If mixing was perfect throughout the reactor, the degree of polymerization would be the same at any height. However, if the (effective) axial diffusion of monomer is slow relative to the propagation reaction rate, the degree of polymerization of the polymer in any layer is distributed in a way which can be described by some convenient function, say a decaying exponential-like function, with parameter (*a*) being the rate of reaction relative to diffusion.

It can be seen from the values presented in Table 2 that even with moderate values of *a* and dp the dispersity index can reach values approaching 10 for the particular model used. Although extreme conditions have been taken in this example, the predicted dispersities are almost an order of magnitude greater than those which might be expected from the broadening effect arising from slow initiation or termination reaction rates in perfectly mixed batch reactors.

TABLE 2. Dispersity index as the function of the parameters *a* and *dp*.

	a	1	10	20
dp	D_n			
100		1.095	3.39	6.4
1000		1.083	3.93	8.7
10 000		1.082	3.99	9.0

The impact of poor mixing of the type described above can be readily demonstrated in a laboratory-scale glass reactor (a full description of the polymerization reactor has been given previously [14]). A number of *sec*-butyllithium initiated polymerization reactions of styrene were performed in dilute solution in cyclohexane (above the θ-temperature) using identical conditions except for the speed at which a simple paddle stirrer was operated in each of the experiments. The molecular weight and molecular weight distribution of the polymer product at 100% conversion of monomer for experiments with stirrer speeds of 20, 200 and 1000 rpm are shown in Table 3.

TABLE 3. Experimental values of dispersity index as a function stirrer speed

Stirrer speed rpm	D_n
20	1.6
200	1.14
1000	1.09

If D_n is taken as a measure of the "livingness" of a reaction, and care has not been taken to ensure good mixing, an incorrect view of the nature of the chemistry taking place will be formed .

In a previous report [55] on the mixing behavior in polymerization reactors, it was suggested that segregation of the initiator (i.e. the production of a uniform distribution of "clumps" of initiator throughout a batch reactor, that are slow to disperse to a system mixed to the molecular level) could lead to narrowing of the observed MWD. In effect, the incomplete or slow physical mixing process produces results akin to those for well-mixed reactions where initiation is very slow. If all the monomer is consumed by reaction with fewer reactive sites, a higher average molecular weight product than that expected would be produced having a value for D_n closer to 1.0 than expected as shown in Table 1.

Broadening Effects. Many factors can conspire, either singularly or on concert, to broaden the observed molecular weight distribution of polymers from living polymerization reactions but there are few that conspire to narrow a distribution. Some factors which give rise to a broadening of MWD are not unique to living systems but are less evident in reactions which give rise to broad distribution products simply because the effect is not manifest in experimentally measured dispersity indices. Broadening of the MWD could stem from the chemistry of the reaction changing with monomer conversion, non-isothermal reaction conditions or a distribution of temperatures in a reaction vessel [60] disturbance in the reactor feed conditions, poor mixing or even deliberate reactor control strategies [58]. Few would question the difference of D_n values quoted as 3.4 and 3.7. However, the same numerical difference between a D_n of 1.07 and 1.47 would not escape attention.

Unsuspected narrowing of the observed D_n of a polymer is a rare phenomenon but might arise because of fractionation during polymerization as a result of phase separation, or because fractionation takes place during product isolation or reprecipitation, or because only a fraction of the initiator is used in a living polymerization, as mentioned earlier.

5. Control of Molecular Weight Distribution

5.1. GENERAL COMMENTS ON LIVING POLYMERIZATIONS

For a truly ideal reaction and reactor system, it is possible to contemplate a fully computer controlled reactor system with which one could 'dial-a-polymer' at will. In the real world, one falls short of perfection for a number of reasons, but the principle of how a simple 'technological' means of control of molecular weight distribution might be achieved is readily demonstrated.

5.2. CONTROL STRATEGY FOR MWD CONTROL; PERTURBED FEEDS TO A TUBULAR REACTOR

If one considers a volume element in a tubular reactor, if the contents of that element remain intact as it moves along the tube, then any reaction taking place in that volume element is occurring in a batch reactor albeit moving along the tube. If an ideal living polymerization process of the type described above were to take place in such an element, when that element reaches the end of the tube and the reaction had gone to complete conversion, then the product from the tube would be exactly the same as that from a batch process. If one arranges for elements along the tube to contain different ratios of monomer to initiator, each element on reaching the end of the tube would discharge polymer with a distinctive molecular weight relating its contents. If the effluent is collected, the polymer in the collection reservoir would have a molecular weight distribution which related to the sum of all the individual components collected with time. Provided the contents of each small segment of the tube can be quantitatively controlled, then the overall molecular weight distribution of the accumulated product can also be controlled.

It is possible to get close to ideal plug flow behavior with a living polymerization process if a dilute monomer solution is used (say 5% solids content - higher final solids contents lead to significant deviations from plug flow behavior when the living polymerization reaches high conversion). To achieve a different monomer-to-initiator ratio in any given small element of the tube, it is only necessary to pump monomer and initiator at different individual rates into the reactor, ensuring very good and fast mixing at the inlet, in some controlled manner. In this case the observed molecular weight distribution of the accumulated product should reflect the flow pattern used for monomer and initiator at the inlet. The input flow strategies for monomer and initiator can be readily correlated as shown in Figures 3-4.

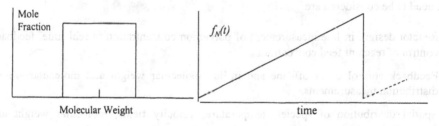

Figure 3. Desired MWD   *Figure 4*. Feed profile (monomer or initiator)

Figure 3 shows a target molecular weight distribution expressed as a plot of mole fraction against molecular weight. This plot can be readily translated into a monomer flow profile at a fixed initiator level as shown in Figure 4. If this flow pattern is introduced in a repeated systematic way, then a cyclical steady-state will be established to give a product which is the accumulation of the material produced over many cycles. Using this basic concept, it is possible to develop reactor feed profiles to give almost any desired molecular weight distribution.

This simple hypothesis has been tested and some typical results are shown Table 4 for the steady state operation of the reactor. It can be seen that very low dispersity indices are obtainable when the molecular weight is less than 50,000. As the molecular weight increases by reducing the initiator concentration, some broadening of the molecular weight distribution is seen indicating deviation from ideal plug flow behavior in the reactor. Under steady state feed input conditions, a polymer with a dispersity index greater than 2.0 is produced when using 5-10 weight percent monomer in the feed solution.

TABLE 4. Variation in D_n with molecular weight for a well behaved living polymerization reaction carried out in a tubular reactor

M_n	D_n
1 600	1.07
1 900	1.06
2 700	1.07
5 800	1.07
16 900	1.10
4 900	1.18
85 000	1.36
>100 000	>1.5

Perhaps sufficient data and arguments were presented in this paper to illustrate the concept, but it is a demanding task to understand what is needed in order to (i) establish a meaningful automatic feedback control system for such a process, and to (ii) understand what the actual relationship is between the input feed profiles and the measured molecular weight distribution of the products and how these might be improved. Some of the areas that need to be considered are:

- Reactor design, in-line measurement of polyanion concentration in real time, feedback control of reactant feed concentration.

- Feedback control using off-line and in-line molecular weight and molecular weight distribution measurements.

- Spatial distribution of species, temperature, velocity fields, molecular weight and dispersity indices.

- Broadening effects on the molecular weight.

- Generic nature of the approach in relation to other living polymerization processes.

However successful the concept, there are significant problems with tubular reactors for the long-term production of polymers. The primary drawback is the build-up on the walls of the reactor of a layer of polymer. This is a direct consequence of the nature of the flow front profile of fluids in a tube. At the walls there will always be a stationary or very slow moving region which, in the case of a living polymerization reaction, leads to the formation of very high molecular weight polymer which settles and eventually causes plugging of the reactor. The problem is exacerbated by the fact that the polymerization reaction is highly exothermic and fast which gives rise to a significant 'hot spot' at some point along the tube, the position depending on many factors including reactor dimension, reagent flow rate and heat removal capabilities of the reactor. Wall build up is much faster at this point and, once the build-up starts, heat transfer out of the reactor becomes more difficult and the on-set of plugging can be quite rapid. Forced oscillations actually help to reduce the rate of wall fouling, but do not eliminate the problem in our experience. It is possible to introduce mechanical stirring or wall scraping devices into a tube if the tube diameters are large, but the technology is not simple.

5.3. CONTROL STRATEGY FOR MWD CONTROL: PERTURBED FEEDS TO A CSTR

Rather than consider complex problems or a real tubular reactor technology, it is perhaps more appropriate to examine the use of a single or multiple stage CSTR reactor system. Some of the issues concerning this type of reactor are briefly considered.

The primary aim here is to identify the main factors affecting the shape of the MWD with the help of our recently developed approximate calculation method [61]. The algorithm utilized avoids the solution of the MWD equations, and only calculates the initiator, monomer and total live ends component balances. The main simplifying step in the method is the assumption that living polymerization gives rise to *truly monodisperse* chains. The length of all chains initiated at a certain time is taken to be uniform with an average length, μ. The rate of change in average chain length is calculated as the ratio of rate of propagation to live ends concentration. On the basis of this simplification it has been shown [61] that the shape of the *reversed* MWD is essentially defined by the initiator concentration profile. In reactor configurations other than batch this profile also has to be weighted with the residence time distribution of the reactor. However, for a batch reactor the approximate NCLD is derived as shown in Equation (17):

$$LE_j = \frac{k_i}{k_p} I(\mu - j)$$

(17)

where μ is the maximum chain length. If the ratio of the rate constants remains unchanged during the process, the NCLD is simply the inverse of the initiator concentration profile.

For a steady state CSTR the approximate NCLD is derived as given in Equation (18):

$$LE_j = e^{-\frac{j}{\tau k_p M}} \frac{k_i}{k_p} I$$

(18)

It is easy to see that Equation (22) is identical to the Schulz-Flory distribution (12), if q is substituted with the exponential term shown in (23).

$$q \sim e^{-\frac{1}{\tau k_p M}} \qquad (19)$$

Since the initiator concentration is constant, it is clear from Equation (18) that the NCLD reflects the residence time distribution curve, i.e. it is a decaying exponential function.

Given a qualitative understanding of what defines the shape of the MWD, the very complex MWD envelopes (see Figure 5) from a single stage CSTR with perturbed feed flow of monomer and/or initiator [14] can be explained. Let the initiator feed function used be a periodic square pulse with a constant overall volumetric flow rate through the reactor, while the monomer feed concentration is maintained constant. The distance between peaks in the MWD can be related to time through the average chain growth, μ. The average increase in chain length per one unit residence time is the distance between the peaks of the WMWD. The MWD is approximately described by the initiator concentration profile against average length of growth with reversed horizontal axis and weighted with the exponential residence time function as shown in Figure 5 (line without markers). The precisely calculated MWD is denoted with markers in Figure 5.

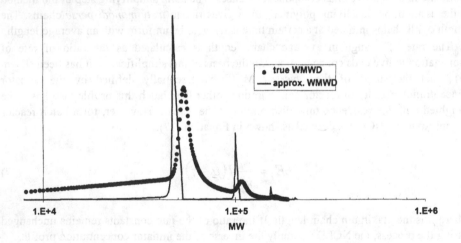

Figure 5. Approximate and rigorously calculated WMWDs for a periodically perturbed CSTR.

The impact of long passed pulse events is extremely small compared to more recent ones as the overall flow of material through the reactor removes material relating to the early events. It is difficult to detect peaks with intensity 0.1% (or much less) of the maximum intensity using a gel permeation chromatograph, hence only a few recent pulses can be resolved, all others becoming lost in the signal noise.

Feedback Control where the Target is a Molecular Weight Distribution. The major problem in tailoring the MWD by control of the feeds to reactors used to living polymerizations is the lack of a simple way of predicting the input perturbations required in order to achieve the desired MWD. The intrinsic difficulty is that the predictive models cannot be run 'backwards', i.e., it is not readily possible to define the required feed profiles to a reactor for any particular desired distribution of molecular weights in the product. It might be possible to circumvent this difficulty with a neural network approach the main advantage being that the impact of any non-idealities relating to the reactor, such as broadening effects, can be accommodated. The ideal approach would be to find an approximate analytical solution to the problem; this work will be described elsewhere.

6. Conclusions

- The modeling and simulation of polymerization processes is always instructive irrespective of whether the information generated is used for real-time control purposes or not. For example, it greatly enhances ones capability of comparing experimental results with kinetic models and to resolve the nature of complex reaction mechanisms.

- Simulated data are only as good as (i) the quality of the model which is used to define the chemistry or overall reactor system, (ii) the kinetic and physical constants used in the simulations, (iii) any assumptions made about the chemistry or any mathematical algorithms used to facilitate calculations, and (iv) in some instances, the rounding error which comes into iterative processes when repeated many times.

- In principle, the purely technological approach of putting feed perturbations of the reactants in living polymerization reactions carried out in tubular reactors allows the prospect of controlling the molecular weight distribution of the polymers fully automatically. In practice there are many limitations to the approach such as broadening effects because of bad mixing, wall fouling and possibly killing reactions.

- In order to achieve any sort of control over a polymerization reaction, it is essential to first understand the precise nature of the reaction mechanism. In order to invoke the assistance to the chemical technology to enhance the overall management of the products of a polymerization reactor it is essential to have a quantitative understanding of the reaction mechanism. Finally, it is vital to understand the impact of physical processes such as reaction medium viscosity, heat transfer and mixing on the reaction mechanism so that these can either be enhanced or minimized to suit ones purpose.

Acknowledgements. The invitation of Dr. J. Puskas and Dr. O. Nuyken to participate in the NATO meeting is greatly appreciated as is the support of NATO which has made the visit to London, Ontario, possible. The EPSRC is gratefully acknowledged for its financial support of the IRC in Polymer Science and Technology. We are also indebted to many research staff, students and technical staff, but, in particular, G. Meira, K. Arnold, and R. G. Gosden.

348

7. References

1. Coughanowr, D.R. and Koppel, L.B. (1965) *Process systems analysis and control*, McGraw Hill, New York.
2. Pollard, A. (1971) *Process Control, Heinemann*, London.
3. Coates, P.D. and Johnson, A.F. (1992) *Encyclopedia of Polymer processing & Applications*, Pergamon Press, Oxford.
4. Luyben (1992) *Simulation and Control for Chemical Engineers*, McGraw Hill, New York.
5. Meszena, R.G. (1998) NATO Meeting, University of Western Ontario, London, Ontario, Canada.
6. Bandermann, F. (1971) *Angew Makromol. Chem.* **18**, 137.
7. Bamford, C.H. and Tompa, H. (1954) *Trans. Faraday. Soc.* 50, 1097.
8. Ray, W.H. (1972) *J. Macromol. Sci. Revs.* **C8**, 1.
9. Cramer H (1946) *Mathematical Methods of Statistics*, Princton University Press.
10. Laurence, R.L., Galvan, R., and Tirrell, M.V. (1994) *Polymer Reaction Engineering*, Ed. C McGreavy, Blackie Academic & Professional, Glasgow.
11. Wulkow, M., (1996) *Macromol Theory Simul.* **5**, 393,
12. Meira, G. and Johnson, A.F. (1981) *Polym. Eng. and Sci.* **21(7)**, 415,
13. Gosden, R.G., Auguste, S., Edwards, H.G.M., Johnson, A.F., Meszena, Z.G., and Mohsin, M.A. (1995) *Polym. Reaction Eng.* **3(4)**, 331,
14. Gosden, R.G., Meszena, Z.G., Auguste, S., Johnson, A. F., and Mohsin, M.A. (1997) *Polym. Reaction Eng.* **5(1&2)**, 45.
15. Gosden, R.G., Meszena, Z.G., Mohsin, M.A:, and Johnson, A.F. (1997) *Polym. Reaction Eng.* **5(4)**, 205.
16. Flory, P.J. (1940) *J. Amer. Chem. Soc.* **62**, 1561.
17. Flory, P.J. (1953) *Principles of Polymer Chemistry*, Cornell University Press.
18. Szwarc, M. (1968) *Carbanions, Living Polymers and Electron Transfer Processes*, Wiley Interscience, New York.
19. Bywater, S. (1975) *Anionic Polymerization*, Progress in Polymer Science Vol 4 Ed. A. D. Jenkins, Pergamon, New York, Chapter 2.
20. Morton, M. (1983) *Anionic Polymerization, Principles and Practice*, Academic Press, New York.
21. Kennedy, J.P. and Ivan, B. (1991) *Designed Polymers by Carbocationic Macromolecular Engineering; Theory and Practice*, Hanser, Munich.
22. Penczek, S., Kubisa, P., and Matyjaszewski, K. (1980), (1985) *Adv. Polym. Sci.*, , **37**, 1, **68/69**, 1.
23. Nuyken, O. and Crivello, J. (1991) *Handbook of Polymer Synthesis*, Ed. H. R. Kercheldorf, Part A, 145.
24. Grubbs, R.H. and Tumas, W. (1986) *Science*, **243**, 907.
25. Gilliom, L.R. and Grubbs, R.H. (1986) *J. Amer. Chem. Soc.*, **108**, 733.
26. 26.Webster, O.W. (1987) in Mark, H.F., Bikales, M.N.C., Overberger, G. and Menges, G., *Encyclopedia of Polymer Science and Engineering*, Vol. 7 Eds., Wiley-Interscience, New York, pp 580.
27. Webster, O.W. and Sogah, D.Y. (1989) in Eastmond, G.C., Ledwith, A., Russo, S., and Sigwalt, P. Pergamon, *Comprehensive Polymer Science*, Vol. 4, Eds., London, pp 163.
28. Sogah, D.Y., Hertler, W.R.O., Webster, W., and Cohen, G.M.J. (1987) *Amer. Chem. Soc.*, **20**, 1473.
29. Hawker, C.J. (1996) *Trends in Polymer Science*, **4(6)**, 183.
30. Kaminsky, W. and Arndt, M. (1996) *Metallocenes for Polymer Catalysis*, Ed., H Ringsdorf, pp 143.
31. Wagner, P. H. (1992) *Chem.& Ind.*, May, pp 330.
32. Sassmannshausen, J., and Bochmann, M., and Rosch J., and Lige D. *J. Organometallic Chem.*, in press.
33. Szwarc, M. (1996) *Ionic Polymerization Fundamentals*, Hanser, Munich.
34. Szwarc, M. and van Beylen, M. (1993) *Ionic Polymerization and Living Polymers*, Chapman & Hall, New York.
35. Szwarc, M., van Beylen, M. and Hoyweghen, V. (1987) *Macromolecules*, **20**, 445.
36. Feng, J. and Yan, D.-Y. (1993) *Polym. Int.*, **30**, 407.
37. Gold, L. (1958) *J. Chem. Phys.*, **28**, 91.
38. Nanda, V.S. and Jain, R.K. (1964) *J. Polymer Sci.*, A2, 4583.
39. Hocker, H. (1972) *Makromol. Chem.*, **157**, 187.
40. Yan, D.-Y. (1984) *J. Chem. Phys.*, **80**, 3434.

41. Yuan, C.-M. and Yan, D.-Y. (1988) *Eur. Polymer J.*, **24**, 729.
42. Cai, G.-F., Yan D.-Y. and Litt, M. (1988) *Macromolecules*, **21**, 578.
43. Feng, J. and Yan, D.-Y. (1993) *Makromol. Chem. Theory Simul.*, **2**, 129.
44. Litvinenko, G.I. and Arest-Yakubovich, A.A. (1992) *Makromol. Chem. Theory Simul.*, **1**, 321.
45. Coleman, B.D. and Fox, T.G. (1963) *J. Amer. Chem. Soc.*, **85**, 1241.
46. Coleman, B.D. and Fox, T.G. (1968) *J. Polym. Sci.*, **4**, 345.
47. Figini, R.V. (1964) *Makromol. Chem.*, **71**, 193.
48. Bhattacharyya, D.N., Lee, C.L., Smid, J., and Szwarc, M. (1965) *J. Phys. Chem.*, **69**, 706.
49. Figini, R.V., Hostalka, H., Hurm, K., Lohr, G., and Schulz, G.V. (1965) *Z. Phys. Chem.*, **45**, 269.
50. Matyjaszewski, K. (1993) *J. Polym. Sci.*, **31**, 995.
51. Matyjaszewski, K., Szymanski, R., and Teodorescu, M. (1994) *Macromolecules*, **27**, 7565.
52. Pepper, D. C. (1995) *Makromol. Chem.*, **196**, 963.
53. Müller, A.H.E. (1994) *Macromolecules*, **27**, 1685.
54. Müller, A.H.E., Zhuang, R. and Litwinenko, G. (1995) *Macromolecules*, **28**, 4326.
55. Tadmore, Z. and Beissenberger, J.A. (1996) *I & EC Fund.*, **5**, 336.
56. Nauman, E.B.J. (1974) *Macromol. Sci.*, Rev. C, **10**, 74.
57. Litt, M. (1962) *J. Polymer Sci.*, **28**, 429.
58. Figini, R.V. (1960) *Z. Physik. Chem.*, **23**, 224.
59. Figini, R.V. and Schulz, G.V. (1960) *Makromol. Chem.*, **41**, 1.
60. Meszena, Z.G. and Johnson, A.F. (1995) *Phoenics J. Comput. Fluid Dynamics & Applic.*, **8**(1), 55.
61. Meszena, Z.G., Viczian, Z., Gosden, R.G., Mohsin, M.A., and Johnson, A.F. (1998) *Polym. Reaction Eng.*, accepted for publication.

CONFERENCE PHOTOS

(LEFT TO RIGHT)

1. Mohan Mathur, Dean of Engineering UWO; Joe Kennedy; Judit Puskas

2. Ingrid Kennedy; Joe Kennedy, Judit Puskas

3. Heinz Greve; Judit Puskas; Ron Commander; Argyrios Margaritis

4. Gerd Langstein; Werner Obrecht; Ron Commander; Doris Langstein

5. Miklos Zsuga; Rudi Faust; Ingrid Kennedy; Gyorgy Deak

6. Oskar Nuyken; Judit Puskas; Joe Kennedy; Mohan Mathur

7. Haihong Peng; Christophe Paulo; Shahzad Barghi; Gabor Kaszas

8. Susan Hoopes, Rob Storey; D. Sinai (Industry Liaison Officer, UWO)

9. Front: P. Hemery; Hung Nguyen; Bernadette Charleux, Rudi Faust;
   Back: Ulrich Schubert; Alain Deffieux; Michel Moreau

10. Nihan Nugay; Turgut Nugay; Gurkan Hizal; Yusuf Yagci

11. Konrad Knoll; Jurgen Ismeier

12. Tony Johnson, Ron Young; Zsolt Meszena

13. Stan Slomkowski; Stan Penczek

14. Timea Marsalko; Gyorgy Deak; Sandor Keki; Miklos Zsuga

15. Karin Weiss, Monika Bauer; Martin Thuring; Sandra Botzenhardt; Markus
    Krombholz

16. Stefan Spange; Joe Kennedy

17. Chris Curry; Judit Puskas; Keri Diamond; Adam Gronowski; Carsten Kreuder

18. Istvan Majoros; Petr Vlcek; Jaroslav Kriz

19. Tim Schaffer

352

20. Diana Hull; Tom Maggio

21. Wouter Reyntjens; Metin Acar

22. Bryan Brister; Metin Acar; Armin Michel

23. Banquet

24. Niagara Falls

1

2

3

4

5

6

7

8

9

10

11

12

13

14

15

16

17

18

19

20

21

22

23

24

CONFERENCE PHOTOS

Having fun